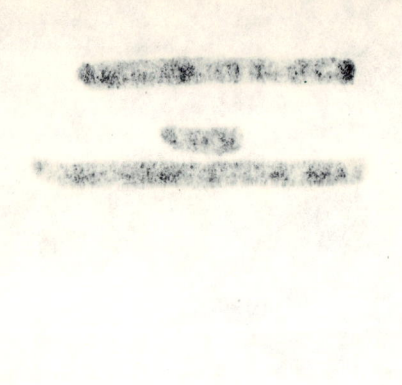

# EXOBIOLOGY:
## A Research Guide

MARTIN H. SABLE
PROFESSOR
THE UNIVERSITY OF WISCONSIN — MILWAUKEE

GREEN OAK PRESS ° Brighton, Michigan ° 1978

EXOBIOLOGY: A RESEARCH GUIDE
by Martin H. Sable

Copyright © 1978 by Martin H. Sable

All rights reserved. No part of this book may be reproduced in any form or by any means without permission in writing from the publisher, except for the inclusion of brief quotations in a review.

---

Library of Congress Cataloging in Publication Data

Sable, Martin Howard, 1924-
   Exobiology: a research guide.

   Includes indexes.
   1. Life on other planets--Bibliography.
2. Flying saucers--Bibliography. I. Title.

Z5154.L5S24   [QB54]   016.574999   78-7287
ISBN 0-931600-00-6

---

Manufactured in the United States of America.

Published by **GREEN OAK PRESS**
9339 Spicer, Brighton, Michigan 48116

THIS BOOK IS DEDICATED TO MY SONS

James Samuel Sable    -    Charles David Sable

TABLE OF CONTENTS

FOREWORD by Dr. J. Allen Hynek .. .. .. .. vii

PREFACE .. .. .. .. .. .. .. .. .. .. ix

BIBLIOGRAPHY SECTION

    Astronomy and Origin of Life and of Planets .. .. 1
    Bibliographies .. .. .. .. .. .. .. .. 10
    Flying Saucer Sightings .. .. .. .. .. .. 13
    Humanoid Contact on Earth .. .. .. .. .. 47
    Intelligent Life on Other Planets .. .. .. 63
    Intelligent Life within Earth .. .. .. .. 91
    Interstellar Communication .. .. .. .. .. 93
    Life on Other Planets in General and in Space
        (Excluding Intelligent Life) .. .. .. .. 106
    Life on Individual Planets
        (Excluding Intelligent Life) .. .. .. .. 131
    Occult Aspects, Including ESP, Mental Telepathy,
        Psychological Physical Effects, etc. .. .. .. 150
    Organizations .. .. .. .. .. .. .. .. 157
    Religious and Philosophical Aspects .. .. .. 160
    The Search for Life .. .. .. .. .. .. .. 169
    Space Flight .. .. .. .. .. .. .. .. 184
    Teleportations .. .. .. .. .. .. .. .. 191
    Unidentified Flying Objects Including Flying Saucers 195

DIRECTORY SECTION

    Organizations .. .. .. .. .. .. .. .. 252
    Periodicals .. .. .. .. .. .. .. .. 260

AUTHOR INDEX .. .. .. .. .. .. .. .. 265

SUBJECT INDEX .. .. .. .. .. .. .. .. 312

INDEX TO DIRECTORY SECTION .. .. .. .. .. 320

ABOUT THE AUTHOR .. .. .. .. .. .. .. 323

# CENTER FOR UFO STUDIES
### 2623 RIDGE AVE. — EVANSTON, ILLINOIS 60201

J. ALLEN HYNEK, Director
PROFESSOR OF ASTRONOMY
NORTHWESTERN UNIVERSITY


The bastions of established science today are being beseiged and battered by new phenomena which, for the most part, by their very nature it seems, should be excluded from the confines of the modern scientific structure. Yet these new empirical observations and phenomena no longer can be summarily dismissed. Even now, some infiltration is taking place. Prominent in these new, but as yet generally unaccepted phenomena are the UFOs and their related topics of Extraterrestrial Life, Interstellar Communication, etc. Most of us, busy with routine, mundane affairs, are hardly aware of the rising tide of these seeming anomalies. UFO literature, which has captured the mind of an increasing number of inquisitive scientists as well as the mind of the mass public through the multi-media exposure it has received, is a problem to the librarian and bibliographer as well as the serious student and researcher.

The present bibliography, monumental in its range, forcefully illustrates the scope of these new subjects and their impact. Entries have been culled from journals in some 20 languages, and over a long span of years to the current epoch. The nature of these subjects is worldwide, and Sable provides the first comprehensive look at these subjects as documented and reported in many different languages. Heretofore obscure references, and serious organizations and individuals known only to the handful of researchers in each field, are now presented to facilitate exchange of information and provide an opportunity for international cooperation and study.

The value of such a compendium can only be truly appreciated by those already aware of the recondite journals and papers cited. There are few men who have both the scholarly ability and the rare temperament to undertake the tedious compilation of such a work. Dr. Sable has performed for us a truly notable task and we must remain long in his debt.

It is the bête-noir of any budding departure from the established belief schemes that the infusion of cranks, unstable individuals, and charlatans find so fertile a ground for their peculiar maneuverings. It is neither the duty nor the perogative of the bibliographer to assess the ultimate value of potential entries, and thus possibly delete many. This must be left to the individual scholar and research worker in these respective fields. The serious researcher must compete with the pseudo-scientists, and the student must sift the documented evidence from spurious conjecture. Bibliographer Sable has provided a tool for this task. This unique work will, I am sure, prove of value for many years to come.

J. Allen Hynek<br>
Professor of Astronomy,<br>
Northwestern University<br>
<br>
Director,<br>
Center for UFO Studies

PREFACE

As a professional Bibliographer, my primary criterion in selecting a topic for compilation centers on the following questions: 1). Will the anticipated bibliography fill a gap in the reference literature? 2). Which categories of users will it serve (i.e., will it fill a real need)?

The answers, in the affirmative, to the preceding questions were set forth in 1966, when I began the compilation of The UFO Guide, 1947-1967 (1967, 100 pages), which is the antecedent of the present work. Exobiology: a Research Guide, is however a decided improvement over my first attempt "to fill the gap," in many respects.

The UFO Guide covered the field for print-format materials in perhaps a dozen languages through the year 1967. Exobiology represents more than text and reference works, pamphlets, government publications, technical reports, theses and dissertations, conference proceedings and periodical articles. It also includes radio and television scripts, films and materials on microfilm. In all, there are 3832 entries for materials published between the years 1648 and 1975, in approximately two dozen languages (entries utilizing non-Roman script appear in transliteration). A further differentiation: whereas The UFO Guide was restricted to treatment of UFO's and of possible intelligent beings in the universe, Exobiology covers all aspects of this topic as well as scientific aspects of life (including the search for life) in all of its forms outside planet Earth, whether biological, chemical, geological or physical, and interdisciplinary grouping of these fields. Other subjects dealt with include interstellar communication, the origin of life (on Earth and elsewhere), the evolution of planets, and the space research activities of the U.S. Government agencies, professional scientific societies and the UFO groups (the last composed mainly of laymen). The final sections of this work are comprised of directories of organizations and of periodicals, international in scope.

Throughout the Bibliography section, periodical articles have been separated from the other types of materials mentioned. The subject arrangement of <u>Exobiology</u> is by broad topic (see Table of Contents); for specific aspects of these topics, <u>it is important to consult the Subject Index</u>. Within each broad topic, the sub-arrangement is <u>chronological</u>: such a sub-arrangement is logical, especially in scientific fields, in which it is advantageous to follow advances made over periods of time. Within any one year, however, should there be two or more entries under "Books," which also includes all forms of materials other than periodical articles, the arrangement becomes alphabetical. In the case of "Periodical Articles," the chronological and alphabetical arrangements apply, for example, even when two or more articles appear in the same month of the same year.

All individual entries in both the Bibliography and Directory sections are numbered, and all the numbers utilized in the Author and Subject Indexes of the Bibliography section, refer to their respective entries. Prices of books (where included), are subject to change, and should be ordered from dealers or publishers.

I am indebted to Dr. J. Allen Hynek, Chairman of the Department of Astronomy at Northwestern University in two respects: first, Dr. Hynek (the acknowledged foremost authority in the world today on the UFO phenomenon) advised me to include the so-called "non-scientific" entries of books and articles dealing with UFO's. Second, it was through Dr. Hynek that I was able to borrow issues of many periodicals of UFO groups, almost none of which are acquired by public or university libraries. I am especially grateful for obtaining access to the foreign titles. As an example of the non-availability of "runs" of popular periodicals, I was able to utilize issues of <u>National Enquirer</u> only by visiting the Library of Congress.

I also wish to thank my sons, Jim and Chuck, to whom this book is dedicated (as well as my dear wife, Minna), for their patience and forbearance with me during the years when I was engaged in this undertaking. If, as Dr. Hynek stated in his Foreword, this work will be of service to scientists and will "prove of value for many years to come," then I shall indeed be rewarded.


Martin H. Sable
March 1978

# BIBLIOGRAPHY SECTION

## ASTRONOMY AND ORIGIN OF LIFE AND OF PLANETS

### BOOKS

**1**
Brahe, T.
 Tychonis Brahe Mathim: Eminent: Dani Opera Omnia, Sive Astronomiae Instauratae Progymnasmata, in Duas Partes Distributa, Quorum Prima de Restitutione Motuum Solis et Lunae, Stellarmus Inerrantium Tractat. Secunda Autem de Mundi Aetherei Recentioribus Phaenomensis Agit. Editio Ultima Nunc cum Indicibus Figuris Prodit. Frankfort, Imprensis I. G. Schonvvetteri, 1648, 470 & 217pp.

**2**
Huygens, C.
 The Celestial Worlds Discovered; or Conjectures Concerning the Inhabitants, Plants and Productions of the Worlds in the Planets. 2d Edition Corrected and Enlarged. Printed for James Knapton, at the Crown in St. Paul's Church Yard, London, 1722, 162p.

**3**
Wright, T.
 An Original Theory or new Hypothesis of the Universe. London, Printed for the Author, 1750, 84p.

**4**
Brahe, T.
 Observationes Cometae Anni 1585 Uraniburgi Havitae a Tychone Brahe. Jussu Serenissimi Regis Daniae Christiani VIII. Edidit H. C. Schumacher. Altonae, Typis Hammericianis & Lesserianis, 1845, 30p.

**5**
Flammarion, Camille
 La Pluralité des Mondes Habitués; Etude ou l'on Expose les Conditions d'Habitabilité des Terres Célestes Discutées au Point de vue de l'Astronomie, de la Physiologie et de la Philosophie Naturelle...28th Edition. Paris, Didier, 1880, 479p.

**6**
Arrhenius, Svante
 Worlds in the Making; the Evolution of the Universe. Trans. by H. Borns. New York, Harper & Brothers, c/o Harper and Row, 1908, 229p.

**7**
Lowell, P.
 The Evolution of Worlds. New York, Macmillan, 1909, 262p.

**8**
Heath, T.
 Aristarchus of Samos, the Ancient Copernicus. ...Oxford, Clarendon Press, 1913, 425p.

NOTES:

**9**
Hughes, T. A.
  Plurality of Worlds, and Other Essays.
London, Longmans, Green, 1927, 276p.

**10**
Jeans, J.
  Astronomy and Cosmogony. 2d ed. London,
Cambridge University Press, 1929, 428pp.

**11**
Stetson, H. T.
  Has Life any Cosmic Significance? New York,
McGraw-Hill, 1930, 221p.

**12**
Sitter, W. de
  Kosmos. Cambridge, Harvard University Press,
1932, 138p.

**13**
Russell, H. N.
  The Solar System and its Origin. New York,
Macmillan, 1935, 144p.

**14**
Kepler, J.
  Das Weltgeheimnis, Mysterium Cosmographicum.
Trans. and edited by Max Casper. Munich,
Oldenbourg, 1936, 147p.

**15**
Copernicus, N.
  On the Revolutions of the Celestial Spheres.
3v. Annapolis, The St. John's Bookstore,
1939.

**16**
Kepler, J.
  Weltharmon, K. Trans. and edited by Max
Casper. Munich, Oldenbourg, 1939, 403p.

**17**
White, George S.
  The Book of Revelations; a True Narrative of
Life on Many Planets; How our Earth was Created
and Inhabited; a Condensed History of the
Earth From its Creation to the Present Time
as Recorded in the Master Book of Life on the
Golden Planet. Los Angeles, George S. White,
1945, 198p.

**18**
Lucretius, Carus T.
  Of the Nature of Things. Trans. by W. H.
Brown. New Brunswick, Rutgers University
Press, 1950, 262p.

**19**
Struve, O.
  Stellar Evolution, an Exploration From the
Observatory. Princeton, Princeton University
Press, 1950, 266p. ($5.00)

**20**
Darwin, C. G.
  The Next Million Years. New York, Doubleday,
1952, 210pp. ($2.75)

**21**
Urey, H. C.
  The Planets, Their Origin and Development.
New Haven, Yale University Press, 1952, 245p.
($5.00)

**22**
Darwin, Charles G.
  The Next Million Years. Garden City, N.Y.,
Doubleday & Co., 1953, 210p.

**23**
Lockyer, J.N.
  Astronomía, la Tierra y el Universo. 2d ed.
Madrid, Eds. Ibéricas, 1953, 58p.

NOTES:

## ASTRONOMY AND ORIGIN OF LIFE AND OF PLANETS

**24**
Nelson, Albert F.J.H.N.
  <u>Life and the Universe</u>.  London, Staples Press, 1953, 258p.

**25**
Dauvillier, A.
  <u>L'Origine des Planètes</u>.  Paris, Presses Universitaires de France, 1955, 224p.

**26**
Hoyle, F.
  <u>Frontiers of Astronomy</u>.  New York, Harper & Brothers, c/o Harper & Row, 1955, 360p.  ($.95)

**27**
Hoyle, F.
  <u>The Nature of the Universe</u>.  Rev. ed. New York, Harper & Brothers, c/o Harper & Row, 1955, 142p.  ($3.50)

**28**
Gatland, Kenneth W., & Derek D. Dempster
  <u>The Inhabited Universe</u>.  London, Alan Wingate, 1957, 182p.

**29**
Oparin, A. I.
  <u>The Origin of Life on the Earth</u>.  Trans. From the Russian by Ann Synge.  New York, Academic Press, 1957, 495p.  ($6.80)

**30**
Henderson, L. J.
  <u>Life and the Cosmos</u>.  (In <u>The Fitness of the Environment</u>.  Boston, Beacon Press, 1958, pp. 274-312).  ($1.95)

**31**
Hutchings, E., editor
  <u>Frontiers in Science, a Survey</u>.  New York, Basic Books, 1958, 362p.  ($6.00)

**32**
Ruggieri, Guido
  <u>Mondi nel Spazio.  I Pianeti nel Sistema Solare e nell Universo Siderale</u>.  Rome, Associazione per le Scienze Astronautiche, 1958, 187p.  (2000 lira)

**33**
Calvin, Melvin
  <u>Origin of Life on Earth and Elsewhere</u>. Report #UCRL-9005.  Livermore, University of California Radiation Laboratory, Dec. 1959.

**34**
Lovell, A. C. B.
  <u>The Individual and the Universe</u>.  New York, New American Library, 1959, 111p.  ($.60)

**35**
Gauroy, Pierre
  <u>Les Mondes du Ciel: Terres Vivantes ou Cimetières?</u>  Paris, Librairie Artheme Fayard, 1960, 303p.

**36**
Kapp, R. O.
  <u>Towards a Unified Cosmology</u>.  New York, Basic Books, 1960, 303p.  ($6.50)

**37**
Oparin, Alexander I.
  <u>The Universe</u>. .2d rev. ed. Moscow, Foreign Languages Pub. House, 1960, 244p.

**38**
Öpik, E. J.
  <u>The Oscillating Universe</u>.  New York, New American Library, 1960, 144p.  ($.60)

**39**
Pons, W.
  <u>Steht uns der Himmel Offen?  Entropie, Ektropie, Ethik; ein Beitrag zur Philosophie des Weltraumzeitalters</u>.  Wiesbaden, O. Krausskopf-Verlag, 1960, 158p.  (11.80 marks)

NOTES:

## ASTRONOMY AND ORIGIN OF LIFE AND OF PLANETS

**40**
Vries, T. E. de
  A la Découverte de l'Univers. Trans. From the Dutch by André M. Hendrickx. Paris, Editions Sequoia, 1960, 155p. (2.25 NF)

**41**
Herrmann, Joachim
  Das Falsche Weltbild. Astronomie und Aberglaube. Eine Kritische Untersuchung über Astrologie, Welteislehre, Hohlwelttheorie, Bewohnbarkeit d. Sonne, Fliegenden Untertassen und and. Astronomische Irreehren. Stuttgart, Franckh, 1962, 162p. (16.80 marks)

**42**
Huang, Su-Shu
  Some Astronomical Aspects of Life in the Universe; Annual Report, 1961: Smithsonian Institution. Washington, 1962, pp. 239-240.

**43**
Rousseau, P.
  L'Univers et les Frontières de la vie. Paris, Hachette et Cie., 1962, 184p.

**44**
Tassi, D.
  The Mind and Time and Space. Philadelphia, Dorrance, 1962, 114p. ($3.00)

**45**
Tocquet, Robert
  Life on the Planets. New York, Grove Press, 1962, 192p.

**46**
Carles, Jules
  The Origins of Life. New York, Walker, 1963, 131p.

**47**
Huang, Su-Shu
  The Problems of Life in the Universe and the Mode of Star Formation. (In Cameron, A. G. W., ed. Interstellar Communication. New York, Benjamin, 1963, pp. 89-92). ($8.50)

**48**
International Space Science Symposium. 3rd, Washington, 1962
  Life Sciences and Space Research; a session edited by R. B. Livingston. New York, Wiley, 1963, 184p. ($7.75)

**49**
Jastrow, Robert, & A. G. W. Cameron, eds.
  Origin of the Solar System. Proceedings of a Conference Held at the Goddard Institute for Space Studies, New York, Jan. 23-24, 1962. New York, Academic Press, 1963, 176p.

**50**
U.S. National Aeronautics & Space Administration. Ames Research Center, Moffett Field, California
  Chemical Evolution and the Origins of Life, by Cyril Ponnamperuma. Condensation of an Address Presented at the Annual Commencement Luncheon, Sigma Xi; and the Phi Beta Kappa, University of Missouri, Columbia, June 3, 1963. NASA TM-X-54008; N64-22754. Washington, 1963, 16p.

**51**
Berrill, N. J.
  Worlds Without End; a Reflection on Planets, Life and Time. New York, Macmillan, 1964, 240 p. ($5.95)

**52**
Keosian, J.
  The Origin of Life. New York, Reinhold, 1964, 118p. ($1.95)

**53**
De Araujo, Hernani Ebecken
  Einstein, Espaco-tempo. Rio de Janeiro, 1965, 89p.

NOTES:

## ASTRONOMY AND ORIGIN OF LIFE AND OF PLANETS

**54** Ehrensvard, G. C. G.
Man on Another World. Trans. by Lennart and Kajsa Roden. Chicago, University of Chicago Press, 1965, 182p. ($5.95)

**55** Ponnamperuma, Cyril
"Chemical Studies on the Origin of Life" (In VPI Proceedings of the Conference on the Exploration of Mars and Venus). Alexandria, Va., NTIS, Aug., 1965, 8p. ($7.00)

**56** Sagan, Carl
Selected Studies in Exobiology, Planetary Environments and Problems Related to the Origin of Life. Semiannual Progress Report #1, 1 Oct. 1965-31 March 1966. (NASA-CR-76106). Alexandria, Va., June 1966, 130p.

**57** Sky and Telescope (periodical)
The Origin of the Solar System, Genesis of the Sun and Planets, and Life on Other Worlds. Ed. by Thornton Page & Lou Williams Page. New York, Macmillan, 1966, 336p. ($7.95)

**58** Firsoff, V. A.
Life, Mind and Galaxies. London, Oliver & Boyd, 1967, 111p.

**59** Murchie, G.
Music of the Spheres. 2v. New York, Dover Pubs., 1967. ($5.00)

**60** Moroz, V.I.
Physics of Planets. Tr. from Russian by NASA. Washington, 1968, 416p. (NASA-TT-F-515). ($3.00) (Available from NTIS).

**61** Mikhailov, A. A., et al
Developments of Astronomy in the USSR. Volume 1: 50 Years of Soviet Science and Technology. (English translation from Russian). (AD-696504; FTD-MT-24-186-68-Vol. 1). Dayton, U.S. Air Force Systems Command, Wright-Patterson Air Force Base, Foreign Technology Div., 1969, 371p.

**62** U.S. Geological Survey
Interagency Report: Astrogeology 19: Strategy for the Geologic Exploration of the Planets, ed. by M.H. Carr. (NASA Order W-12650). (NASA-CR-108783). Washington, 1969, 110p. Available from NTIS.

**63** Glasby, J. S.
Boundaries of the Universe. Cambridge, Harvard University Press, 1971, 305p.

**64** Rasool, S.I., ed.
Physics of the Solar System. NASA-SP-300. Washington, G.P.O., 1972, 519p. (Available on microfiche from NTIS).

**65** Soviet Bloc Research in Geophysics, Astronomy and Space; # 275. Arlington, Va., Joint Publications Information Service, 1972, 94p. ($6.75) (Available from NTIS).

**66** U.S. Joint Publications Research Service
Soviet Bloc Publications in Geophysics, Astronomy, and Space, #287. JPRS-57392. Tr. into English From Various Soviet-bloc Publications. Arlington, Va., 1972, 60p. ($5.00) Available From NTIS.

NOTES:

ASTRONOMY AND ORIGIN OF LIFE AND OF PLANETS

## PERIODICAL ARTICLES

**67**
Merriman, C. C.
 "Creation of Worlds." Popular Science Monthly 10, 1876, p. 701.

**68**
Vaughn, D.
 "Astronomical History of Worlds." Popular Science Monthly 13, 1878, p. 571.

**69**
Arrhenius, S.
 "The Transmission of Life From Star to Star." Scientific American 96, March 2, 1907, p. 196.

**70**
Dean, J. C.
 "The Transmission of Life From Star to Star." Scientific American 96, April 20, 1907, p. 331.

**71**
Lowell, P.
 "Evolution of Life." Century 75, Feb. 1908, pp. 499-510.

**72**
Berget, Alphonse
 "The Appearance of Life on Worlds and the Hypothesis of Arrhenius." (In The Smithsonian Institution. Annual Report of the Board of Regents for the Year Ending June 30, 1912. Washington, U.S. Govt. Printing Office, 1913, pp. 543-551).

**73**
"Did Life Reach Earth on a Comet's Bright Tail?" Newsweek 6, Dec. 14, 1935, p. 39.

**74**
Lighthall, W. D.
 "The law of Cosmic Evolutionary Adaptation: an Interpretation of Recent Thought." Ottawa, Transactions of the Royal Society of Canada, Series 3, #34, Section 2, 1940, pp. 135-141.

**75**
Banerji, A. C.
 "The Origin of the Solar System." Calcutta, Science and Culture (10:8), 1945, pp. 317-323.

**76**
Kavanau, J. L.
 "Some Physico-chemical Aspects of Life and Evolution in Relation to the Living State." American Naturalist (81:798), May 1947, pp. 161-183.

**77**
Wald, G.
 "The Origin of Life." Scientific American (191:2), Aug. 1954, pp. 44-53.

**78**
Blum, H. F.
 "Perspectives in Evolution." American Scientist (43:4), Oct. 1955, pp. 595-610.

**79**
Horowitz, Norman H.
 "The Origin of Life." Engineering Science (20: 2), Nov. 1956, pp. 21-25.

**80**
Shapley, Harlow
 "Coming to Terms With the Cosmos; Excerpt From Of Stars and men." Saturday Review 41, Sept. 6, 1958, pp. 51-54.

**81**
Wilson, A. G.
 "Introduction to Problems Common to the Field of Astronomy and Biology, a Symposium." Publications of the Astronomical Society of the Pacific (70: 42), Feb. 1958, p. 41.

NOTES:

**82** Heard, J. F.
"The Physical Nature of the Planets and Their Probable Course of Evolution." Ottawa, Canadian Aeronautical Journal (5:5), 1959, pp. 184-186.

**83** Briggs, Michael H.
"Meteorites and the Origin of Life." London, Spaceflight (2:2), April 1959, pp. 39-44.

**84** Huang, S. S.
"The Problem of Life in the Universe and the Mode of Star Formation." American Scientist (47:3), Sept. 1959, pp. 397-402.

**85** Gaffron, H.
"The Origin of Life." Perspectives in Biology and Medicine 3, 1960, p. 163.

**86** Gilbert, D. L.
"Speculation on the Relationship Between Organic and Atmospheric Evolution." Perspectives in Biology and Medicine 4, 1960, pp. 58-71.

**87** Muller, H. J.
"Issues in Evolution." (In Fox, S., & C. Collender, eds. Evolution After Darwin. Chicago, University of Chicago Press, 1960, pp. 69-105).

**88** Shapley, Harlow, ed.
"Stellar Evolution," sec. X (In his Source Book in Astronomy, 1900-1950. Cambridge, Harvard University Press, 1960).

**89** "Life Without end." Time 75, Jan. 4, 1960, p. 54.

**90** Fox, Sidney W.
"How did Life Begin?" Science 132, July 22, 1960, p. 200.

**91** Struve, O.
"Astronomers in Turmoil." Physics Today (13:9), Sept. 1960, pp. 18-23.

**92** Calvin, Melvin
"Origin of Life on Earth and Elsewhere." Armed Forces Chemical Journal 15, March-April 1961, pp. 26-35.

**93** Calvin, Melvin
"Origin of Life on Earth and Elsewhere." Annals of Internal Medicine 54, May 1961, pp. 954-976.

**94** Cramp, Leonard G.
"The Cosmos--Expanding or Orbital?" London, Flying Saucer Review 7, May-June 1961, pp. 8-12.

**95** Sagan, Carl
"On the Origin and Planetary Distribution of Life." Radiation Research 15, Aug. 1961, pp. 174-192.

**96** "Ecosphere may Shape Life on Distant Planets." Science Newsletter 80, Oct. 21, 1961, p. 272.

**97** Oparin, A. I., & V. G. Fesenkov
"Le Cosmos vu par les Savants Sovietiques." La vie et son Apparition, pt. 2. Lucerne, l'Homme et l'Espace 8-9, Nov.-Dec. 1961, pp. 16-23.

NOTES:

**98**
Horowitz, N. H., & S. L. Miller
 "Current Theories on the Origin of Life." Vienna, Fortschritte der Chemie Organischer Naturstoffe 20, 1962, pp. 423-455.

**99**
Wald, G.
 "The Origin of Life." (In Moment, G. B., ed. Frontiers of Modern Biology. Boston, Houghton Mifflin, 1962, pp. 185-192).

**100**
Wood, J. A.
 "Chrondrules and the Origin of the Terrestrial Planets." Nature 194, April 14, 1962, pp. 127-130.

**101**
Cameron, A. G. W.
 "Studies on the Origin of the Solar System." Yale Scientific Magazine 38, April 1963, pp. 12-14.

**102**
Imshenetskii, A. A.
 Life and Space. (In Florkin, M., & A. Dollfus, eds. Life Sciences and Space Research II: International Space Science Symposium. 4th, Warsaw, June 3-12, 1963. New York, Interscience, 1964, pp. 25-34).

**103**
Pryor, H.
 "This View of Life: the World of an Evolutionist, by G. G. Simpson: Review." Science Digest 55, May 1964, p. 42.

**104**
Oparin, A. I.
 "The Origin of Life in Space." Space Science Reviews 3, July 1964, pp. 5-26.

**105**
Ponnamperuma, Cyril
 "Primordial Organic Chemistry and the Origin of Life" (In Southwest Research Institute. Bioastronautics and the Exploration of Space. (AD-627686). Alexandria, Va., NTIS, 1965, pp. 117-128). ($9.40)

**106**
Sagan, Carl
 "The Quest for Life Beyond the Earth." (In Smithsonian Institution. Annual Report of the Board of Regents for the Year Ending June 30, 1964. Washington, U.S. Govt. Printing Office, 1965, pp. 297-306).

**107**
Smith, S. L.
 "The Problems of Exobiology. Part I: the Origin of Life." London, BUFORA Journal and Bulletin 1, Winter 1965, pp. 13-16.

**108**
"Ames Exobiologists Provide Major Link in Understanding Origin of Life on Earth." Missiles and Rockets 16, Aug. 16, 1965, pp. 38-39.

**109**
Frisch, B. H.
 "Obituary of a Planet." Science Digest 58, Oct. 1965, pp. 8-9.

**110**
Baur, Franz
 "Kosmische Aspekte der Organischen Evolution; eine Ergänzung zum Beitrag von Prof. G. Simpson." Stuttgart, Naturwissenschaftliche Rundschau 22, April 1969, pp. 167-168.

**111**
Keller, Eugenia
 "The Origin of Life." Chemistry 42, April 1969, pp. 8-13.

NOTES:

## ASTRONOMY AND ORIGIN OF LIFE AND OF PLANETS

**112** Ambartsumyan, V.
"Science Expands its Horizons: the Mysterious World of the Galaxies." Tr. into English from Pravda (USSR), Feb. 1, 1970, p. 3. (AD-706651; FTD-HT-23-121-70). Dayton, U.S. Air Force Systems Command, Wright-Patterson Air Force Base, Foreign Technology Div., Feb. 27, 1970, 10p. Available on microfilm from NTIS.

**113** Ponnamperuma, Cyril
"Chemical Evolution and the Origin of Life." New York State Journal of Medicine 70, May 15, 1970, pp. 1169-1175.

**114** Otrotchenko, V. A., & L. M. Mukhin
"Extraterrestrial Life Study. Problem of its Origin and Evolution." (In Chemical Evolution and the Origin of Life. Proceedings of the Third International Conference, Pont-à-Mousson, France, April 19-25, 1970. v. 1. Amsterdam, North-Holland Pub., 1971, pp. 516-522).

**115** Sagan, Carl
"The Solar System Beyond Mars, an Exobiological Survey." Space Science Reviews 11, March 1971, pp. 827-866.

**116** "Astronomy" (In Soviet Bloc Research in Geophysics, Astronomy and Space #251, April 28, 1971, pp. 1-5. Washington, Joint Publications Research Service. Available from NTIS).

**117** Buhl, D., & Cyril Ponnamperuma
"Interstellar Molecules and the Origin of Life." Dordrecht, Holland, Space Life Sciences 3, Dec. 1971, pp. 157-164.

**118** "Origin of Life." London, Spaceflight 13, Dec. 1971, pp. 458-459.

**119** Kumar, S. K.
"Planetary Systems." (In The Emerging Universe: Essays on Contemporary Astronomy. Charolottesville, University of Virginia Press, 1972, pp. 25-34).

**120** Young, R. S.
"Space Exploration and the Origin of Life." (In Molecular Evolution, Prebiological and Biological. New York, Plenum Press, 1972, pp. 43-65).

**121** Ehricke, Krafft A.
"Astrogenic Environments: the Effect of Stellar Spectral Classes on the Evolutionary Pace of Life." London, Spaceflight 14, Jan. 1972, pp. 2-14.

**122** Wilhelm, John
"Cosmic Connection, by Carl Sagan (Review)." Time 103, Jan. 21, 1974, p. 74.

NOTES:

# BIBLIOGRAPHIES

123
National Investigations Committee on Aerial Phenomena
  UFO Newspaper Clipping File, 1947-1966. Washington, 1947-,(looseleaf).

124
Briggs, Michael H., et al, comps.
  A Bibliography of Exobiology: (NASA-CR-52484). Pasadena, Calif., Jet Propulsion Lab., California Institute of Technology, 1962, 109p. (Available from NASA: $9.10).

125
Ordway, Frederick I.
  Annotated Bibliography of Space Science and Technology, With an Astronomical Supplement. A History of Astronautical Book Literature, 1931 Through 1961. 3d ed. Washington, ARFOR Publications, 1962, 77p.

126
U.S. Air Force Systems Command, Kirtland AFB, New Mexico
  Bibliography of Extraterrestrial Research, by Robert W. Henry. AFWL-RTL-TDR-63-3025; N63-18778. Kirtland Air Force Base, 1963, 104p.

127
Henry, Robert W.
  Bibliography of Extraterrestrial Research. (AFWL-RTD-TDR-63-3025). Kirtland Air Force Base, N.M., Air Force Systems Command, June 1963, 104p.

128
Spiegler, Paul E.
  Bibliography of Bioregenerating Systems for Extraterrestrial Habitation. Technical Documentation Report (15 Aug. 1962-15 Sept. 1963). (AMRL-TDR-63-121; AD-430814). Dayton, Wright-Patterson Air Force Base, Biomedical Lab, Nov. 1963, 99p. (Available from NTIS: $2.50).

129
George Washington University, Washington, D.C.
  Exobiology, Annotated Bibliography, 1951-1964, Compiled by Joe W. Tyson & Ruby W. Moats. NASA-CR-53806; N64-23393. Washington, 1964, 77p.

130
Magnolia, L. R., et al
  Exobiology: a Bibliography. Redondo Beach, Calif., TRW Space Technology Laboratories, 1964, 166p. (STL Technical Library. Advanced Research Group. Research bibliography no. 52)

131
U.S. National Aeronautics & Space Administration. Scientific & Technical Information Facility.
  Extraterrestrial Life, a Bibliography. Part 1: Report Literature: a Select Listing of Annotated References to Unclassified Scientific and Technical Reports, 1952-1964. Bethesda, 1964, Unpaged.

NOTES:

# BIBLIOGRAPHIES

**132**
U.S. National Aeronautics & Space Administration
   Extraterrestrial Life, Bibliography, Part 1: Report Literature, Selected Listing of Annotated References to Unclassified Scientific and Technical Reports, 1952-64; NASA SP-7015. For Sale by U.S. Government Printing Office, Washington, 1965, 76p. ($.45)

**133**
U.S. National Aeronautics & Space Administration
   Extraterrestrial Life-a Bibliography. Pt. II. Published Literature, 1900-1964. (NASA-SP-7015). Washington, U.S. Gov't. Printing Office, 1965, 345p. ($2.00)

**134**
Shneour, Elie A., & Eric A. Ottesen, comp
   Extraterrestrial Life: an Anthology and Bibliography. (Publication 1296A). Washington, National Research Council, 1966, 478p.

**135**
Jain, Sushil K., & Christine Horswell, comps.
   20 Years of Flying Saucers: a Select List of Interesting Books and Periodical Articles Published During 1947-1966; Part 2, Periodical Articles. Tenderden (Kent.), Sushil Jain Pubs., 1967, 20p.

**136**
Sable, Martin H.
   UFO Guide: 1947-1967; containing international lists of books and magazine articles on UFO's, flying saucers, and about life on other planets; world-wide directories of flying saucer organizations, professional groups and research centers concerned with space research and astronautics, a partial list of sightings, and an international directory of flying saucer magazines. 1st ed. Beverly Hills, Calif., Rainbow Press Co., 1967, 100p. ($2.95)

**137**
U.S. Air Force Cambridge Research Laboratories. Space Physics Lab
   Bibliography of Lunar and Planetary Research. Supplement #2-1966, ed. by J.W. Salisbury, et al. (AFCRL-67-0518; AD-662876). Bedford, Mass., Sept. 1967, 235p.

**138**
Page, Henrietta M.
   Flying Saucers: a Bibliography. (n.p.), 1968, 17p.

**139**
Salisbury, John W., ed.
   Bibliography of Lunar and Planetary Research; Supplement #3, 1967. (AD-679995; AFCRL-68-0533). Bedford, Mass., Air Force Cambridge Research Laboratories, Space Physics Lab., Oct. 1968, 311p. (Available from NTIS)

**140**
U.S. Defense Documentation Center
   The Planet Mars, v. 1. Report Bibliography, Jan. 1953-Dec. 1968. (AD-685800; DDC-TAS-69-38-1-Vol-1). Alexandria, Va., May 1969, 116p. (Available from NTIS).

**141**
Catoe, Lynne
   UFO's and Related Subjects: an Annotated Bibliography. Washington, Science & Technology Div., Library of Congress, July 1969, 415p. (AD-688332; AFOSR-68-1656). Available from NTIS: $3.50.

**142**
U.S. Air Force Cambridge Research Laboratories. Space Physics Laboratory
   Bibliography of Lunar and Planetary Research, Supplement #4, 1968, ed. by J.W. Salisbury, et al. (AD-700051; AFCRL-Sr-92; AFCRL-69-0486; Report-92). Bedford, Mass., Oct. 1969, 346p. Available from NTIS.

NOTES:

## BIBLIOGRAPHIES

**143**
Beard, Robert B.
  Flying Saucers. UFO's and Extraterrestrial Life: a Bibliography of British Books. 1950-1970. Swindon, R. Beard, 1971, 5p.

**144**
Brennan, Norman
  Flying Saucer Books and Pamphlets in English; a Bibliographical Checklist. Buffalo? 1971? 94p.

**145**
Unesco
  Annotated Bibliography of Unesco Publications and Documents Dealing with Space Communication, 1953-1970. Paris, 1971, 44p.

**146**
Netto, Paulo Coelho
  Bibliografia Brasileira Sôbre os Discos Voadores. Rio de Janeiro, Libraria São José, 1972, 25p.

**147**
Wegner, Willy
  Dansk UFO-Litteratur 1946-1970. En Bibliografi. Copenhagen, University of Copenhagen, Institut for Folkemindevidenskab, 1972, 253p.

NOTES:

# FLYING SAUCER SIGHTINGS

## BOOKS

**148**
Last, Cecil E.
　Man in the Universe.  London, Werner Laurie, 1954, 166p.

**149**
Anchor
　Transvaal Episode; a UFO Lands in Africa. Corpus Christi, Essene Press, 1958, 48p.

**150**
Zinsstag, Lou, & Theodor Allemann
　UFO-Sichtungen über der Schweiz, 1949-1958.  Zürich, UFO-Verlag, 1958, 48p.
(3.20 Swiss francs)

**151**
Unidentified Flying Object Research Committee
　The Fitzgerald Report:  a Complete and Detailed Account of the Sighting of an Unidentified Flying Object. Sheffield Lake, Ohio, Sept. 21, 1958.  Akron? 1959, 19p.

**152**
Unidentified Flying Objects Research Committee
　Report on Unidentified Objects Observed Feb. 24, 1959, by American and United Airlines Pilots.  Akron, 1960, 22p.

**153**
Gonzáles Ganteaume, Horacio
　Platillos Voladores Sobre Venezuela. Caracas, 1961, 250p.

**154**
Quarnström, Gunnar
　Dikten och den nya Vetenskapen.  Lund, Sweden, C. W. K. Gleerup, 1961, 300p.

**155**
Ribera, Antonio
　Objetos Desconocidos en el Cielo.  Barcelona, Librería Editorial Argos, 1961, 289p.

**156**
Perego, Alberto
　L'aviazione di Altri Pianeti Opera tra noi; Rapporto agli Italiane (1943-1963).  Rome, Edizioni del Centro Italiano Studi Aviazione Elettromagnetica, 1963, 563p.

**157**
Holledge, James
　Flying Saucers Over Australia.  Sidney, Horwitz, Pubs., Inc., 1965, 130p.

**158**
Perego, Alberto
　The Monguzzi Case.  Basle, 1965, 2p.

NOTES:

## FLYING SAUCER SIGHTINGS

**159**
Babcock, Edward J., and Timothy G. Beckley, eds.
  UFO's Around the World. (n.p.) Interplanetary News Service, 1966, 64p.

**160**
Fry, Daniel W.
  The White Sands Incident. Louisville, Ky., Best Books, Inc., 1966, 120p.

**161**
Keel, John A.
  Project 'B' 1966. New York, 1967, 5p.

**162**
Bloecher, Ted
  Report on the UFO Wave of 1947. Introd. by Dr. James E. McDonald. Washington, 1967, variously paged.

**163**
Flying Saucers Pictorial. Tucson, Arizill Realty and Pub. Co., 1967.

**164**
Fontes, Olavo T.
  Trinidade Observationerne-der Fremtvang Officiel Brasiliansk Anerkendelse af UFO-ernes Eksistens. Randers, Denmark, UFO-NYTS Forlag, 1967, 30p.

**165**
Sherwood, John C.
  Flying Saucers are Watching you. Clarksburg, W. Va., Saucerian Books, 1967, 78p.

**166**
Lorenzen, Jim, & Coral Lorenzen
  UFO's Over the Americas. New York, New American Library, 1968, 254p.

**167**
Akademski Astronomsko-Astronautitski Klub, Sarajevo, Yugoslavia
  (Identifikacija Leteceg Objekta od 18 Oktobra 1968 Godine). In Serbo-Croatian. Sarajevo, 1969, 42p. (Available from NTIS).

**168**
Ribera Jorda, Antonio, & Rafael Farriols
  Un Caso Perfecto; Platillos Volantes Sobre España. Barcelona, Editorial Pomaire, 1969, 196p.

**169**
Armand, John, & Bjorn Holm-Hansen
  UFO'er over Norge. Oslo, Gyldendal, 1971, 100p.

**170**
Uriondo, Oscar A.
  Los Aterrizajes de OVNI en la Argentina. Catálogo Sistemático y Descriptivo de Avistamientos Tipo-I (Años 1943-1971). Buenos Aires, Centro de Estudios de Fenómenos Aéreos Inusuales, 1972, 55p.

**171**
Banchs, Roberto E.
  Fenómenos Aéreos Inusuales; Cuadro General de Observaciones OVNI en la Argentina. Buenos Aires, Centro de Estudios de Fenómenos Aéreos Inusuales, 1973, 61p.

**172**
Hobana, Ion, & Louis J. Weverbergh
  UFO's from Behind the Iron Curtain. Tr. from Dutch. London, Souvenir Press, 1974, 309p.

**173**
McWane, Glenn, & David Graham
  The New UFO Sightings. New York, Warner Paperback Library, 1974, 173p.

**174**
Salisbury, Frank B.
  The Utah UFO Display: a Biologist's Report. New York, Devin-Adair, 1974.

NOTES:

## PERIODICAL ARTICLES

**175**
Luckiesh, Matthew
  "Seeing is Deceiving." Science Illustrated 1, June 1946, pp. 86-87.

**176**
"Signs, Portents, and Flying Saucers." Newsweek 30, July 1947, pp. 19-20.

**177**
"Speaking of Pictures: a Rash of Flying Disks Breaks out Over the U.S." Life, July 1947, pp. 14-16.

**178**
"U.S. Army Air Force Drops Inquiry into Reports of Strange Objects Flying at 1200 Miles per Hour Over Western U.S.; Denies Saucers are new Aerial Missiles." New York Times, July 4, 1947, p. 26, col. 2.

**179**
"Flying Saucer Spots Before Their Eyes." Newsweek 30, July 14, 1947, p. 19.

**180**
"Flying Disks Break Out Over the U.S." Life 23, July 21, 1947, pp. 14-16.

**181**
"Illusions of Nature." Science Illustrated 2, Oct. 1947, pp. 42-44.

**182**
"One of 3 Kentucky National Guard Planes Crashes in Investigating new Reports of Flying Saucers in Franklin, Ky. Area..." New York Times, Jan. 9, 1948, p. 11, col. 1.

**183**
Arnold, Kenneth A.
  "I did see the Flying Disks." Fate, Spring 1948, pp. 4-10.

**184**
Arnold, Kenneth A.
  "Mystery of the Flying Disks." Fate, Spring 1948, pp. 19-48.

**185**
Ross, John C.
  "What Were the Doughnuts?" Fate, Spring 1948, pp. 12-14.

**186**
"Flying Saucers Definitely Proved." Fate 1, Winter 1949, pp. 47-51.

**187**
Hansen, L. T.
  "The Mystery Ship." Fate 1, Winter 1949, pp. 30-33.

**188**
Moorehouse, Frederick G.
  "The Case of the Flying Saucers." Argosy 329, July 1949, pp. 22-24, 92.

**189**
Janssen, John H.
  "My Encounter With the Flying Discs." Fate 2, Sept. 1949, pp. 12-16.

**190**
"Tribal Memories of the Flying Saucers." Fate 2, Sept. 1949, pp. 17-21.

**191**
McLaughlin, Robert B.
  "How Scientists Tracked a Flying Saucer." True, March 1950.

NOTES:

# FLYING SAUCER SIGHTINGS

192
Waithman, Robert
  "These Flying Saucers." London, The Spectator 184, April 1950, pp. 489-490.

193
"Pies in the sky in Mexico City." Time 55, April 3, 1950, p. 36.

194
"Farmer Trent's Flying Saucer." Life 28, June 26, 1950, p. 40.

195
"The Camera Sees Flying Saucers." True 27, July 1950, pp. 44-45, 82.

196
Taylor, H. J.
  "Flying Saucer is Good News." Reader's Digest 57, July 1950, pp. 14-16.

197
Keyhoe, Donald E.
  "Flight 117 and the Flying Saucer." True 27, Aug. 1950, pp. 24-25, 75.

198
Palmer, Ray
  "New Report on the Flying Saucers." Fate 4, Jan. 1951, pp. 63-81.

199
Adamski, George
  "I Photographed Space Ships." Fate 4, July 1951, pp. 64-74.

200
"What Were the Flying Saucers?" Popular Science 159, Aug. 1951, pp. 74-75, 228.

201
Dixon, William A.
  "Saucers or Illusions?" Air Facts 14, Sept. 1951, pp. 37-43.

202
Webster, Robert N.
  "Let's get up to Date on the Flying Saucer." Fate 5, Jan. 1952, pp. 4-8.

203
Kaempffert, Waldemar
  "Expert Sees Flying Object--Saucer or Balloon." Science Digest 31, Feb. 1952, p. 74.

204
"Korean Saucers." Newsweek 39, March 3, 1952, p. 44.

205
"More Saucers Over North Korea." Time 59, March 3, 1952, p. 92.

206
"Blips on the Scopes." Time 60, Aug. 4, 1952, p. 40.

207
"Washington's Blips." Life 33, Aug. 4, 1952, pp. 39-40.

208
"What's Going on in the Skies." U.S. News & World Report 33, Aug. 8, 1952, pp. 13-15.

209
"Saucer Season." Newsweek 40, Aug. 11, 1952, p. 56.

210
"Something in the air." Time 60, Aug. 11, 1952, p. 58.

211
"New Saucer Epidemic." New Republic 127, Aug. 18, 1952, p. 7.

NOTES:

**212**
Swezey, K.
"How to see Flying Saucers." *Popular Science* 161, Sept. 1952, pp. 167-170.

**213**
"The Wind is up in Kansas." *Time* 60, Sept. 8, 1952, p. 86.

**214**
Nash, William B., & William H. Fortenberry
"We Flew Above Flying Saucers." *True*, October 1952.

**215**
Sheridan, H.
"Logging Time." *Flying* 51, Nov. 1952, p. 51.

**216**
Keyhoe, Donald E.
"What Radar Tells About Flying Saucers." *True*, December 1952.

**217**
"Temperature Inversions Cause 'Flying Saucers'." *Science News Letter* 62, Dec. 1952, p. 388.

**218**
Mauer, E. F.
"Of Spots Before the Eyes." *Science* 116, Dec. 19, 1952, p. 693.

**219**
Steiner, R.
"How to Expose Flying Saucers: Air Force's Project for Photographing Them." *Popular Science* 162, Jan. 1953, pp. 226-229.

**220**
Perret, Jacques
"Barbu, Marc de Cafe, Hallebardes." Paris, *Mercure de France*, March 1953, pp. 408-427.

**221**
Liddel, Urner
"Phantasmagoria or Unusual Observations in the Atmosphere." *Journal of the Optical Society of America* 43, April 1953, pp. 314-317.

**222**
Menzel, Donald H.
"Saucers on Radar?" *Popular Science* 162, April 1953, pp. 168-171.

**223**
"Air Training (U.S. Air Force Publication) Reports Over 1000 Reports of Saucers Received in 1952 at Wright-Patterson Field, Ohio; 20% of Unknown Origin." *New York Times*, April 19, 1953, sec. IV, p. 9, col. 5.

**224**
Gibbons, Russell W.
"Ohio Northern Investigates the Saucers." *Fate* 6, Oct. 1953, pp. 18-20.

**225**
Hickman, Warren, & Eric Turner
"Report From Ohio Northern." *Fate* 6, Oct. 1953, pp. 21-25.

**226**
Ross, John C.
"Fate's Report on the Flying Saucers." *Fate* 6, Oct. 1953, pp. 6-13.

**227**
"Flying Saucer Controversy; Meteorological Balloons, and Weather Conditions Which may Provide Explanations of the Phenomenon." London, *Illustrated London News* 223, Dec. 5, 1953, pp. 936-937.

**228**
"Saucer Sightings Fall Sharply." *Air Force Times* 14, Dec. 5, 1953, p. 2.

NOTES:

FLYING SAUCER SIGHTINGS

229
"Plenty Going on in the Skies." U.S. News & World Report 36, Jan. 1, 1954, pp. 27-28.

230
Griffin, E.
"Canada's Flying Saucer Lookout Lab." Science Digest 35, Feb. 1954, p. 36.

231
"On the Flying Saucer Trail." American Magazine 157, April 1954, p. 56.

232
Fuller, Curtis
"Fate's Report on Flying Saucers." Fate 7, May 1954, pp. 16-22.

233
"Reports From Everywhere." Fate 7, May 1954, pp. 23-31.

234
Ross, John C.
"The Lights That Failed." Fate 7, May 1954, pp. 37-39.

235
Towner, Larry E.
"The Night of August 9th." Fate 7, May 1954, pp. 32-36.

236
Wilson, Harlan
"There are Meteors After all." Fate 7, May 1954, pp. 40-43.

237
Knight, C.
"Report on our Flying Saucer Balloons." Collier's 133, June 11, 1954, p. 50.

238
"U.S. Air Force Reports Sightings off Sharply Since 1952; 87 Reports Thus far in 1954." New York Times, June 20, 1954, p. 28, col. 3.

239
Bessor, John P.
"Some Strange Meteors." Fate 7, July 1954, pp. 78-79.

240
"Observations Françaises Plus Anciennes." Paris, Phénomènes Spatiaux, Nov. 1954, pp. 26-36.

241
"Now They're in Italy: Astral Intruders." Life 37, Nov. 29, 1954, pp. 133-134.

242
Dowling, J.
"Flying Saucer Over Durban." Bloemfontein, South Africa, Outspan (55:1449), Dec. 3, 1954, pp. 14-17.

243
Hamilton, Jared
"Saucers Over Italy." Fate 8, March 1955, pp. 14-19.

244
Serpas, Paul F.
"The Saucer That got Away." Fate 8, March 1955, pp. 34-37.

245
King, R. S.
"Flying Saucers Over South Africa." Johannesburg, South Africa, Spotlight, April 15, 1955, p. 24.

NOTES:

**246**
Jones, S. H.
  "The Swedish Ghost Rockets." Flying Saucer News, Summer 1955, pp. 10-12.

**247**
"U.S. Earth Satellite Program Held not Responsible for Saucer Rumors." New York Times, July 30, 1955, p. 30, col. 7.

**248**
Woodley, Morris
  "The Gyrating UFO of Cobalt." Fate 8, Aug. 1955, pp. 34-35.

**249**
"Dr. J. Huxley Holds he Cannot Dismiss Reports." New York Times, Aug. 2, 1955, p. 21, col. 1.

**250**
Constance, Arthur
  "What Blazed Over Britain?" Flying Saucer News, Autumn 1955, pp. 8-15.

**251**
Shanklin, H. A.
  "Flying Saucers I've Seen." Flying 57, Sept. 1955, p. 38.

**252**
Volkman, Frank
  "Mystery Lights Over the Orient." Fate 8, Dec. 1955, pp. 13-16.

**253**
"Strange Shapes Seen in the sky." Life 39, Dec. 5, 1955, pp. 177-178.

**254**
Edwards, Frank
  "To see or not to see." In his My First 10,000,000 Sponsors. New York, Ballantine Books, 1956, pp. 110-125.

**255**
Norman, Samuel
  "Recent UFO's Over Japan." Fate 9, June 1956, pp. 22-24.

**256**
Michel, Aimé
  "The Mystery at Marignane." Fate 9, Dec. 1956, pp. 22-30.

**257**
Walling, Theodore L.
  "Saucer Sighting." Fate 9, Dec. 1956, pp. 35-37.

**258**
Caldbeck, David
  "Flying Saucers Over South Africa." Moorabin, Australia, Australian Saucer Record 3, First Quarter 1957, pp. 3-8.

**259**
"Retired Rear Admiral Fahrney Says Reports Indicate High-Speed, Directed Objects are Entering Earth's Atmosphere." New York Times, Jan. 17, 1957, p. 31, col. 8.

**260**
Towner, Cliff R.
  "Silver Chaff From the sky." Fate 10, March 1957, pp. 94-98.

**261**
Goble, H. C.
  "Did Jones Chart an Unknown World?" Fate 10, April 1957, pp. 68-70.

**262**
Mann, Mary, & Amey Hoag
  "We saw a Flying Saucer." Fate 10, April 1957, pp. 54-56.

**263**
Fuller, Curtis
  "Report From New Zealand." Fate 10, June 1957, pp. 15-16.

NOTES:

# FLYING SAUCER SIGHTINGS

264
"Saucers Over Paris." *Flying Saucers From Other Worlds*, June 1957, pp. 76-77.

265
"Sightings by Scientists." *Flying Saucers From Other Worlds*, June 1957, 82-85.

266
Sondy, Dominic
"Space Ship Over Detroit." *Fate* 10, June 1957, pp. 86-89.

267
"I saw a Flying Saucer." *Flying Saucers From Other Worlds*, June 1957, pp. 44-50; Aug. 1957, pp. 74-80.

268
Michel, Aimé
"Flying Saucers in Europe: the Crisis of Autumn, 1954." *Fate* 10, Aug. 1957, pp. 28-35.

269
Roberts, August C.
"The Skywatch Tower Case." *Flying Saucers From Other Worlds*, Aug. 1957, pp. 8-15.

270
Michel, Aimé
"Flying Saucers in Europe: Italian Flying Saucer." *Fate* 10, Oct. 1957, pp. 19-24.

271
Edwards, Frank
"Frank Edwards' Report: Flying Saucers Stopped the Ball Game!" *Fate* 10, Nov. 1957, pp. 35-41.

272
"U.S. Air Force Intelligence Center Reports That Investigation of 5,700 Sightings From 1947-57 Reveals no Physical Evidence of Anything not Identifiable." *New York Times*, Nov. 7, 1957, p. 24, col. 5.

273
"If You're Seeing Things in the sky." *U.S. News & World Report* 43, Nov. 15, 1957, p. 122.

274
"U.S. Air Force Describes Saucer Reports as Hoaxes." *New York Times*, Nov. 16, 1957, p. 3, col. 4.

275
"Seeing Things." *Newsweek* 50, Nov. 18, 1957, p. 41.

276
Edwards, Frank
"Frank Edwards' Report: How to Fake a Flying Saucer." *Fate* 10, Dec. 1957, pp. 46-54.

277
Michel, Aimé
"Flying Saucers Over Europe, French Flying Saucer." *Fate* 10, Dec. 1957, pp. 33-38.

278
Michel, Aimé
"Flying Saucers in Europe: Saucers--or Delusions?" *Fate* 11, Jan. 1958, pp. 73-79.

279
Berry, Bruce D.
"Flying Saucer Over the Golden Gate." *Fate* 11, Feb. 1958, pp. 52-54.

280
"Flying Saucer Roundup: Saucer Report no. 1." *Fate* 11, Feb. 1958, pp. 29-37.

281
Miller, Max B.
"UFO's Invade Australian Skies." *Fate* 11, Feb. 1958, pp. 48-51.

282
Ross, John C.
"The big, fat UFO: Saucer Report no. 2." *Fate* 11, Feb. 1958, pp. 38-43.

NOTES:

283 "Slim Chance for Saucer Sightings." *Science Digest* 43, Feb. 1958, p. 60.

284 Wilson, Harlan
"The Saucer That Made Tracks." *Fate* 11, Feb. 1958, pp. 44-47.

285 "Brief Saucer Review." *Fate* 11, March 1958, pp. 14-17.

286 Edwards, Frank
"Frank Edwards' Report: UFO Sightings and Alibis." *Fate* 11, March 1958, pp. 27-34.

287 Sklarewitz, Norman
"Tokyo Space Ship." *Fate* 11, March 1958, pp. 46-48.

288 Ley, Willy
"The Mighty Invaders From Outer Space." *Catholic Digest* 22, April 1958, pp. 25-29.

289 Mollohan, Hank
"The Holly River Sighting, as told to Gray Barker." *Flying Saucers From Other Worlds*, May 1958, pp. 36-41.

290 Roberts, August C., & Dominick Lucchesi
"Saucers in the wee Hours: the 'Long John' Party Line. *Flying Saucers From Other Worlds*, May 1958, pp. 14-19.

291 "Saucers in the News." *Flying Saucers From Other Worlds*, May 1958, pp. 76-77.

292 Everett, Eldon K.
"Saucers Over Puget Sound." *Flying Saucers*, July-Aug. 1958, pp. 52-60.

293 "Saucers in the News." *Flying Saucers*, July-Aug. 1958, pp. 61-63.

294 Lang, Daniel
"Something in the sky." (In *From Hiroshima to the Moon*). New York, Simon and Schuster, 1959, pp. 320-346.

295 Comella, Tom
"UFO's: Problems in Perception." *Fate* 12, Jan. 1959, pp. 92-96.

296 Barker, Gray
"Chasing the Flying Saucers." *Flying Saucers From Other Worlds*, June 1957, pp. 28-38; Aug. 1957, pp. 29-40; Dec. 1958, pp. 43-56; Feb. 1959, pp. 29-34.

297 Edwards, Frank
"UFO Buzzes Train." *Fate* 12, Feb. 1959, pp. 25-30.

298 "The Night of September 29." *Fate* 12, Feb. 1959, pp. 31-38.

299 Allen, W. Gordon
"Spacecraft Over Mexico." London, *Flying Saucer Review* 5, March-April 1959, pp. 16-19.

---

NOTES:

# FLYING SAUCER SIGHTINGS

**300**
Saunders, Alex
   "Flying Saucer Scrap Book." <u>Search</u>, April 1959, pp. 22-39, 53.

**301**
Barker, Gray
   "Chasing the Flying Saucers." <u>Flying Saucers From Other Worlds</u>, May 1958, pp. 20-35, 80; July-Aug. 1958, p. 2035; May 1959, pp. 19-43.

**302**
Ribera, Antonio
   "UFO Waves Follow a Certain Pattern." London, <u>Flying Saucer Review</u> 5, May-June 1959, pp. 12-14.

**303**
Fogl, T.
   "Saucer Photographed at sea." <u>Flying Saucers</u>, July 1959, pp. 6-9.

**304**
Fuller, Curtis
   "Saucers Trail Airliner." <u>Fate</u> 12, Aug. 1959, pp. 25-31.

**305**
Fontes, Olavo T.
   "Project Argus and the 'Anonymous' Satellite." <u>Flying Saucers From Other Worlds</u>, Oct. 1959, pp. 8-12.

**306**
Morgan, Dean
   "The Red Bud, Illinois Photo." <u>Flying Saucers</u>, Oct. 1959, pp. 5-7.

**307**
Maney, Charles A.
   "An Evaluation of Aimé Michel's Study of the Straight Line Mystery." London, <u>Flying Saucer Review</u> 5, Nov.-Dec. 1959, pp. 10-14.

**308**
Barker, Gray
   "Chasing the Flying Saucers." <u>Flying Saucers From Other Worlds</u>, Dec. 1959, pp. 22-29.

**309**
Dagenis, Arleigh J.
   "Do you Believe in Flying Saucers?" <u>Michigan Technic</u>, Jan. 1960, pp. 16-17, 50-51.

**310**
Miller, Max B.
   "Recent Dramatic Saucer Reports." <u>Fate</u> 13, Jan. 1960, pp. 29-32.

**311**
Barker, Gray
   "Chasing the Flying Saucers." <u>Flying Saucers</u>, July 1959, pp. 24-37; Feb. 1960, pp. 15-24.

**312**
Kowalezewski, Stanislaw
   "U.F.O. Photographed Over Poland." <u>Flying Saucers</u>, Feb. 1960, pp. 27-28.

**313**
Posin, Dan Q.
   "An eye on Space." <u>Popular Mechanics</u> 113, Feb. 1960, p. 103.

**314**
"Alaska Air Command Probes Reports of Silvery Objects Trailing Flames, in Western Alaska; North American Air Defense Command Confirms Reports of 2 Objects Sighted Over Alaska." <u>New York Times</u>, Feb. 17, 1960, p. 14, col. 4.

**315**
"National Investigations Commission on Aerial Phenomena Reveals U. S. Air Force Issued Directive in Dec. 1959, Warning Commands to Treat Sightings Seriously; U. S. Air Force Confirms Report; Commission Scores U. S. Air Force Reports Belittling Sightings..." <u>New York Times</u>, Feb. 28, 1960, col. 4.

NOTES:

# FLYING SAUCER SIGHTINGS

**316**
Ross, John C.
  "UFO's Over New Guinea." <u>Fate</u> 13, March 1960, pp. 44-52.

**317**
"Aerial Phenomena Investigations Committee Urges Secretary Sharp Clarify Report on Flame-Spouting Object that Reportedly Escaped Jet Pursuit over Oregon, Sept. 1959; Holds FAA Report Proves U. S. Air Force Conceals Facts...U. S. Air Force Holds Sighting Data Insufficient." <u>New York Times</u>, March 20, 1960, p. 20, col. 7.

**318**
Fontes, Olavo T.
  "Brazil Under UFO Survey." London, <u>Flying Saucer Review</u> 7, March-April 1960, pp. 10-14.

**319**
"The Mystery Satellite: a Study in Confusion." London, <u>Flying Saucer Review</u> 6, May-June 1960, pp. 25-26.

**320**
"Saucers and the Iron Curtain: a Report From Czechoslovakia." London, <u>Flying Saucer Review</u> 6, July-Aug. 1960, pp. 31-32.

**321**
"Adler Planetarium Director Johnson and Aide Recall Tracking Reddish Object Moving over Chicago Aug. 26; Reports National Investigations Commission on Aerial Phenomena Informed him Georgetown University Astronomers saw it 2 Days Earlier." <u>New York Times</u>, Aug. 31, 1960, p. 4, col. 4.

**322**
Edwards, Frank
  "Flying Saucers Over Brazil." <u>Fate</u> 13, Sept. 1960, pp. 42-49.

**323**
Wilson, Harlan
  "Photographers Analyze UFO Picture." <u>Fate</u> 13, Oct. 1960, pp. 70-73.

**324**
Cruttwell, Norman E. G.
  "What Happened in Papua in 1959?" London, <u>Flying Saucer Review</u> 6, Nov.-Dec. 1960, pp. 3-7.

**325**
Ross, John C.
  "State Cops Race 'Flying Saucer'." <u>Fate</u> 13, Dec. 1960, pp. 44-47.

**326**
"Two USSR Newspapers Score Russians Believing Reports of Flying Saucers." <u>New York Times</u>, Jan. 9, 1961, p. 7, col. 7.

**327**
Barker, Gray
  "Chasing the Flying Saucers." <u>Flying Saucers</u>, Feb. 1961, pp. 22-24, 58.

**328**
Fontes, Olavo T.
  "The Brazilian Navy Sighting at the Island of Trinidade." <u>Flying Saucers</u>, Feb. 1961, pp. 27-54.

**329**
Fuller, Curtis
  "The November 23 UFO." <u>Fate</u> 14, March 1961, pp. 46-51.

**330**
"Russians say That Flying Saucers Exist." <u>National Enquirer</u> 42, March 1961, pp. 1-3.

**331**
Buelta, Eduardo
  "Investigaciones Estadísticas." Barcelona, <u>Boletín del Centro de Estudios Interplanetarios</u> (3:9), Oct. 1961.

**332**
Gregg, Doris D.
  "I am Being Watched by a UFO." <u>Fate</u> 14, Nov. 1961, pp. 27-33.

NOTES:

## FLYING SAUCER SIGHTINGS

**333**
Ribera, Antonio
 "Spanish Orthotenies in 1950." London, Flying Saucer Review 7, Nov.-Dec. 1961, pp. 9-11.

**334**
Fuller, Curtis
 "The Boys who 'Caught' a Flying Saucer." Fate 15, Jan. 1962, pp. 36-42.

**335**
Edwards, Frank
 "My First UFO." Fate 15, Feb. 1962, pp. 27-31.

**336**
Fawcett, George D.
 "A Camera's eye Analysis of 411 Flying Saucers From 281 Photographs Taken Around the World." Fate 15, Feb. 1962, pp. 67-87.

**337**
Vallée, Jacques
 "Towards a Generalisation of Orthoteny and its Application to the North African Sightings." London, Flying Saucer Review 8, March-April 1962, pp. 3-7.

**338**
Cappa, Fidel A.
 "Saucers Over the Argentine." London, Flying Saucer Review 8, May-June 1962, pp. 29-30.

**339**
Maney, Charles A.
 "The Campinas Sighting." London, Flying Saucer Review 8, May-June 1962, pp. 3-6.

**340**
Creighton, Gordon
 "Saucers and South Africa." London, Flying Saucer Review 8, July-Aug. 1962, pp. 18-21.

**341**
Ribera, Antonio
 "Two More Facts for the UFO File." London, Flying Saucer Review 8, July-Aug. 1962, pp. 14-15.

**342**
"X-15 Pilot Shows his Film." London, Flying Saucer Review 8, July-Aug. 1962, pp. 3-4, 13.

**343**
Edwards, Frank
 "Mystery Blast Over Nevada." Fate 15, Aug. 1962, pp. 68-74.

**344**
Ross, John C.
 "UFO's and the Record Flight of the X-15." Fate 15, Aug. 1962, pp. 38-44.

**345**
Lorenzen, Coral
 "Rocket-Shooting Saucers Over Tucson." Fate 15, Oct. 1962, pp. 36-43.

**346**
"An Airline Captain Speaks out." The UFO Investigator 2, Oct.-Nov. 1962, pp. 5-6.

**347**
Creighton, Gordon W.
 "Amazing News From Russia." London, Flying Saucer Review 8, Nov.-Dec. 1962, pp. 27-28.

**348**
"Sheffield's Sensational Week." London, Flying Saucer Review 8, Nov.-Dec. 1962, pp. 6-9.

**349**
Birch, A.
 "Flying Saucers Photographed in England." Fate 16, Jan. 1963, pp. 26-27.

NOTES:

**350**
Ribera, Antonio
  "UFO Survey of Spain: More Evidence." London, Flying Saucer Review 9, Jan.-Feb. 1963, pp. 14-17.

**351**
Pennington, J., & R. Brissenden
  "Visual Capability in Rendez-Vous." Astronautics, Feb. 1963, pp. 96-99.

**352**
"U.S. Air Force Reports no Evidence of Flying Saucers Found in 483 Investigations." New York Times, Feb. 8, 1963, p. 12, col. 8.

**353**
"The Lowestoft Sighting: Object Observed for an Hour." London, Flying Saucer Review 9, March-April 1963, pp. 17-18.

**354**
Liddell, Urner
  "Phantasmagoria or Unusual Observations in the Atmosphere." Journal of the Optical Society of America XLIII, April 1963, pp. 314-317.

**355**
Michel, Aimé
  "Global Orthoteny." London, Flying Saucer Review 9, May-June 1963, pp. 3-7.

**356**
"Two Classic Sightings." London, Flying Saucer Review 9, May-June 1963, pp. 11-12.

**357**
"The Vauriat Sighting." London, Flying Saucer Review 9, July-Aug. 1963, pp. 3-5.

**358**
Girvan, Waveney
  "The Wiltshire Crater Mystery, the Meteor That Never was." London, Flying Saucer Review 9, Sept.-Oct. 1963, pp. 3-7.

**359**
Ribera, Antonio
  "'Bavic' in the Iberian Peninsula." London, Flying Saucer Review 9, Sept.-Oct. 1963, pp. 30-32.

**360**
Liss, Jeffrey G.
  "The Light That Chased a car." Fate 16, Nov. 1963, pp. 26-35.

**361**
Haythornthwaite, P. K.
  "Bavic Plotted as a World Circle Line." London, Flying Saucer Review 9, Nov.-Dec. 1963, pp. 17-18.

**362**
Vallée, Jacques
  "Recent Developments in Orthotenic Research." London, Flying Saucer Review 9, Nov.-Dec. 1963, pp. 3-6.

**363**
Edwards, Frank
  "UFO's, Natural Satellites, or Meteors?" Fate 16, Dec. 1963, pp. 49-54.

**364**
Johnson, R. D.
  "The Priest and the Saucer." Fate 17, Jan. 1964, pp. 26-31.

**365**
Menzel, Donald H.
  "Do Flying Saucers Move in Straight Lines?" London, Flying Saucer Review 10, March-April 1964, pp. 3-7.

NOTES:

## FLYING SAUCER SIGHTINGS

**366**
Lorenzen, Coral
"Diving for Lost UFO." Fate 17, May 1964, pp. 62-65.

**367**
Lorenzen, Coral
"Besieged by UFO's." Fate 17, June 1964, pp. 34-38.

**368**
Ibbotson, B.
"I saw a Flying Saucer." Bloemfontein, South Africa, Personality, June 25, 1964, p. 127.

**369**
Creighton, Gordon W.
"Argentina 1962." London, Flying Saucer Review 10, July-Aug. 1964, pp. 10-13.

**370**
Menzel, Donald H.
"Global Orthoteny." London, Flying Saucer Review 10, July-Aug. 1964, pp. 3-4.

**371**
"Physical Evidence Landing Reports." The UFO Investigator 2, July-Aug. 1964, pp. 4-6.

**372**
Vallée, Jacques
"Ghost Rockets: a Moment of History." London, Flying Saucer Review 10, July-Aug. 1964, pp. 30-32.

**373**
Vallée, Jacques
"The Menzel-Michel Controversy." London, Flying Saucer Review 10, July-Aug. 1964, pp. 4-6.

**374**
Couten, François
"Curieux et Secret Contact Interplanétair Survenu en Argentine le 2 Janvier 1953." Paris, Phénomènes Spatiaux, Nov. 1964, pp. 5-7.

**375**
Creighton, Gordon W.
"The Mysterious Templeton Photograph." London, Flying Saucer Review 10, Nov.-Dec. 1964, pp. 11-12.

**376**
"The New Zealand 'Flap' of 1909." London, Flying Saucer Review 10, Nov.-Dec. 1964, pp. 32-33.

**377**
Keyhoe, Donald E.
"The Air Force Censorship of UFO Sightings." True, January 1965.

**378**
Bowen, Charles
"A South American Trio." London, Flying Saucer Review 11, Jan.-Feb. 1965, pp. 19-21.

**379**
Schönherr, Luis
"Spindles in the sky." London, Flying Saucer Review 11, Jan.-Feb. 1965, pp. 9-11.

**380**
"UFO's? No! Lens Flare? YES!" London, Flying Saucer Review 11, Jan.-Feb. 1965, pp. 7-9.

**381**
"Observations Etrangères." Paris, Phénomènes Spatiaux, Feb. 1965, pp. 32-33.

**382**
Lorenzen, Coral
"Brazilian Official Report on the Trinidade UFO." Fate 18, March 1965, pp. 38-48.

**383**
"532 UFO Sightings Checked During 1964; 16 Remain Unidentified." Air Force Times 25, March 17, 1965, p. 7.

NOTES:

**384**
Kelly, Peter J.
 "Another Southampton Flap." London, <u>Flying Saucer Review</u> 11, March-April 1965, pp. 3-4, 20.

**385**
Seevior, Peter M.
 "Foundations of Orthoteny." London, <u>Flying Saucer Review</u> 11, March-April 1965, pp. 10-12.

**386**
Maney, Charles A.
 "Donald Menzel and the Newport News UFO: a Critical Report." <u>Fate</u> 18, April 1965, pp. 64-75.

**387**
Duchêne, J. L.
 "La Repartition des Atterrissages de Soucoupes Volantes en France." Paris, <u>Phénomènes Spatiaux</u>, May 1965, pp. 21-25.

**388**
"Observations Etrangères." Paris, <u>Phénomènes Spatiaux</u>, Nov. 1954, pp. 30-31; May 1965, pp. 37-47.

**389**
Bowen, Charles
 "Crash-landed UFO Near Mendoza." London, <u>Flying Saucer Review</u> 11, May-June 1965, pp. 7-9.

**390**
Farish, Lucius
 "An 1880 UFO." London, <u>Flying Saucer Review</u> 11, May-June 1965, pp. 34-35.

**391**
"Opposition Flap 1965." London, <u>Flying Saucer Review</u> 11, May-June 1965, pp. 3-6.

**392**
"Australian Scene, 1963-64." Sydney, <u>Australian Flying Saucer Review</u> 8, June 1965, pp. 8-12.

**393**
Bowen, Charles
 "Who Hoaxes who?" London, <u>Flying Saucer Review</u> 11, July-Aug. 1965, pp. 6-9.

**394**
Haythornthwaite, P. K.
 "Bavic as a Permanent Alignment." London, <u>BUFORA Journal and Bulletin</u> 1, Autumn 1965, pp. 16-18.

**395**
Leiber, Fritz
 "Homes for men in the Stars." <u>Science Digest</u> 58, Sept. 1965, pp. 53-57.

**396**
"Sur les Solitudes Glacées de l'Antarctique." Paris, <u>Phenomenes Spatiaux</u>, Sept. 1965, pp. 25-31.

**397**
"R. Heflin Releases 3 Photos of Saucer-Shaped Flying Object Taken Aug. 3 near El Toro Marine Airfield, Calif...Says 2-way Radio in Auto was Inoperable While Object was in Area." <u>New York Times</u>, Sept. 22, 1965, p. 1, col. 3.

**398**
Bowen, Charles
 "A Significant Report From France." London, <u>Flying Saucer Review</u> 11, Sept.-Oct. 1965, pp. 9-11.

**399**
Lloyd, Dan
 "Things are Hotting up in the Antarctic." London, <u>Flying Saucer Review</u> 11, Sept.-Oct. 1965, pp. 4-5.

NOTES:

# FLYING SAUCER SIGHTINGS

**400**
Magee, Judith
"UFO Activity Along the North-east Coast of Australia." London, Flying Saucer Review 11, Sept.-Oct. 1965, pp. 12-13.

**401**
"The Significant Report From France." London, Flying Saucer Review 11, Nov.-Dec. 1965, pp. 5-6.

**402**
"Enquêtes et Observations Diverses." Paris, Phénomènes Spatiaux, Dec. 1965, pp. 39-43.

**403**
"Que S'est-il Passé à Plestan?" Paris, Phénomènes Spatiaux, Dec. 1965, pp. 28-32.

**404**
Lorenzen, Coral
"UFO's Blanket South America." Fate 19, Jan. 1966, pp. 51-56.

**405**
Michel, Aimé
"New Thoughts on Orthoteny." London, Flying Saucer Review 12, Jan.-Feb. 1966, p. 19.

**406**
Rathbun, Mabel
"Flying Saucer Over San Salvador." Fate 19, Feb. 1966, pp. 41-43.

**407**
Thompson, T. A.
"Means of Detecting UFO's." London, BUFORA Journal and Bulletin 1, Spring 1966, pp. 6-8.

**408**
La 'Soucoupe' Carrée de Bolazec ou le Tracteur Volant." Paris, Phénomènes Spatiaux, March 1966, pp. 17-20.

**409**
Sanderson, Ivan T.
"'Something' Landed in Pennsylvania." Fate 19, March 1966, pp. 33-35.

**410**
"At Least 40 Report Flying Object and 4 Sister Ships Landed in Swamp near Ann Arbor, Michigan, 3 Times in Week..." New York Times, March 22, 1966, p. 18, col. 5.

**411**
Hanlon, Donald B.
"Virginia 1965 Flap." London, Flying Saucer Review 12, March-April 1966, pp. 14-16.

**412**
Hunt, Richard
"Canadian Fireballs." London, Flying Saucer Review 12, March-April 1966, pp. 33-34.

**413**
"Police Chase low Flying UFO." The UFO Investigator 3, March-April 1966, p. 1.

**414**
"Pi in the sky." Newsweek 67, April 1966, pp. 22, 27.

**415**
Veestraeten, Door J.
"Ze Zien ze Weer Vliegen." Haarlem, Netherlands, Panorama, April 1966, pp. 56, 57.

**416**
"Fatous Season: Ann Arbor and Hillsdale Sightings." Time 87, April 1, 1966, p. 25B.

**417**
"Well-witnessed Invasion, by Something: Australia to Michigan, With Report by P. O'Neil." Life 60, April 1, 1966, pp. 24-31.

NOTES:

**418**
"Flying Saucers: Illusions or Reality? Unidentified Flying Objects in Michigan." U.S. News & World Report 60, April 4, 1966, p. 14.

**419**
"Sightings Spur Review; UFO Probe Methods, Findings Studied." Air Force Times 26, April 6, 1966, p. 5.

**420**
"Marsh Gas in Michigan: Latest UFO Incident." America 114, April 9, 1966, p. 473.

**421**
"Notes and Comment: Saucer Flap." New Yorker 42, April 9, 1966, pp. 32-34.

**422**
Fuller, John G.
"Trade Winds: U.S. Air Force Reactions to Recent Sightings." Saturday Review 49, April 16, 1966, p. 10.

**423**
Bowen, Charles
"Michigan Furor." London, Flying Saucer Review 12, May-June 1966, pp. 4-6.

**424**
Creighton, Gordon W.
"Argentina 1963-1964." London, Flying Saucer Review 11, Nov.-Dec. 1965, pp. 14-17; 12, Jan.-Feb. 1966, pp. 23-26; 12, March-April 1966, pp. 24-26; 12, May-June 1966, pp. 25-28.

**425**
Ribera, Antonio
"The Madrid Landing." London, Flying Saucer Review 12, May-June 1966, pp. 28-31.

**426**
"Ann Arbor, Hillsdale et Autres Lieux." Paris, Phénomènes Spatiaux, June 1966, pp. 17-25.

**427**
Dunn, William J., Jr.
"An Analysis of the 1965 Brookville Landing Case." Saucer News 13, June 1966, pp. 6-8.

**428**
Fouéré, René
"Observations d'un Astronome Argentin." Paris, Phénomènes Spatiaux, June 1966, pp. 3-11.

**429**
Roussel, Robert
"La Roue d'Attigneville." Paris, Phénomènes Spatiaux, June 1966, pp. 25-31.

**430**
"Soucoupes Carrées Avant Bolazec." Paris, Phénomènes Spatiaux, June 1966, pp. 14-16.

**431**
Strauch, Arthur A.
"I Photographed a UFO." Fate 19, June 1966, pp. 67-72.

**432**
Carpenter, Mark
"The Great UFO Flap at Ann Arbor." Fate 19, July 1966, pp. 50-58.

**433**
Clark, Jerome
"The Strange Case of the 1897 Airship." London, Flying Saucer Review 12, July-Aug. 1966, pp. 10-17.

**434**
Anderson, Dave
"The Saucer That Terrorized a Town." Saga, Aug. 1966, pp. 12-15, 71-72.

**435**
Duplantier, Gene
"Ontario's Spring Flood of UFO's." Flying Saucers, Aug. 1966, pp. 22-24.

NOTES:

# FLYING SAUCER SIGHTINGS

**436**
Palmer, Ray
"1914 UFO Photo Puzzles Experts." *Flying Saucers*, Aug. 1966, pp. 7-9.

**437**
"What the British Press Reports on Flying Saucers." *Flying Saucers*, Aug. 1966, pp. 29-33.

**438**
Caputo, Livio
"Anche gli Astronauti Hanno Visto i Dischi Volanti." Milan, *Epoca* LXIV, Aug. 28, 1966, pp. 16-23.

**439**
Shuttlewood, Arthur
"Warminster UFO's Puzzling Behavior." London, *BUFORA Journal and Bulletin* 1, Autumn 1966, pp. 14-17.

**440**
Smith, Stephen L.
"The Bent Beams Case." London, *BUFORA Journal and Bulletin* 1, Autumn 1966, pp. 4-5.

**441**
Schang, Casimiro A.
"Recents Incidents." Paris, *Phénomènes Spatiaux*, Sept. 1966, pp. 25-28.

**442**
"Some 'Saucers' may be Electrical." London, *New Scientist* 31, Sept 1966, p. 463.

**443**
Zulli, Alphonse
"Glassboro UFO Landing Reviewed." *Fate* 19, Sept. 1966, pp. 32-39.

**444**
"Great Balls of Fire: Philip Klass Theory." *Newsweek* 68, Sept. 5, 1966, p. 78.

**445**
Bowen, Charles
"Cross Country cog Wheels." London, *Flying Saucer Review* 12, Sept.-Oct. 1966, pp. 16-17.

**446**
Galindez, Oscar A.
"Argentine Astronomer Observes UFO Buzzing Echo II." London, *Flying Saucer Review* 12, Sept.-Oct. 1966, p. 31.

**447**
Hanlon, Donald B.
"Texas Odyssey of 1897." London, *Flying Saucer Review* 12, Sept.-Oct. 1966, pp. 8-11.

**448**
Smith, Stephen L.
"The Bent Beams Case." London, *Flying Saucer Review* 12, Sept.-Oct. 1966, 111.

**449**
Babcock, Edward J., & Timothy G. Beckley
"UFO Plagues N. J. Reservoir." *Fate* 19, Oct. 1966, pp. 34-44.

**450**
Clark, Jerome
"UFO Landings in South America." *Flying Saucers*, Oct. 1966, pp. 24-25.

**451**
"Did UFO Bring Death?" *Flying Saucers*, Oct. 1966, pp. 11-14.

**452**
"Do Flying Saucers Ever Land?" *Flying Saucers*, Oct. 1966, p. 30.

**453**
Heath, David
"The Kingsford Heights, Indiana, Sighting." *Flying Saucers*, Oct. 1966, pp. 16-17.

NOTES:

# FLYING SAUCER SIGHTINGS

**454**
Palmer, Ray
"Navy Claps Saucer Sighters in Psychiatric Ward." Flying Saucers, Oct. 1966, pp. 7-9.

**455**
Sacksteder, Fred V.
"Horned UFO Sighted at La Porte." Flying Saucers, Oct. 1966, p. 15.

**456**
"Flying Something Touches Down in Brazil." Life 61, Oct. 28, 1966, pp. 40-40A.

**457**
"After Tully." Moorabin, Australia, Australian Flying Saucer Review 9, Nov. 1966, pp. 18-21.

**458**
"Australian Scene 1965." Moorabin, Australia, Australian Flying Saucer Review 9, Nov. 1966, pp. 8-11.

**459**
Drury, Neville
"Flying Ships in 'Oahspe'." Moorabin, Australia, Australian Flying Saucer Review 9, Nov. 1966, pp. 40-41.

**460**
Farish, Lucius
"Unidentified 'Airships' of the Gay Nineties." Fate 19, Nov. 1966, pp. 94-104.

**461**
Maney, Charles A.
"The Antarctica Sighting." Moorabin, Australia, Australian Flying Saucer Review 9, Nov. 1966, pp. 24-25.

**462**
Rankow, Ralph
"The Martin B-57 and the Changing UFO." Fate 19, Nov. 1966, pp. 36-45.

**463**
"Saucer Sightings Rise During Year." Air Force Times 27, Nov. 16, 1966, p. 6.

**464**
Clark, Jerome
"The Greatest Flap yet?" London, Flying Saucer Review 12, Jan.-Feb. 1966, pp. 27-30; March-April 1966, pp. 8-11; May-June 1966, pp. 13-15; Nov.-Dec. 1966, pp. 9-13.

**465**
Evans, Gordon H.
"Do the UFO's use a Paralysis ray?" Fate 19, Dec. 1966, pp. 101-105.

**466**
"Le Vaudriat II" Paris, Phénomènes Spatiaux, Dec. 1966, pp. 26-27.

**467**
"Observations Françaises Récentes." Paris, Phénomènes Spatiaux, Feb. 1965, pp. 27-28; May 1965, pp. 35-37; Nov. 1964, pp. 16-25; Dec. 1966, pp. 30-32.

**468**
"Observations Hors Presse." Paris, Phénomènes Spatiaux, Dec. 1966, pp. 27-30.

**469**
Seaman, E. A.
"A 1953 Sighting." Science 154, Dec. 1966, p. 1118.

**470**
Steiger, Brad
"Flying Saucers on the Attack." Saga 34, 1967, pp. 32-35, 76, 78-81, 83-84.

**471**
"Flygande Tefat Over Lappland." Stockholm, Sökaren (4:3), 1967, p. 3.

---

NOTES:

## FLYING SAUCER SIGHTINGS

**472**
Keyhoe, Donald E.
"Tefatens Hemlighet: Antigravitation."
Stockholm, Sökaren (4:6), 1967, pp. 7-10.

**473**
Bergendahl, P. E.
"UFO Förföljde Bilist." Stockholm, Sökaren 4, #8, 1967, p. 8.

**474**
Bowen, Charles
"The Russell Photograph." London, Flying Saucer Review 13, Jan.-Feb. 1967, p. 29.

**475**
Galindez, Oscar A.
"Unusual Photographs From Argentina."
London, Flying Saucer Review 13, Jan.-Feb. 1967, pp. 8-9.

**476**
Hanlon, Donald B., & Jacques Vallée
"Airships Over Texas." London, Flying Saucer Review 13, Jan.-Feb. 1967, pp. 20-25.

**477**
"Immense Triangular Object Over Majorca."
London, Flying Saucer Review 13, Jan.-Feb. 1967, pp. 19, 33.

**478**
Pastorino, Luiz P.
"The Flying Discs and the U.S.S.R." London, BUFORA Journal and Bulletin 1, Spring 1967, pp. 4-6.

**479**
Beckley, Timothy G.
"On the Trail of the Flying Saucers." Flying Saucers 51, March 1967, pp. 22-26.

**480**
Chaloupek, Henri
"Une Observation en Moravie en 1913." Paris, Phénomènes Spatiaux, March 1967, p. 28.

**481**
Ferriere, Joseph L.
"We Photographed UFO's." Fate 20, March 1967, pp. 52-55.

**482**
"Fort-Lamy, 27 Mars 1955." Paris, Phénomènes Spatiaux, March 1967, pp. 25-27.

**483**
Fouéré, René
"Une Lumière qui Traverse les Murs." Paris, Phénomènes Spatiaux, March 1967, pp. 23-24.

**484**
"Poursuites Dans le Ciel." Paris, Phénomènes Spatiaux, March 1967, pp. 29-30.

**485**
Trouble, Michel, & René Fouéré
"La Grande Panne USA-Canada et son Explication Technique." Paris, Phénomènes Spatiaux, March 1967, pp. 8-13.

**486**
Piel, J.
"UFO-watcher Watcher." Newsweek 69, March 20, 1967, p. 111.

**487**
Rogers, W.
"Flying Saucers: Sightings and Study of UFO's." Look 31, March 21, 1967, pp. 76-80.

**488**
Keel, John A.
"North America 1966: Development of a Great Wave." London, Flying Saucer Review 13, March-April 1967, pp. 3-9.

**489**
Durham, Anthony, & Keith Watkins
"Visual Perception of UFO's." London, Flying Saucer Review 13, May-June 1967, pp. 27-29.

NOTES:

**490**
Lloyd, Dan
"Crawling Lights--a new Development." London, Flying Saucer Review 13, May-June 1967, pp. 29-30.

**491**
"UFO Activity in Brazil During 1965." London, BUFORA Journal and Bulletin 2, Summer 1967, pp. 14-15.

**492**
"An Account of the Michigan Incident Through the Experts and Witnesses." Flying Saucers, June 1967, pp. 9-11.

**493**
Darden, Mona
"One UFO for the Road." Fate 20, June 1967, pp. 66-70.

**494**
Fouéré, René
"Rencontre Avec Eugène Coquil; Nouvelles Soucoupes Quadrangulaires; un Mode de Sustenation Mystérieux." Paris, Phénomènes Spatiaux, June 1967, pp. 13-17.

**495**
Hewes, Hayden C.
"The day the Flying Saucers Came to Oklahoma." Flying Saucers, June 1967, pp. 7-8.

**496**
"Que S'est-il Passé à Marliens." Paris, Phénomènes Spatiaux, June 1967, pp. 24-30.

**497**
Toulet, François
"Mathematic de L'orthotenie." Paris, Phénomènes Spatiaux, June 1967, pp. 7-11.

**498**
Hughes, F. P.
"Trained eye on UFO's; Letter." Science 156, June 9, 1967, pp. 1311-1312.

**499**
Clark, Jerome
"More on 1897." London, Flying Saucer Review 13, July-Aug. 1967, pp. 22-23.

**500**
Carson, Anthony
"Cigars Over Epping." London, New Statesman 74, July 7, 1967, p. 13.

**501**
Durham, Anthony, & Keith Watkins
"Visual Perception of UFO's: Part II." London, Flying Saucer Review 13, July-Aug. 1967, pp. 24-26.

**502**
Green, Gabriel, & Warren Smith
"UFO Raids Inside Russia." Saga 34, Aug. 1967, pp. 34-36, 76, 78-81, 83-84.

**503**
"More About Marliens." London, Flying Saucer Review 13, Sept.-Oct. 1967, pp. 14-15.

**504**
Rifat, Alain
"Was it a Landing at Marliens?" London, Flying Saucer Review 13, Sept.-Oct. 1967, pp. 11-13.

**505**
"Happening at Hoogdal: an Unidentified Beeping Object." Look 31, Nov. 1967, pp. 42-43.

**506**
Hinman, Grace
"How to Take a UFO Photograph." Fate 20, Nov. 1967, pp. 78-80.

NOTES:

# FLYING SAUCER SIGHTINGS

**507**
Beauchamp, T.
"Happening at Hoogdal: an Unidentified Beeping Object; UFO, or Midget Saw-whet owl near Sedro-Wooley, Washington." Look 31, Nov. 14, 1967, pp. 42-43.

**508**
Bowen, Charles
"Britan's Busiest UFO Days." London, Flying Saucer Review 13, Nov.-Dec. 1967, pp. 3-4.

**509**
Ribera, Antonio
"Midsummer Sightings Over Andorra." London, Flying Saucer Review 13, Nov.-Dec. 1967, p. 25.

**510**
Vallée, Jacques, & Aleksandr Kasantsev
"What is it That is Flying in our Skies?" London, Flying Saucer Review 13, Nov.-Dec. 1967, pp. 11-12.

**511**
Watson, W. H.
"19th Century Paraglider?" London, Flying Saucer Review 13, Nov.-Dec. 1967, p. 21.

**512**
Mesnard, Joel
"Quatre Enquêtes." Paris, Phénomènes Spatiaux, Dec. 1967, pp. 18-24.

**513**
Ribera, Antonio
"Cortège de Soucoupes Volantes sur Andorre-la Vieille." Paris, Phénomènes Spatiaux, Dec. 1967, pp. 29-30.

**514**
Senelier, Jean
"Observations sur le Rapport D'analyse de la Terre de Marliens; Remarques sur de Preténdus Débris D'O.V.N.I." Paris, Phénomènes Spatiaux, Dec. 1967, pp. 12-15.

**515**
Cramp, Leonard G.
"Report on UFO Sighting and Ground Effect at Whippingham, Isle of Wright." London, BUFORA Journal and Bulletin 2, Winter 1967-1968, pp. 5-9.

**516**
Page, Thornton
"Photographic Sky Coverage for the Detection of UFO's." Science, (new series), (160:3833), 1968, pp. 1258-1260.

**517**
Powers, William T.
"UFO in 1800?" Science (new series), (160:3833), 1968, p. 1260.

**518**
Stanford, Rex G.
"Brev Från Rex Stanford." Stockholm, Sökaren (5:3), 1968, p. 16.

**519**
Brooks, Angus
"Remarkable Sighting Near Dorset Coast." London, Flying Saucer Review 14, Jan.-Feb. 1968, pp. 3-4.

**520**
Creighton, Gordon
"UFO's in the South Atlantic." London, Flying Saucer Review 14, Jan.-Feb. 1968, p. 13.

**521**
Powers, W. T.
"Photographic Surveillance for UFO's: Is it Feasible?" London, Flying Saucer Review 14, Jan.-Feb. 1968, pp. 14, 17.

**522**
Rankow, Ralph
"The Heflin Photographs." London, Flying Saucer Review 14, Jan.-Feb. 1968, pp. 21-24.

NOTES:

## FLYING SAUCER SIGHTINGS

**523**
Winder, R. H. B.
"Comment on the Angus Brooks Sighting."
London, Flying Saucer Review 14, Jan.-Feb. 1968, pp. 4-5.

**524**
Chaloupek, Henri
"Observations en Tchécoslovaquie et en Bulgarie." Paris, Phénomènes Spatiaux, March 1968, pp. 20-24.

**525**
Fink, Herschel P.
"A Tale of 2 Saucers, or Which UFO Hangs on a String?" True, March 1968.

**526**
Hynek, J. A.
"How to Photograph a UFO." Popular Photography 62, March 1968, p. 69.

**527**
"La Nuit du 17 au 18 Juillet 1967." Paris, Phénomènes Spatiaux, March 1968, pp. 11-18.

**528**
Ribera, Antonio
"L'Etrange Affaire de Nuria." Paris, Phénomènes Spatiaux, March 1968, pp. 27-30.

**529**
"Atterrissages de 'M.O.C.' en 1967." Paris, Lumières Dans la Nuit 93, March-April 1968, pp. 17-18.

**530**
Bowen, Charles, & Gordon Creighton
"The Storrington Reports, Landings in Sussex?" London, Flying Saucer Review 14, March-April 1968, pp. 4-6.

**531**
Creighton, Gordon
"A Cigar-shaped UFO Over Antarctica." London, Flying Saucer Review 14, March-April 1968, pp. 20-22.

**532**
Dufour, Jean C.
"Pigeon Shoot at the Col d'Aspin." London, Flying Saucer Review 14, March-April 1968, p. 29.

**533**
Hugill, Joanna
"On the Road From Sydney to Melbourne." London, Flying Saucer Review 14, March-April 1968, pp. 3, 11.

**534**
Jonsson, Ake
"Reports From Sweden." London, Flying Saucer Review 14, March-April 1968, pp. 12-16.

**535**
Fawcett, George D.
"Flying Saucers, Explosive Situation for 1968." Flying Saucers, April 1968, pp. 22-25.

**536**
"Priest Astronomer's Report: Observations of an Argentine Astronomer." Flying Saucers, April 1968, pp. 15-17.

**537**
Bryant, Larry W.
"The UFO Cover-up at Langley Air Force Base." Flying Saucers From Other Worlds, June 1968, pp. 11-14.

**538**
Duplantier, Gene
"Alien Crafts Curious About our Cars and Occupants." Flying Saucers, June 1968, pp. 22-24.

**539**
Vezina, Allan K.
"Canada 1967--a big Year for UFO Research." Flying Saucers, June 1968, pp. 8-10.

**540**
"Russian UFO's." Soviet Science in the News 5, July 1968, pp. 1-2.

NOTES:

## FLYING SAUCER SIGHTINGS

**541**
Wright, Eric
"UFO Lands Near Pond and the Water Freezes in 50 Degree Weather." National Enquirer, July 1968, p. 32.

**542**
Walker, Sydney
"The Applied Assessment of Central Nervous System Integrity: a Method for Establishing the Credibility of Eye Witnesses and Other Observers" (In Symposium on UFO's. Hearings Before the Committee on Science and Astronautics, U. S. House of Representatives, 89th Congress, July 29, 1968. Washington, 1968, pp. 152-176, 185-189).

**543**
Roberts, August C.
"The Nicholson Photos." Flying Saucers, July-Aug. 1968, pp. 84-85.

**544**
"UFO's Over Washington: Call for Renewed Scientific and Governmental Saucer Research." Newsweek 72, Aug. 12, 1968, p. 72.

**545**
Drake, Walter R., tr.
"A Flying Saucer Discovered Near the Coasts of North Germany." Flying Saucers 60, Oct. 1968, p. 24.

**546**
Glemser, Kurt
"A Roundup on the Bender Mystery." Flying Saucers 61, Dec. 1968, pp. 16-20.

**547**
Mesnard, Joel, & René Fouéré
"Enquêtes dans le Nivernais et le Morvan." Paris, Phénomènes Spatiaux 18, Dec. 1968, pp. 22-26.

**548**
Lorenzen, Coral
"UFO in California." Fate 228, March 1969, pp. 74-79.

**549**
Lyon, Bill
"UFO Eludes Spanish Jets." Fate 229, April 1969, pp. 36-41.

**550**
Boyer, Samuel H.
"UFO Encounter Brings Elation." Fate 230, May 1969, pp. 40-43.

**551**
Butcher, Lee
"Florida Reports UFO Swarm." Fate 230, May 1969, pp. 44-50.

**552**
Remaley, Sally
"Luminous Objects on Arctic Ice." Fate 231, June 1969, pp. 66-72.

**553**
Sanderson, Jay
"UFO's Panic Veracruz." Fate 232, July 1969, pp. 51-56.

**554**
Saalsaa, Geneva
"UFO Sighting in Lafayette County." Flying Saucers 65, Aug. 1969, p. 7.

**555**
Smith, Gordon W.
"More Australia UFO's." Flying Saucers 65, Aug. 1969, pp. 23-28.

**556**
Locke, L. J.
"UFO's in Western Australia." Flying Saucers 66, Oct. 1969, pp. 7-13.

**557**
Gheorgita, Florin
"Flying Saucer over Cluj, Romania." London, Flying Saucer Review (15: 6), Nov.-Dec. 1969, pp. 12-16.

**558**
"Un Article d'Antonio Ribera." Paris, Phénomènes Spatiaux 22, Dec. 1969, pp. 17-26.

NOTES:

**559**
Raynes, Brent
  "The Lemon Grove UFO." Flying Saucers 67, Dec. 1969, p. 16.

**560**
"Unidentified Aerial Phenomena" (In U.S. Air Force Academy. Introductory Space Science, v. 2. Washington, 1970, 7p.).

**561**
Bemelmans, Hans
  "Reports From Ibiuna." London, Flying Saucer Review (16:1), Jan.-Feb. 1970, pp. 15-19.

**562**
Perego, Alberto
  "40 Flying Saucers in a Cross Formation Over the Vatican City." Flying Saucers 68, March 1970, pp. 17-21.

**563**
Hynek, J. Allen
  "21 Years of UFO Reports, 1." London, Flying Saucer Review (16: 1), Jan.-Feb. 1970, pp. 3-6; 2 (16: 2), March-April 1970, pp. 6-8, 22.

**564**
McDonald, James E.
  "UFO's Over Lakenheath in 1956." London, Flying Saucer Review (16: 2), March-April 1970, pp. 9-17, 29.

**565**
McDonald, James E.
  "The 1957 Gulf Coast RB-47 Incident." London, Flying Saucer Review (16: 1), May-June 1970, pp. 2-6.

**566**
Schönherr, Luis
  "Observations of a Sceptical Believer." London, Flying Saucer Review (16: 3), May-June 1970, pp. 16-21.

**567**
Raynes, Brent
  "Flying Saucer Landings and UFO Kidnaping Reports." Flying Saucers 69, June 1970, p. 32.

**568**
Ribera, Antonio
  "Le cas de San Marti de Tous." Paris, Phénomènes Spatiaux 24, June 1970, pp. 25-28.

**569**
"UFO Sightings at sea." Flying Saucers 69, June 1970, pp. 7-8.

**570**
Keel, John A.
  "Mystery Aeroplanes of the 1930's." London, Flying Saucer Review (16: 3), May-June 1970, pp. 10-13; (16: 4), July-Aug. 1970, pp. 9-14.

**571**
Fawcett, George D.
  "Three-Quarters of a Century of Florida Flying Saucers." Flying Saucers 70, Sept. 1970, pp. 2-8.

**572**
Henry, Dick
  "1970 UFO Over Nebraska." Fate 246, Sept. 1970, pp. 82-87.

**573**
Hynek, J. Allen
  "C'est un Oiseau, un Avion, c'est..." Paris, Phénomènes Spatiaux 25, Sept. 1970, pp. 4-7.

**574**
Warren, Donald I.
  "Status Inconsistency Theory and Flying Saucer Sightings." Science 170, Nov. 6, 1970, pp. 599-603.

NOTES:

## FLYING SAUCER SIGHTINGS

575
Grove, Carl
"The Airship Wave of 1909: a Preliminary Survey." London, Flying Saucer Review (16: 6), Nov.-Dec. 1970, pp. 9-11.

576
Cowgill, G. L.
"People who see Flying Saucers." Science 171, 1971, pp. 956-957.

577
Warren, Donald I.
"People who see Flying Saucers: Reply." Science 171, 1971, p. 857.

578
Templin, K. W.
"Flying Saucer Sightings." Science 173, 1971, p. 1083.

579
Fredrickson, Sven-Olaf
"A Landing Near Lake Anten?" London, Flying Saucer Review (17:1), Jan.-Feb. 1971, pp. 13-17.

580
Grove, Carl
"The Airship Wave of 1909-Part 2." London, Flying Saucer Review (17:1), Jan.-Feb. 1971, pp. 17-19.

581
Lagarde, Fernand
"The Aveyron Enquiry-3." London, Flying Saucer Review (17:1), Jan.-Feb. 1971, pp. 3-9.

582
Fawcett, George
"UFO's: 1971, a new Start Toward Further Respect and Evidence." Flying Saucers 72, March 1971, pp. 28-29.

583
"Incident en Norvège." Paris, Phénomènes Spatiaux 27, March 1971, pp. 14-18.

584
Jensen, Erling
"UFO in Denmark." Flying Saucers 72, March 1971, pp. 17-19.

585
"Singulière Observation Russe en 1663." Paris, Phénomènes Spatiaux 27, March 1971, pp. 9-14.

586
Clark, Jerome
"Indian Prophecy and the Prescott UFO's." Fate 253, April 1971, pp. 54-61.

587
Vallée, Jacques F.
"The Landings of 1970." Ben Lomond, Calif., Data-Net (5:5), May 1971.

588
Buhler, Dr. W.
"UFO on the Sea Near Rio." London, Flying Saucer Review (17:3), May- June 1971, pp. 3-7.

589
Galíndez, Oscar A.
"Trancas, After Seven Years." London, Flying Saucer Review (17:3), May-June 1971, pp. 14-20+.

590
Schwartz, Berthold E.
"The Port Monmouth Landing." London, Flying Saucer Review (17:3), May-June 1971, pp. 21-27.

591
Vallée, Jacques
"UFO Activity in Relation to Night-of-the-Week." London, Flying Saucer Review (17:3), May-June 1971, pp. 8-10.

592
Phillips. Z. T.
"The Landing Traces Found at Alleged UFO Landing Sites." Ben Lomond, Calif., Data-Net (5:5), June 1971.

NOTES:

**593**
Sibol, R. F.
 "Ballet in the sky." Fate 255, June 1971, pp. 68-70.

**594**
Ballester-Olmos, Viente J., & Jacques F. Vallée
 "Estudio de 100 Aterrizajes de OVNI en la Península Ibérica." Barcelona, Stendek, #1 (extra), July 1971.

**595**
Clark, J. J.
 "A Survey of 322 United States UFO Reports." Ben Lomond, Calif., Data-Net (5:7), July 1971.

**596**
Glemser, Kurt
 "Astronauts Spot UFO's." Search 98, July 1971, pp. 77-78.

**597**
Ballester Olmos, Vicente J.
 "Landings in Spain in 1958." London, Flying Saucer Review (17:4), July-Aug. 1971, pp. 10-12.

**598**
Buckle, Eileen
 "Defenestration at Kempsey: Bizarre levitation effect during UFO flap in New South Wales." London, Flying Saucer Review (17:4), July-Aug. 1971, pp. 20-22.

**599**
Creighton, Gordon
 "Follow-up on the Morro Do Vintem Mystery." London, Flying Saucer Review (17:4), July-Aug. 1971, pp. 6-9.

**600**
Saunders, David R.
 "Is Bavic Remarkable?" London, Flying Saucer Review (17:4), July-Aug. 1971, pp. 13-16+.

**601**
Ballester Olmos, Vicente J.
 "Survey of Iberian Landings." London, Flying Saucer Review (special issue no. 4), August 1971, pp. 46-56.

**602**
Ballester Olmos, Vicente J., & Jacques Vallee
 "Type-1 Phenomena in Spain and Portugal-1." London, Flying Saucer Review (Special Issue no. 4), August 1971, pp. 40-45.

**603**
Ballester Olmos, Vicente J., & Jacques Vallée
 "Type-1 Phenomena in Spain and Portugal-2." London, Flying Saucer Review (special issue no. 4), August 1971, pp. 57-64.

**604**
Cruttwell, Norman E. G.
 "Flying Saucers over Papua." London, Flying Saucer Review (special issue no. 4), August 1971, pp. 3-38.

**605**
Lorenzen, Coral
 "Brazilian Shoot-outs With UFO's." Fate 258, Sept. 1971, pp. 91-98.

**606**
Clark, Josephine, and Jacques Vallée
 "Researching the American Landings." London, Flying Saucer Review (17:5), Sept.-Oct. 1971, pp. 3-8.

**607**
Fredrickson, Sven-Olaf
 "Recent Observations over Southern Sweden." London, Flying Saucer Review (17:5), Sept.-Oct. 1971, pp. 9-11.

**608**
Karavieri, Ahti
 "The Saapunki UFO." London, Flying Saucer Review (17:5), Sept.-Oct. 1971, pp. 23-26.

**609**
Keel, John A.
 "Mystery Aeroplanes of the 1930's-Part IV." London, Flying Saucer Review (17:5), Sept.-Oct. 1971, pp. 20-22+.

**610**
Dick, William, & David Klein
 "UFO Sighting Mystifies England." National Enquirer, Oct. 17, 1971, p.2.

---

NOTES:

## FLYING SAUCER SIGHTINGS

**611**
James, Michael
"British Government Investigating Reports of UFO Sighted by Hundreds." <u>National Enquirer</u>, Nov. 28, 1971, back page.

**612**
Clark, Josephine J.
"UFO Observed During Californian Blackout." London, <u>Flying Saucer Review</u> (17:6), Nov.-Dec. 1971, pp. 21-23+.

**613**
Clark, Josephine, and Jacques Vallée
"Researching the American Landings-Part 1." London, <u>Flying Saucer Review</u> (17:6), Nov.-Dec. 1971, pp. 10-14.

**614**
Creighton, Gordon
"Uproar in Brazil." London, <u>Flying Saucer Review</u> (17:6), Nov.-Dec. 1971, pp. 24-29.

**615**
Prytz, John
"Flaps, a new Approach." <u>Flying Saucers</u> 75, Dec. 1971, pp. 33-34.

**616**
Saunders, David R.
"The Shapes of UFO Waves." Ben Lomond, Calif., <u>Data-Net</u> (5:12), Dec. 1971.

**617**
Vallee, Jacques F., & Vicente J. Ballester Olmos
"Sociología de los Aterrizajes Ibéricos." Barcelona, <u>Stendek</u> (3:7), Dec. 1971.

**618**
Schwartz, Berthold E.
"Stella Lansing's UFO Motion Pictures." London, <u>Flying Saucer Review</u> (18:1), Jan.-Feb. 1972, pp. 3-12.

**619**
Pye, Fred
"Members of TV Crew Film UFO Flying over Southern England." <u>National Enquirer</u>, Feb. 13, 1972, p.20.

**620**
Ballester Olmos, Vicente J., & Carlos Orlando
"Notas Estadísticas Sobre la Oleada de 1950 en España y Portugal." Barcelona, <u>Stendek</u> (3:8), March 1972.

**621**
Larson, Kenneth
"The Super Spaceship, American UFO Report Sites Outline on Ground..." <u>Flying Saucers</u> 76, March 1972, pp. 6 10.

**622**
"Lueurs Singulières dans l'Ardèche. Compte Rendu d'Enquête." Paris, <u>Phénomènes Spatiaux</u> 31, March 1972, pp. 13-22.

**623**
McDonald, James
"Some Pennsylvania UFO Cases and Their Bearing on the Condon Report." <u>Flying Saucers</u> 76, March 1972, pp. 2-5.

**624**
Mesnard, Joel, & Maryvonne Eveno
"Observations Récentes dans l'Oise." Paris, <u>Phénomènes Spatiaux</u> 31, March 1972, pp. 25-29.

**625**
Saunders, David R.
"Actividad OVNI en Relación con los Días de la Semana." Barcelona, <u>Stendek</u> (3:8), March 1972.

**626**
Saunders, David R.
"Bavic est-il Remarquable?" Paris, <u>Phénomènes Spatiaux</u> 31, March 1972, pp. 4-11.

NOTES:

# FLYING SAUCER SIGHTINGS

**627**
"Sightings in Finland." *Flying Saucers* 76, March 1972, pp. 24-25.

**628**
"UFO Sighting Reports Over the British Isles." *Flying Saucers* 76, March 1972, pp. 15-17.

**629**
Vallée, Jacques F.
"Actividad OVNI en Relación con las Noches de los Días de la Semana." Barcelona, *Stendek* (3:8), March 1972.

**630**
Dick, William
"UFO Terrorizes Girl Scouts at Campground." *National Enquirer*, March 5, 1972, p.2.

**631**
Fredrickson, Sven-Olaf
"The Ängelholm Landing Report." London, *Flying Saucer Review* (18:2), March-April 1972, pp. 15-17.

**632**
Creighton, Gordon
"Brazil Learns at Last About A.V.B." London, *Flying Saucer Review* (18:3), May-June 1972, pp. 9-12.

**633**
Creighton, Gordon
"UFO's with Multiple Beams of Light." London, *Flying Saucer Review* (18:3), May-June 1972, pp. 4-8.

**634**
"Du Côte de la mer." Paris, *Phénomènes Spatiaux* 32, June 1972, pp. 15-21.

**635**
Fawcett, George D.
"Quarter Century of UFO's in North Carolina." *Flying Saucers* 77, June 1972, pp. 2-7.

**636**
Fawcett, George D.
"UFO's Show Dramatic Worldwide Increase." *Fate* 267, June 1972, p. 59.

**637**
Larson, Kenneth
"Analysis of the Liberty Saucers." *Flying Saucers* 77, June 1972, pp. 27-29.

**638**
"Sightings in Britain." *Flying Saucers* 77, June 1972, pp. 8-14.

**639**
"Article on UFO's Notes for Most of Past 25 Years 'Hardly a Month has Gone by Without a Deluge' of new Sightings of Mysterious Flying Craft; Says There Have Been Reports of at Least 50,000 Sightings, Perhaps 100,000, Since First Report in June 1947; Organizations Studying UFO Phenomenon Discussed; Illustration of Mrs. C. Lorenzen With Model of UFO She Maintains she saw." *New York Times*, June 25, 1972, p. 36, col. 4.

**640**
Kelley, Jack
"Small English Town Records over 1,000 UFO Sightings in 5 Months." *National Enquirer*, July 23, 1972, p.2.

**641**
Ballester Olmos, Vicente J.
"Record and Analysis of the Spanish 'Negative' Landings." London, *Flying Saucer Review* (18:4), July-Aug. 1972, pp. 31-iv.

**642**
Malthaner, Hubert
"Mystery Flying Object Rolls Along a German Road." London, *Flying Saucer Review* (18:4), July-Aug. 1972, pp. 15-17.

NOTES:

# FLYING SAUCER SIGHTINGS

**643**
Thomas, Jane
"Recent Reports from Agentina (sic) and Peru."  London, Flying Saucer Review (18:4), July-Aug. 1972, pp. 21-24.

**644**
Orlando, Carlos, & Vicente J. Ballester Olmos
"Ampliación al Ensayo Sobre 1950." Barcelona, Stendek, (3:9), Aug. 1972.

**645**
Friedman, Stanton, & B. A. Slate
"Hundreds see UFO in Ceylon."  Fate 270, Sept. 1972, pp. 84-89.

**646**
Schalin, Sven
"Space Visit, or What?"  Flying Saucers 78, Sept. 1972, pp. 18-25.

**647**
Bowen, Charles
"South African Mini-Wave, 1972-Part 1." London, Flying Saucer Review (18:5), Sept.-Oct. 1972, pp. 5-8+.

**648**
Conti, Sergio
"The Cennina Landing of 1954."  London, Flying Saucer Review (18:5), Sept.-Oct. 1972, pp. 11-15.

**649**
Creighton, Gordon
"Brazil Once More."  London, Flying Saucer Review (18:5), Sept.-Oct. 1972, pp. 24-26.

**650**
Ballester-Olmos, Vicente J., & J. Bonabot
"The Worldwide Wave of 1950: Further Inquiries."  Ben Lomond, Calif., Data-Net (6:10) Oct. 1972.

**651**
Balfour, Malcolm
"Police and Farmer Shoot at Terrifying UFO-45 Feet Away."  National Enquirer, Oct. 15, 1972, back page.

**652**
Burt, Bill
"Amazing UFO Sightings in Argentina." National Enquirer, Oct. 29, 1972, p.4.

**653**
Burt, Bill
"Soccer Fans Tell of UFO's Over Stadium in Brazil."  National Enquirer, Nov. 5, 1972, p.25.

**654**
Burt, Bill
"UFO's in South America, an Enquirer Special Investigation."  National Enquirer, Nov. 12, 1972, p.5.

**655**
Dick, William
"UFO Tracked on Radar Screens in Florida." National Enquirer, Nov. 26, 1972, p.3.

**656**
Bowen, Charles
"South African Mini-Wave, 1972- Part 2." London, Flying Saucer Review (18:6), Nov.-Dec. 1972, pp. 14-17.

**657**
Magee, Judith
"UFO Over the Mooraduc Road."  London, Flying Saucer Review (18:6), Nov.-Dec. 1972, pp. 3-5.

**658**
Wiersema, G.S.
"Landing on the Leusderheide."  London, Flying Saucer Review (18:6), Nov.-Dec. 1972, pp. 10-13.

**659**
"Sightings in Britain."  Flying Saucers 79, Dec. 1972, pp. 35-42.

**660**
"UFO Global Landings Show Dramatic Increase, E-M Effects Noted."  Flying Saucers 79, Dec. 1972, pp. 2-7.

NOTES:

**661**
Besset, Henry J., et al
  "A Taize, le 12 Août 1972." Paris, Phénomènes Spatiaux 35, March 1973, pp. 11-21.

**662**
"Radio Uganda Reports Pres. Amin Sees UFO Near Kampala." Los Angeles Times, March 5, 1973, sec. 1, p. 2, col. 1.

**663**
Dick, William
  "Trained Observer Sights UFO's on 6 Different Nights in State of Washington." National Enquirer, March 11, 1973, p.3.

**664**
"Youths Tell of Seeing 'Glowing Sphere' at Pt. Dume, Calif." Los Angeles Times, March 25, 1973, sec. 1, p. 26, col. 1.

**665**
Vance, Adrian
  "The Oregon Photo-Using Photography to Tackle a Mystery." London, Flying Saucer Review (19:2), March-April 1973, pp. 3-6.

**666**
Slate, B. A.
  "Kansas UFO Leaves Hard Evidence." Fate 277, April 1973, pp. 88-96.

**667**
"Editorial: Presence of UFO's Near Midwest Lead Fields Viewed." Chicago Tribune, April 10, 1973, sec. 1, p. 10, col. 1.

**668**
"Aurora, Texas Site of 1897 UFO Crash Investigated." New Orleans Times-Picayune, April 16, 1973, sec. 3, p. 7, col. 5.

**669**
"Student, Ben Baron, Takes Picture of Piedmont, Mo., UFO." Chicago Tribune, April 22, 1973, sec. 1, p. 4, col. 2.

**670**
"9 UFO's Sighted in Suburbs North of Chicago." Chicago Tribune, May 11, 1973, sec. 1, p. 1, col. 4.

**671**
"Coachella Valley Residents Report UFO's in Riverside, Cal." Los Angeles Times, May 15, 1973, sec. 1, p. 3, col. 3.

**672**
"Riverside, Calif. UFO Said to be Meteor." Los Angeles Times, May 18, 1973, sec. 1, p. 2, col. 6.

**673**
"Case of Dallas, Texas 'Blob' Tied to UFO Sighting." Washington Post, May 26, 1973, sec. A, p. 4, col. 6.

**674**
Dick, William
  "Enquirer Awards Kansas Family $5,000 for Supplying the Most Scientifically Valuable UFO Evidence in 1972." National Enquirer, May 27, 1973, p.2.

**675**
"Texas Woman Recalls Crash of 'Spaceship' in 1897." Chicago Tribune, May 31, 1973, sec. 1, p. 3, col. 6.

**676**
"UFO's and UFOnauts, Landings More Frequent in 1973." Flying Saucers 81, Summer 1973, pp. 30-36.

**677**
Villela, Rubens Junqueira
  "Montauroux." Paris, Phénomènes Spatiaux 36, June 1973, pp. 13-22.

**678**
"A Look at Alleged 1897 UFO Crash in Aurora, Texas." Chicago Tribune, June 3, 1973, sec. 1, p. 24, col. 2.

NOTES:

679
"S. Nixon, NICAP Researcher, Eyes UFO's Seen by Astronauts." Los Angeles Times, June 5, 1973, sec. 1A, p. 8, col. 1.

680
"L. J. Lorenzen Calls Aurora, Texas Story 'Hoax'." Chicago Tribune, June 16, 1973, sec. 1, p. 8, col. 1.

681
"Green UFO Sighting in New Orleans Said to be Advertizing Gimmick." New Orleans Times-Picayune, June 22, 1973, sec. 4, p. 7, col 3.

682
Dick, William
"Hundreds see UFO in Alabama." National Enquirer, June 24, 1973, back page.

683
Williams, Frances (Mathis)
"UFO's in the Florida Swamp." Fate 281, Aug. 1973, pp. 74-76.

684
"A List of Sightings by Astronomers." Flying Saucers 83, Fall 1973, pp. 53-62.

685
Evans, Livio
"Les OVNIS Envahissent le Chili." Paris, Phénomènes Spatiaux 37, Sept. 1973, pp. 3-8.

686
Osuna Llemente, Manuel
"Cazalla de la Sierra, Printemps 1973." Paris, Phénomènes Spatiaux 37, Sept. 1973, pp. 22-27.

687
Techter, David
"UFO Makes its Mark in South Africa." Fate 282, Sept. 1973, pp. 62-69.

688
"Saucerers: UFO Watchers Report Eerie Events." Milwaukee, Milwaukee Journal, Oct. 17, 1973, pp. 1, 2.

689
"Ring of Lights set up in Texas as UFO Bait". Milwaukee, Milwaukee Journal, Oct.18, 1973, p. 13.

690
"W. Sullivan on Rash of...Sightings of UFO's over Various Areas of U. S...." New York Times, Oct. 21, 1973, p. 65, col. 1.

691
Pratt, Bob
"Mysterious Saucer-Shaped Depressions Surrounded by Scorched Crops Found on 2 Iowa Farms." National Enquirer, Oct. 28, 1973, p.6.

692
"Stardust and Moonshine: UFO Sightings." Newsweek 82, Oct. 29, 1973, p. 31.

693
Benedict, W. R.
"Quebec UFO Flap." Fate 284, Nov. 1973, p. 104.

694
"Mysterious Wave of UFO's Reported in Mass Sighting." National Enquirer, Nov. 4, 1973, p.41.

695
Ingle, Bob
"Scientist Sees Spaceship Where Ezekiel saw Wheel." Milwaukee Journal (Green Sheet), Nov. 22, 1973, p. 1.

NOTES:

**696**
"UFO's Again?" *Public Conference Program*, Sponsored by Milwaukee Public Library. ABC-TV Channel 6, Milwaukee. Sunday, Nov. 25, 1973, 12 noon (Discussion by John Wallace Spencer, Author of *Limbo of the Lost*).

**697**
Fouéré, René
"Que s'est-il Passé à Turin-Caselle et en d'Autres Lieux d'Italie?" Paris, *Phénomènes Spatiaux* (38: 4), Dec. 1973, pp. 15-18.

**698**
Godefroy, Roland
"Le Mystérieux Objet Lumineux de Carteret." Paris, *Phénomènes Spatiaux* (38: 4), Dec. 1973, pp. 19-24.

**699**
Mesnard, Joel
"Un Automobiliste Passe sous un OVNI?" Paris, *Phénomènes Spatiaux* (38:4), Dec. 1973, pp. 30-31.

**700**
Dickson, Stewart
"UFO's Buzz Tennessee Town." *National Enquirer*, Dec. 9, 1973, p.10.

**701**
Hunt, Gerry
"UFO's in the Night Sky." *National Enquirer*, Dec. 23, 1973, p.5.

**702**
Huguet, Casas
"Observación OVNI en la Comarca del Girones." Barcelona, *Stendek* (V: 15), Dec. 1973 - March 1974, pp. 3-10.

**703**
Cresson, J.
"Erikoinen Laskeutumistapaus Dinanissa." Turku, Finland, *Argos* 1974, pp. 9-12.

**704**
Zullo, Allan
"Enquirer Awards $5,000 to Helicopter Crew for 1973's Most Valuable UFO Evidence." *National Enquirer*, 23, 1974, p.2.

**705**
Wells, Jeff
"UFO Startles Thousands." *National Enquirer*, Feb. 10, 1974, back page.

**706**
Smith, Robert G.
"Los Angeles Policemen Chase UFO for 3 Miles." *National Enquirer*, Feb. 24, 1974, p.4.

**707**
Madden, Ruth S.
"The Silver UFO." *Fate* 288, March 1974, pp. 58-60.

**708**
Camlin, Edward B.
"UFO Plays Strange Beam of Light on Ground as Amazed Pennsylvania State Troopers Look on." *National Enquirer*, March 31, 1974, p.10.

**709**
"Huaco: Los Foo-Fighters en la Argentina." Buenos Aires, *Cuarta Dimensión* 8, April 1974, pp. 33-39.

**710**
"Le Nostre Analisi: Inchiesta UFO in Calabria." Milan, *Notiziario UFO* 62, April-June 1974, pp. 8-13.

**711**
Banchs, Roberto E.
"Estudio Comparativo de las Observaciones de Chile y Uruguay." Buenos Aires, *CEFAI Revista* (3: 2), May 1974, pp. 9-13.

NOTES:

## FLYING SAUCER SIGHTINGS

**712**
Clark, Jerome
"The Brushy Creek UFO Scare." *Fate* 290, May 1974, pp. 96-106.

**713**
Redon, Père, & Maria del Carmen Tamayo
"Sobre la Oleada Ibérica de 1974." Barcelona, *Stendek* (5:16), June 1974, pp. 3-8.

**714**
Smith, Robert G.
"Huge, Glimmering UFO Seen Near New York City One of Most Valuable Sightings Ever." *National Enquirer*, June 30, 1974, p.46.

**715**
Fredrickson, Sven-Olof
"Aterrizaje Cerca del Lago Anten?" Tr. from English by James Thomas. Córdoba, Argentina, *OVNIS, un Desafío a la Ciencia* (I:2), July-Aug. 1974, pp. 8-11.

**716**
Krmelj, Milos
"OVNIS en Países Socialistas: Descenso en Yugoslavia." Córdoba, Arg., *OVNIS, un Desafío a la Ciencia* (I:2), July-Aug. 1974, pp. 25-27.

**717**
Philips, Ted
"OVNI Desciende en Delphos." Córdoba, Arg., *OVNIS, un Desafío a la Ciencia* (I:2), July-Aug. 1974, pp. 22-25.

**718**
Guasp, Miguel
"Análisis Procesal de las Direcciones de los OVNI." Barcelona, *Stendek* (5: 17), Sept. 1974, pp. 11-18.

**719**
Zullo, Allan
"Policeman and Farmer Followed UFO as Big as a Bus for 20 Minutes." *National Enquirer*, Sept. 1, 1974, p.40.

**720**
Toogood, Granville
"Earth is Being Visited by UFO's that may be Using Isolated Area of Utah as a Base." *National Enquirer*, Sept. 8, 1974, p.16.

**721**
Phillips, Ted
"OVNI Desciende en Delphos: Análisis del Suelo." Córdoba, *OVNIS* (I: 3), Sept.-Oct. 1974, pp. 28-33.

**722**
"Puerto Velaz: Plato Volador." Buenos Aires, *La Razón*, Oct. 18, 1974.

**723**
Galindez, Oscar A.
"Los Asombrosos Fenómenos de Trancas." Córdoba, Arg., *OVNIS* (I: 4), Nov.-Dec. 1974, pp. 2-12.

NOTES:

# HUMANOID CONTACT ON EARTH

## BOOKS

**724**
Bethurum, Truman
  The Voice of the Planet Clarion. Prescott, Ariz., 195-, 88p.

**725**
Grant, Walter V.
  Men From the Moon in America: Did They Come in a Russian Satellite? Dallas, Faith Clinic, 195-, 31p.

**726**
Grant, Walter V.
  Men in the Flying Saucers Identified: not a Mystery! Dallas, Faith Clinic, 195-, 32p. ($0.50)

**727**
Pelley, William D.
  Star Guests...Design for Mortality. Noblesville, Ind., Soulcraft Press, 1950, 318p.

**728**
Leslie, Desmond, & George Adamski
  Flying Saucers Have Landed. London, Werner Laurie, 1953, 232p.

**729**
Williamson, George H.
  Other Tongues--Other Flesh. Amherst, Wis., Amherst Press, 1953, 448p.

**730**
Fry, D. W.
  Alan's Message: to Men of Earth. Los Angeles, New Age Pub. Co., 1954, 41p.

**731**
Angelucci, Orfeo M.
  The Secret of the Saucers. Amherst, Wis., Amherst Press, 1955, 167p.

**732**
Buelta, Eduardo
  Astronaves Sobre la Tierra. Barcelona, Oromi, 1955.

**733**
Crandall, L.
  The Venusians. Los Angeles, New Age Pub. Co., 1955, 76p.

**734**
Thomas, Franklin
  We Come in Peace. Los Angeles, New Age Pub. Co., 1955, 53p.

---

NOTES:

## HUMANOID CONTACT ON EARTH

**735**
Marshall, James S., ed.
  The Planet Mars and its Inhabitants, by Eros Urides (a Martian). Chico, Calif., James E. Marshall, 1956, 37p.

**736**
Barton, Michel X.
  Flying Saucer Revelations. Los Angeles, Futura Press, 1957, 38p.

**737**
Barton, Michael X.
  Venusian Secret-science. Los Angeles, Futura Press, 1958, 76p.

**738**
Van Tassel, George
  The Council of Seven Lights. Los Angeles, De Vorss & Co., 1958, 156p.

**739**
Angelucci, Orfeo M.
  Son of the sun. Los Angeles, DeVorss & Co., 1959, 211p.

**740**
Barton, Michael X.
  Venusian Health Magic. Los Angeles, Futura Press, 1959, 59p.

**741**
Mitchell, Helen, & Betty Mitchell
  We met the Space People. Clarksburg, W. Va., Saucerian Pubs., 1959, 17p.

**742**
Faria, J. E.
  Discos Voadores; Contatos com Seres de Outros Planetas. São Paulo, Edições Melhoramentos, 1960, 105p.

**743**
Freudenthal, Hans
  Lincos, Design of a Language for Cosmic Intercourse. Amsterdam, North-Holland Pub., 1960-

**744**
Schmidt, Reinhold
  The Reinhold Schmidt Story. Los Angeles, Amalgamated Flying Saucer Clubs of America, 1960, 18p.

**745**
Nebel, Long John
  The way Out World. Englewood Cliffs, N. J., Prentice-Hall, 1961, 225p.

**746**
Veit, Karl
  Planeten-menschen. Wiesbaden-Schierstein, Ventla-Verlag, 1961, 223p.

**747**
Bender, Albert K.
  Flying Saucers and the Three men. Clarksburg, W. Va., Saucerian Books, 1962, 194p.

**748**
Carrougès, M.
  Les Apparitions des Martiens. Paris, Fayard Librairie, 1963, 287p.

**749**
Haley, Andrew G.
  Metalaw; Reassessment in the Light of Certain Views Expressed by the Chief Justice of the U.S. Paper Presented at the American Institute of Aeronautics and Astronautics, Summer Meeting, Los Angeles, Calif., June 17-20, 1963. Paper 63-279. Tarzana, Calif., 1963, 11p. ($1.00).

**750**
Knight, Oscar F.
  Wolverton Trail Event, a Visitor From Venus. Strathmore, Calif., 1963, 11p.

**751**
Layne, Meade
  The Coming of the Guardians. Vista, Calif., Borderland Sciences Research Associates Foundation, Inc., 1964, 72p.

NOTES:

## HUMANOID CONTACT ON EARTH

752
Mundo, Laura
 Pied Piper From Outer Space. Los Angeles, The Planetary Space Center Working Committee, 1964, 152p.

753
Barker, Gray
 Gray Barker's Book of Saucers. Clarksburg, W. Va., Saucerian Books, 1965, 77p.

754
Fry, Daniel W.
 The White Sands Incident. (Best Books, Inc., 1900 Bashford Manor Lane, Louisville, Ky. 40218), 1966, 120p. ($3.95)

755
Fuller, J. G.
 Incident at Exeter: the Story of Unidentified Flying Objects Over America Today. New York, G. P. Putnam's Sons, 1966, 251p. ($5.95)

756
Fuller, John G.
 The Interrupted Journey. New York, Dial Press, 1966, 302p.

757
Owens, Ted
 Flying Saucer Intelligences Speak. New Brunswick, Interplanetary News Service, 1966, 7p.

758
Barker, Gray, ed.
 Gray Barker's Book of Adamski. Clarksburg, W. Va., Saucerian Books, 1967, 78p.

759
Hudson, Jan
 Those Sexy Saucer People. Canterbury, N. H., Greenleaf Classics, 1967, 176p.

760
Lorenzen, Coral, & Jim Lorenzen
 Flying Saucer Occupants; a Selection of Reports About Unidentified Flying Objects Seen on the Ground and Their Occupants, From Aerial Phenomena Research Organization. New York, New American Library, 1967, 215p. ($0.95)

761
Saenz, Manuel, & Willy Wolf
 Los sin Nombre. Santiago, Chile, Editorial ORBE, 1967, 173p.

762
Shuttlewood, Arthur
 The Warminster Mystery. London, Neville Spearman, 1967, 265p. (25 shillings)

763
Stranges, Frank E.
 Stranger at the Pentagon. Van Nuys, Calif., International Evangelism Crusades, 1967, 201p.

764
Sykes, Egerton
 The Extra Terrestrials. London, Markham House P. (31 Kings Rd.), 1967, 24p. (8 pence)

765
Aberg Cobo, Axel
 Kosmokratores. Protectores del Espacio. Buenos Aires, Kier, 1968, 45p.

766
Easley, Robert S., & Rick R. Hilberg
 MIB, a Report on the Mysterious men in Black who Have Terrorized UFO Witnesses and Investigators in all Parts of the Nation. Cleveland, UFO Magazines Pubs., 1968, 24p.

767
Luna, Walter F.
 Los UFO's y su Posible Misión en la Tierra. Montevideo, 1968, 15p.

NOTES:

## HUMANOID CONTACT ON EARTH

**768**
Puccetti, Roland
  Persons: a Study of Possible Moral Agents in the Universe. London, Macmillan, 1968, 145p. (50 shillings)

**769**
Shuttlewood, Arthur
  Warnings from Flying Friends. Warminster, Partway P., 1968, 266p. (26 shillings)

**770**
Vidal, Franco
  Cuando...Extraterrestres en la Tierra. Barcelona, Editorial Linosa, 1968, 223p.

**771**
Michel, Aimé, et al
  The Humanoids. London, Spearman, 1969, 256p. (30 shillings).

**772**
Sesma, Fernando
  La Lógica del Visitante del Espacio. Madrid, Ed. Tesoro, 1969, 146p.

**773**
Smith, Wilbert B.
  The Boys From Topside. Ed. by Timothy Green Beckley. Clarksburg, W. Va., Saucerian Books, 1969, 96p.

**774**
Le Poer-Trench, Brinsley
  Operation Earth. London, Spearman, 1969, 128p. (30 shillings).

**775**
Bergier, Jacques
  Les Extra-Terrestres dans l'Histoire. Paris, Eds. J'ai lu, 1970, 187p. (3 francs)

**776**
Bowen, Charles, ed.
  The Humanoids. Chicago, Regnery, 1970, 256p. ($5.95)

**777**
Cardinale, Quixe
  Il Ritorno Delle Civiltá Perdute. Viaggio Alla Scoperta Delle Colonie Venusiane sul Nostro Pianeta. Rome, Newton Compton Italiana, 1970, 213p. (1200 lira)

**778**
Glemser, Kurt
  Men in Black; Startling New Evidence. Kitchener, Canada, 1970, 45p.

**779**
Anderson, Kenneth V.
  The Morphology and Physiology of UFO Occupants. (Unpublished paper for delivery at the APRO UFO Symposium, Tucson, Arizona, November 22-23, 1971).

**780**
Derenberger, Woodrow W.
  Visitors From Lanulos. As related by Woodrow W. Derenberger to the author, Harold W. Hubbard. New York, Vantage Press, 1971, 111p. ($3.75)

**781**
Keel, John A.
  Our Haunted Planet. Greenwich, Conn., Fawcett Pubs., 1971, 222p.

**782**
Bergier, Jacques, & INFO
  Le Livre de l'Inexplicable. Paris, Michel, 1972, 244p.

**783**
Dutta, Rex
  The Flying Saucer Message. London, Pelham Books, 1972, 117p.

**784**
Moore, Patrick
  Can You Speak Venusian? A Guide to Independent Thinkers. Newton Abbot, England, David & Charles, 1972, 176p.

NOTES:

**785**
Rossi, Dora Nelly V. de
   La Verdad del Ser y el Fenómeno OVNI.
Buenos Aires, Methopress, 1972, 91p.

**786**
Vallée, Jacques
   Chronique des Apparitions Extraterrestres.
Tr. into French by Dominique Pascoli.
Paris, Eds. Ep. Denoel, 1972, 448p. (29 francs)

**787**
Von Krueger, Frederick
   The Transparent People. New York,
Vantage Press, 1972, 33p.

**788**
Beere, D. C.
   USP; a Physics for Flying Saucers;
an Interpretation from Memory of a
Communication from Atos Xetrov, Visitor.
Del Mar, Calif., USP Press, 1973, 54p.

**789**
Ford, Brian J.
   The Earth Watchers. London, Leslie
Frewin, 1973, 189p.

**790**
Hewes, Hayden C.
   Earthprobe. The Complete Story of the "Piedmont Unexplained." Statements Taken From Over 200 Eyewitnesses, Complete With Illustrations and Unexplained Photos. Oklahoma City, International UFO Bureau, Inc., (P.O. Box 281), 1973, 26p.

**791**
Kolosimo, Peter
   This Timeless Earth. Tr. from Italian
by Paul Stevenson. London, Garnstone
Press, 1973, 270p.

**792**
Lob, Jacques, & Robert Gigi
   Ceux Venus d'Ailleurs. Neuilly, Dargaud,
1973, 64p.

**793**
Pinotti, Roberto
   Visitatori Dallo Spazio. Milan,
Armenia Ed., 1973, 277p.

**794**
Romaniuk, Pedro
   La Tierra Está Temblando...Causas y
sus Efectos. Fenómenos Geológicas y su
Relación con las Explosiones Atómicas.
2d ed. Buenos Aires, Larin, 1973 87p.

**795**
Schneider, Adolf
   Besucher aus dem All: das Geheimnis d.
Unbekannten Flugobjekte. Freiburg im
Breisgau, Bauer, 1973, 364p.

**796**
Stewart, Edward A.
   The Apollyon; a Guide in the Observation
and Reporting of the Unidentified Flying
Objects (UFO). Bordentown, N. J., 1973,
17p.

**797**
Bowen, Charles
   En Quête des Humanoïdes. Pari, Eds.
J'ai Lu, 1974, 308p.

**798**
Mark-Age
   Visitors from Other Planets. Miami,
Fla. 33137 (Mark-Age Meta Center, Inc.,
327 N.E. 20th Terrace), 1974, 334p.

NOTES:

## PERIODICAL ARTICLES

**799**
"Visitors From Venus; Flying Saucer Yarn." *Time* 55, Jan. 9, 1950, p. 49.

**800**
Carson, C.
"Those Little Men From Venus. Reply to R. Gelatt." *Saturday Review of Literature* 33, Oct. 21, 1950, p. 25.

**801**
"Steep Rock Flying Saucer." *Fate* 5, Feb.-March 1952, pp. 68-72.

**802**
Darrach, H. B., & R. E. Ginna
"Have we Visitors From Space?" *Life* 32, April 7, 1952, pp. 80-82.

**803**
Cahn, J. P.
"Flying Saucers and the Mysterious Little men." *True* 32, Sept. 1952, pp. 17-19, 102-112.

**804**
Hogben, Lancelot
"Astraglossa, or First Steps in Celestial Syntax." London, *Journal of the British Interplanetary Society* (11: 6), Nov. 1952, p. 258.

**805**
Barker, Gray
"The Monster and the Saucer." *Fate* 6, Jan. 1953, pp. 12-17.

**806**
Eisely, Loren C.
"Little Men and Flying Saucers." *Harper's Magazine* 206, March 1953, pp. 86-91.

**807**
Barker, Gray
"Visitors From Space." *Catholic Digest* 17, April 1953, pp. 7-10.

**808**
Thompson, William C.
"Houston bat man." *Fate* 6, Oct. 1953, pp. 26-27.

**809**
Fuller, Curtis
"The men who Ride in Saucers." *Fate* 7, May 1954, pp. 44-47.

**810**
Kunkel, Wallace
"The Little man who Wasn't There." *Fate* 7, May 1954, pp. 48-52.

**811**
"Martians Over France." *Time* 64, Oct. 25, 1954, p. 71.

**812**
Edwards, Frank
"The Spies From Outer Space." *Real* 5, Nov. 1954, pp. 20-21, 58-60.

**813**
Green, Vaughn M.
"Flying Monsters." *Fate* 8, Jan. 1955, pp. 112-114.

**814**
Fuller, Curtis
"Little men From all Over." *Fate* 8, March 1955, pp. 30-33.

**815**
"Waiting for the Little men." *Newsweek* 45, March 28, 1955, p. 64.

**816**
Ibson, Jack
"Did a Space Visitor Land in Bradford?" *Flying Saucer News*, Autumn 1955, pp. 4-5.

NOTES:

# HUMANOID CONTACT ON EARTH

**817** Bessor, John P.
"Are the Saucers Space Animals?" Fate 8, Dec. 1955, pp. 6-12.

**818** "Space Visitors: Examples of Mysterious and Well-authenticated Unidentified Flying Objects, by 'Theorist'." London, Practical Mechanics 24, Dec. 1956, pp. 138-141.

**819** "Meeting Extraterrestrials." Science News Letter 70, Dec. 15, 1956, p. 382.

**820** "Metalaw." New Yorker 32, Dec. 29, 1956, p. 19.

**821** "Elizabeth Klarer's Flying Saucer." Flying Saucers From Other Worlds, June 1957, pp. 65-69, 75.

**822** Ormond, Ron
"I Found a Little Green man." Flying Saucers From Other Worlds, Aug. 1957, pp. 22-26.

**823** Edwards, Frank
"Frank Edwards' Report: What do Flying Saucers Want?" Fate 10, Sept. 1957, pp. 19-27.

**824** Michel, Aimé
"Flying Saucers in Europe, Meeting With the Martian." Fate 10, Sept. 1957, pp. 43-46.

**825** Michel, Aimé
"Flying Saucers in Europe, the Little men." Fate 10, Nov. 1957, pp. 72-76.

**826** Helland, Albert E.
"They Caught a Spaceman." Fate 11, March 1958, p. 62.

**827** Menger, Howard
"The Howard Menger Story." London, Flying Saucer Review 4, July-Aug. 1958, pp. 10-12, 111.

**828** Edwards, Frank
"Frank Edwards' Report, Spacemen-or Monsters." Fate 11, Sept. 1958, pp. 76-83.

**829** Lorenzen, Coral
"The Reality of the Little men." Flying Saucers, Dec. 1958, pp. 26-34.

**830** Barker, Gray
"The Case for Non-Human Space Visitors." Flying Saucers, Feb. 1959, pp. 18-23.

**831** "A Saucer, two men, and 'Little Creatures'." Flying Saucers, May 1959, pp. 57-60.

**832** Schmidt, Reinhold
"The Kearney Incident." Flying Saucers From Other Worlds, Oct. 1959, pp. 31-45.

**833** Lemaitre, Jules
"Angels or Monsters?" London, Flying Saucer Review 5, Nov.-Dec. 1959, pp. 3-5.

**834** James, Trevor
"Scientists, Contactees and Equilibrium." London, Flying Saucer Review 6, Jan.-Feb. 1960, pp. 19-21.

NOTES:

## HUMANOID CONTACT ON EARTH

835
Fontes, Olavo T.
 "Brazilian top Secret Report Revealed."
 *Flying Saucers*, Feb. 1960, pp. 6-14, 24.

836
Miller, Max B.
 "The men who Ride in Saucers." *Fate* 13,
Feb. 1960, pp. 32-38.

837
James, Trevor
 "Space Animals--a Fact of Life." London,
*Flying Saucer Review* 6, July-Aug. 1960, pp.
3-7.

838
Draper, H.
 "Afternoon With the Space People." *Harper's Magazine* 221, Sept. 1960, pp. 37-40.

839
Golomb, Solomon W.
 "Extraterrestrial Linguistics."
*Astronautics* (6: 5), May 1961, p. 46.

840
Oakley, C. O.
 "Math, our Link With Space People." *Science Digest* 49, June 1961, pp. 7-13.

841
"A Brazilian Contact Claim." London, *Flying Saucer Review* 7, Sept.-Oct. 1961, pp. 18-20.

842
James, Trevor
 "The Case for Contact." London, *Flying Saucer Review* 8, Jan.-Feb. 1962, pp. 9-11.

843
Miller, E. C.
 "Correspondence: Ethics and Space Travel."
London, *Spaceflight* 4, July 1962, p. 139.

844
"Luciano Galli's Contact Claim." London,
*Flying Saucer Review* 8, Sept.-Oct. 1962, pp.
29-30.

845
Silvano, Ceccarelli
 "Mario Zuccala's Strange Encounter." London,
*Flying Saucer Review* 8, July-Aug. 1962, pp.
5-7.

846
Albanesi, Renato
 "The Italian Scene, Signor Siragusa's Message." London, *Flying Saucer Review* 9,
Jan.-Feb. 1963, pp. 3-5.

847
Burr, Frank
 "Visitors From Afar." London, *Flying Saucer Review* 9, March-April 1963, pp. 19-20.

848
Creighton, Gordon W.
 "The Italian Scene, Bruno Ghibaudi's Contact Claim." London, *Flying Saucer Review* 9, May-June 1963, pp. 18-20.

849
"Another Speech by Wilbert B. Smith." London,
 *Flying Saucer Review* 9, Nov.-Dec. 1963, pp.
11-14.

850
Vallée, Jacques
 "A Descriptive Study of the Entities Associated With the Type 1 Sighting." London,
*Flying Saucer Review* 10, Jan.-Feb. 1964, pp.
6-12.

851
Ribera, Antonio
 "What Happened at Fatima?" London, *Flying Saucer Review* 10, March-April 1964, pp. 12-14.

852
"Entities Associated With Type I Sightings.
Part two, the Scientific Interpretation."
London, *Flying Saucer Review* 10, May-June
1964, pp. 3-5.

NOTES:

**853**
Fouéré, René
"Adamski's Last Chance." London, <u>Flying Saucer Review</u> 10, Sept.-Oct. 1964, pp. 27-29.

**854**
Avignon, André
"Des Animaux-machines?" Paris, <u>Phénomènes Spatiaux</u>, Nov. 1964, pp. 13-14.

**855**
Bowen, Charles
"Mystery Animals." London, <u>Flying Saucer Review</u> 10, Nov.-Dec. 1964, pp. 15-17.

**856**
Carr, Aidan M.
"Take me to Your Leader." <u>Homiletic and Pastoral Review</u> 65, Dec. 1964, pp. 255-256.

**857**
Le Poer-Trench, Brinsley
"The Three W's." <u>Saucer News</u> 11, Dec. 1964, pp. 7-10.

**858**
Clark, Jerome
"A Contact Claim." London, <u>Flying Saucer Review</u> 11, Jan.-Feb. 1965, pp. 30-32.

**859**
Roberts, Keith
"Reconsidering the Mysterious 'Little men'." <u>Saucer News</u> 12, March 1965, pp. 7-9.

**860**
Duchêne, J. L.
"La Repartition des Attérrissages de Soucoupes Volantes en France." Paris, <u>Phénomènes Spatiaux</u>, May 1965, pp. 21-25.

**861**
Bowen, Charles
"Crash-landed UFO Near Mendoza." London, <u>Flying Saucer Review</u> 11, May-June 1965, pp. 7-9.

**862**
Clark, Jerome
"Two new Contact Claims." London, <u>Flying Saucer Review</u> 11, May-June 1965, pp. 20-23.

**863**
Leslie, Desmond
"George Adamski." London, <u>Flying Saucer Review</u> 11, July-Aug. 1965, pp. 18-19.

**864**
"The Warminster Phenomenon." London, <u>Flying Saucer Review</u> 11, July-Aug. 1965, pp. 3, 9.

**865**
Cleary-Baker, John
"Report on Warminster." London, <u>BUFORA Journal and Bulletin</u> 1, Autumn 1965, pp. 6-8.

**866**
Cleary-Baker, John
"The Scoriton Affair." London, <u>BUFORA Journal and Bulletin</u> 1, Autumn 1965, pp. 10-11.

**867**
"L'affaire de Valensole." Paris, <u>Phénomènes Spatiaux</u>, Sept. 1965, pp. 5-24.

**868**
"Enquête à Valensole." Paris, <u>Phénomènes Spatiaux</u>, Sept. 1965, pp. 42-46.

**869**
Clark, Jerome
"The Meaning of Contact." London, <u>Flying Saucer Review</u> 11, Sept.-Oct. 1965, pp. 28-29.

**870**
Fuller, John G.
"Trade Winds: Report of an Unidentified Flying Object in Exeter, N.H." <u>Saturday Review</u> 48, Oct. 2, 1965, p. 10.

NOTES:

## HUMANOID CONTACT ON EARTH

**871**
Michel, Aimé
"The Valensole Affair." London, <u>Flying Saucer Review</u> 11, Nov.-Dec. 1965, pp. 7-9.

**872**
"Retour sur Valensole, les Conclusions de Notre Enquêteur." Paris, <u>Phénomènes Spatiaux</u>, Dec. 1965, pp. 11-16.

**873**
Fuller, John G.
"Trade Winds: Exeter People Give Accounts of Observations." <u>Saturday Review</u> 49, Jan. 22, 1966, p. 14.

**874**
Jansen, Clare J.
"Little tin men in Minnesota." <u>Fate</u> 19, Feb. 1966, pp. 36-40.

**875**
Fuller, John G.
"Outer Space Ghost Story: Excerpt From Incident at Exeter." <u>Look</u> 30, Feb. 23, 1966, p. 36.

**876**
Michel, Aimé
"Valensole." Paris, <u>Phénomènes Spatiaux</u>, March 1966, pp. 21-26.

**877**
Adamski, George
"How to Know a Spaceman, if you see one." <u>Probe</u> 3, March-April 1966, pp. 5-6.

**878**
Farish, Lucius
"Cattle Rustling by UFO." <u>Fate</u> 19, April 1966, pp. 42-45.

**879**
Crenshaw, James
"The Great Venusian Mystery." <u>Fate</u> 19, June 1966, pp. 32-39.

**880**
Fouèrè, René
"Seraient-ils des Revenants du Futur?" Paris, <u>Phénomènes Spatiaux</u>, June 1966, pp. 11-14.

**881**
Fuller, John G.
"Incident at Exeter." Review by Oscar Handlin. <u>The Atlantic</u> 218, Aug. 1966, pp. 116-117.

**882**
Robinson, Jack, & Mary Robinson
"The Case for Extraterrestrial Little men." <u>Saucer News</u> 13, Fall 1966, pp. 7-9.

**883**
Caputo, Livio
"Rapporto sui Dischi Volanti, Qualcuno ha Parlato con 'Loro'." Milan, <u>Epoca</u> LXIV, Sept. 11, 1966, pp. 30-38.

**884**
"The Humanoids." London, <u>Flying Saucer Review</u>, Oct.-Nov. 1966, 32p.

**885**
Fuller, John G.
"Aboard a Flying Saucer." <u>Look</u> 20, Oct. 4, 1966, pp. 44-48, 53-56.

**886**
Drury, Neville
"Flying Ships in 'Oahspe'." Moorabin, <u>Australian Flying Saucer Review</u> 9, Nov. 1966, pp. 40-41.

**887**
Keel, John A.
"New Landing and Creature Reports." London, <u>Flying Saucer Review</u> 12, Nov.-Dec. 1966, pp. 5-8.

**888**
"Etranges Créatures." Paris, <u>Phénomènes Spatiaux</u>, Dec. 1966, pp. 32-34.

NOTES:

**889**
"Retour sur Attigneville, L'incident de Xertigney." Paris, Phénomènes Spatiaux, Dec. 1966, pp. 13-17.

**890**
Creighton, Gordon W.
"The Villa Santina Case." London, Flying Saucer Review 13, Jan.-Feb. 1967, pp. 3-7.

**891**
Cohen, D.
"Have we Been Visited by Spacemen?" Science Digest 61, Feb. 1967, pp. 84-85.

**892**
Keel, John A.
"Never Mind the Saucer! Did you see the Guys who Were Driving?" True 48, Feb. 1967, pp. 36-37, 78, 80-83.

**893**
Keel, John A.
"The UFO Kidnappers." Saga 33, Feb. 1967, pp. 10-14, 50, 52-54, 56-60, 62.

**894**
Goupil, Jean
"L'hypothèse du Champ Répulsif, Essaie D'explications des Incidents de Kelly." Paris, Phénomènes Spatiaux, March 1967, pp. 18-23.

**895**
Keel, John A.
"The 'Silencers' at Work." London, Flying Saucer Review 13, March-April 1967, p. 10.

**896**
Giles, Gordon A.
"The UFO's Have Taken Over the Earth." Monsieur 9, April 1967, pp. 10-13, 54-55.

**897**
Sheldon, Jean
"I was Seduced by a Flying Saucer." The National Tattler 6, April 1967, pp. 1, 8-9.

**898**
Brandt, Ivan
"The Problem of the Frankensteins." London, Flying Saucer Review 13, May-June 1967, pp. 16-19.

**899**
Creighton, Gordon W.
"Three More Brazilian Cases." London, Flying Saucer Review 13, May-June 1967, pp. 5-8.

**900**
Keel, John A.
"From my Ohio Valley Note Book." London, Flying Saucer Review 13, May-June 1967, pp. 3-5.

**901**
Bowen, Charles
"Fantasy or Truth?" London, Flying Saucer Review 13, July-Aug. 1967, pp. 11-14.

**902**
Winder, R. B. H.
"The Little Blue man on Studham Common." London, Flying Saucer Review 13, July-Aug. 1967, pp. 3-4.

**903**
Creighton, Gordon
"The Extraordinary Happenings at Casa Blanca." London, Flying Saucer Review 13, Sept.-Oct. 1967, pp. 16-18.

**904**
Fontes, Olavo T.
"Dying Girl Saved by Humanoid Surgeons." London, Flying Saucer Review 13, Sept.-Oct. 1967, pp. 5-6.

**905**
Fuller, Jean
"The Exeter Incidents." London, Flying Saucer Review 13, Sept.-Oct. 1967, pp. 25-27.

**906**
Harney, John, & Alan W. Sharp
"Report on a Visit to Warminister." London, Flying Saucer Review 13, Sept.-Oct. 1967, pp. 3-4.

NOTES:

## HUMANOID CONTACT ON EARTH

**907**
Keel, John A.
"UFO 'Agents of Terror'." Saga 33, Oct. 1967, pp. 29-31, 72-74, 76-79, 81.

**908**
Bessor, John P.
"UFO's: Animal or Mineral?" Fate 20, Nov. 1967, pp. 32-39.

**909**
Keel, John
"Strange Messages From Flying Saucers." Saga 35, Jan. 1968, pp. 22-25, 69-70, 72-74.

**910**
Keel, John
"An Unusual Contact Claim From Ohio." London, Flying Saucer Review 14, Jan.-Feb. 1968, pp. 25-26.

**911**
Michel, Aimé, & Charles Bowen
"A Visit to Valensole." London, Flying Saucer Review 14, Jan.-Feb. 1968, pp. 6-12.

**912**
Steiger, Brad, & Joan Whritenour
"Abominable Spacemen." Saga 35, Feb. 1968, pp. 34-35, 58-60, 62, 64.

**913**
Finch, Bernard E.
"Can They see us?" London, Flying Saucer Review 14, March-April 1968, p 31.

**914**
Hugill, Joanna
"On the Road From Sydney to Melbourne." London, Flying Saucer Review 14, March-April 1968, pp. 3, 11.

**915**
Keel, John A.
"The Little man of Gaffney." London, Flying Saucer Review 14, March-April 1968, pp. 17-19.

**916**
"'M.O.C.' et Leurs Occupants vus au sol en 1967." Paris, Lumières Dans la Nuit 93, March-April 1968, pp. 14-16.

**917**
Keel, John A.
"UFO Report, the Sinister men in Black." Fate 21, April 1968, pp. 32-39.

**918**
Knowlson, James R.
"Note on Bishop Godwin's Man in the Moon: the East Indies Trade Route and a Language of Musical Notes. Modern Philology 65, May 1968, pp. 357-361.

**919**
Steiger, Brad
"What Price Silence?" Flying Saucers, June 1968, p. 31.

**920**
Prytz, John
"Legal Aspects of Exobiology." Flying Saucers 60, Oct. 1968, pp. 14-15.

**921**
Keel, John A.
"The 'Little Man' of North Carolina." London, Flying Saucer Review (15: 1), Jan.-Feb. 1969, pp. 15-16.

**922**
Bowen, Charles
"A Fatal Encounter." London, Flying Saucer Review (15: 2), March-April 1969, pp. 13-14.

**923**
Harney, John
"The 'Men in Black' Reports." London, Flying Saucer Review (15: 2), March-April 1969, pp. 9-11, 20.

NOTES:

**924**
Fawcett, George D.
   "UFO With Occupants." *Flying Saucers* 64, June 1969, p. 13.

**925**
Robinson, G. S.
   "Ecological Foundations of Haley's Metalaw." London, *Journal of the British Interplanetary Society* 22, Aug. 1969, pp. 266-274.

**926**
Creighton, Gordon
   "The 'One-Eyed Entities' of Bêlo Horizonte." London, *Flying Saucer Review*, Special issue #3, Sept. 1969, pp. 28-32.

**927**
Michel, Aimé
   "The Strange Case of Dr. X." London, *Flying Saucer Review*, special issue #3, Sept. 1969, pp. 3-16.

**928**
Schwarz, Berthold
   "Gary Wilcox and the Ufonauts." London, *Flying Saucer Review*, special issue #3, Sept. 1969, pp. 20-27.

**929**
Schwarz, Berthold
   "UFO Occupants: Fact or Fantasy?" London, *Flying Saucer Review* (15: 5), Sept.-Oct. 1969, pp. 14-18.

**930**
Verplaetse, Juliaan
   "Verstärktes Analogon des Irdischen Rechts im Weltraum?" Cologne, *Zeitschrift für Luftrecht und Weltraumrechtsfragen* 19, Jan. 1, 1970, pp. 140-145.

**931**
Glemser, Kurt
   "Little men From Saucers." *Search* 90, March 1970, pp. 44-48.

**932**
Ratliff, Buffard
   "A Fossilized Alien Spaceship and its Occupants." *Flying Saucers* 68, March 1970, pp. 6-7.

**933**
Edwards, P. M. H.
   "Speech of the Aliens." London, *Flying Saucer Review* (16: 1), Jan.-Feb. 1970, pp. 11-12, 14; (16: 1), March-April 1970, pp. 23-25.

**934**
Pereira, Jader U.
   "Les Extra-terrestres." Paris, *Phénomènes Spatiaux* 24, June 1970-.

**935**
Stamey, Dennis
   "Roundup of UFO Occupant Sightings." *Flying Saucers* 69, June 1970, p. 36.

**936**
McDonald, James E.
   "Y a-t-il des Preuves de l'Existence d'un Péril ou d'Hostilité dans le Phénomène UFO." Paris, *Phénomènes Spatiaux* 25, Sept. 1970, pp. 10-14.

**937**
Saunders, Alex
   "Are Martians Living on Earth?" *Search* 93, Sept. 1970, pp. 9-13.

**938**
Fredrickson, Sven O.
   "A Humanoid was Seen at Imjarvi." London, *Flying Saucer Review* (16: 5), Sept.-Oct. 1970, pp. 14-18.

**939**
Liljegren, Anders
   "Mariannelund UFO and Occupants." London, *Flying Saucer Review* (16: 6), Nov.-Dec. 1970, pp. 14-17.

NOTES:

## HUMANOID CONTACT ON EARTH

**940**
Rimes, Nigel
"Muzio's Contacts." London, Flying Saucer Review (17:1), Jan.-Feb. 1971, pp. 24-27.

**941**
Saunders, Alex
"Communicating With Aliens and Non-Human Terrestrials." Flying Saucers 72, March 1971, pp. 12-15.

**942**
Schmidt, Robert A.
"Callery UFO and Occupants: A new report from Pennsylvania." London, Flying Saucer Review (17:4), July-Aug. 1971, pp. 3-5.

**943**
Friedman, Stanton T.
"Top Nuclear Physicist and UFO Expert Claims Evidence Shows Beings From Other Worlds are Exploring Earth." National Enquirer, Aug. 22, 1971, p.5.

**944**
Aggen, Erich
"UFO Mystery men." Flying Saucers 74, Sept. 1971, p. 34.

**945**
Galindez, Oscar A.
"Réflexions sur le Phénomène Humanoïde." Paris, Phénomènes Spatiaux 28, June 1971, pp. 10-17; 29, Sept. 1971, pp. 14-17.

**946**
Pereira, J. U.
"Les Extra-Terrestres." Paris, Phénomènes Spatiaux #24, June 1970, pp. 14-20; #25, Sept. 1970, pp. 21-28; #27, March 1971, pp. 25-31; #28, June 1971, pp. 28-33, #29, Sept. 1971, pp. 18-28.

**947**
Wentworth, Jim
"The Hidden Aspect of Alien Visitation." Flying Saucers 74, Sept. 1971, pp. 2-5.

**948**
Holiday, F.W.
"The Possible Polarisation of Monster Phenomena." London, Flying Saucer Review (17:6), Nov.-Dec. 1971, pp. 18-20.

**949**
Vanquelef, G.
"Les Occupantes des MOC et Leur Comportement." Le Chambon-sur-Lignon, France, Lumières dans la Nuit (14:115), Dec. 1971, pp. 7-11; (15:116), Feb. 1972, pp. 4-7.

**950**
"Entrainés pas des Etres de l'Espace." Paris, Phénomènes Spatiaux 31, March 1972, pp. 29-33.

**951**
Fawcett, George D.
"UFOnauts in the 1970's." Flying Saucers 76, March 1972, pp. 11-13.

**952**
Michel, Aimé
"An Enigmatic Figure of the XVIIth Century." London, Flying Saucer Review (18:2), March-April 1972, pp. 3-6.

**953**
Johannes, Lewis W.
"Shaver, Saucers and the 3 men in Black." Flying Saucers 77, June 1972, pp. 16-18.

**954**
Saunders, Alex
"Are we Never to Confront an Alien?" Flying Saucers 77, June 1972, pp. 25-26.

**955**
Bowen, Charles
"Landings and Humanoids Reported in Cape Province: Speculation on the cause of the damage to Rosmead Tennis court." London, Flying Saucer Review (19:1), Jan.-Feb. 1973, pp. 12-14.

NOTES:

**956**
Clark, Jerome, and Loren Coleman
"Anthropoids, Monsters and UFO's." London, Flying Saucer Review (19:1), Jan.-Feb. 1973, pp. 18-24.

**957**
Vallée, Jacques
"Occupant Symbolism in Phoenician Mythology: Speculation concerning UFO-related rituals in ancient amulets." London, Flying Saucer Review (19:1), Jan.-Feb. 1973, pp. 7-9.

**958**
Ballester Olmos, Vicente J.
"Données Biométriques dan 19 cas d'Occupants d'UFO's." Paris, Phénomènes Spatiaux 35, March 1973, pp. 3-10.

**959**
Glemser, Kurt
"UFO's and Dead Animals." Search 108, March 1973, pp. 29-31.

**960**
Ballester Olmos, Vicente J.
"Biometric Data in 19 UFO Occupant Cases." London, Flying Saucer Review (19:3) May-June 1973, pp. 19-23.

**961**
Prytz, John
"The Sociology of UFOlogy" (Part 2). Flying Saucers 81, Summer 1973, pp. 49-51.

**962**
Saunders, Alex
"Would man Benefit by Alien Contact?" Flying Saucers 81, Summer 1973, pp. 22-27.

**963**
Dutuit, J. M.
"Le Problème des Motivations..." Paris, Phénomènes Spatiaux 36, June 1973, pp. 3-12.

**964**
Buckle, Eileen
"Aurora Spaceman-R.I.P.?" London, Flying Saucer Review (19:4), July-Aug. 1973, pp. 7-9.

**965**
Tyrode, J.
"Taizé: A Case Right Out of the Ordinary." London, Flying Saucer Review (19:4), July-Aug. 1973, pp. 16-21.

**966**
"Le Fantastique Incident de Catanduva." Paris, Phénomènes Spatiaux 37, Sept. 1973, pp. 13-19.

**967**
"Pascagoula ou les Pêcheurs Pêchés: un Drôle d'Occupant." Phénomènes Spatiaux (38:4), Dec. 1973, pp. 32-33.

**968**
Stokes, Bill
"Could Your Next-Door Neighbor be From Mars." Milwaukee Journal, Accent sec., Dec. 18, 1973, p. 4.

**969**
Techter, David
"Mississippi Report: Terror Aboard a UFO." Fate 287, Feb. 1974, pp. 36-42.

**970**
Galindez, Oscar A.
"Argentina: los Fenómenos Antropomorfos de Santa Isabel." Barcelona, Stendek (V:15), Dec. 1972-March 1974, pp. 15-30.

**971**
"ONIFE Investiga Caso Carcaraña." Buenos Aires, Cuarta Dimensión 10, June 1974, pp. 26-27.

NOTES:

## HUMANOID CONTACT ON EARTH

**972**
Zerpa, Fabio
 "Caso Tandil del 14 de Abril de 1974." Buenos Aires, Cuarta Dimensión 10, June 1974, pp. 8-11.

**973**
"Bahía Blanca; l'Extraordinaire Rencontre du Camionneur Dionisio Llanca." Paris, Phénomènes Spatiaux 40-41-42, June-Sept.-Dec. 1974, pp. 18-26.

**974**
Pouèrè, René
 "Un Mort Mystérieuse dont on Reparle Celle de João Prestes Filho." Paris, Phénomènes Spatiaux 40-41-42, June-Sept.-Dec. 1974, pp.26-31.

**975**
Galíndez, Oscar A.
 "Los Fenómenos Antropomorfos de Santa Isabel." Córdoba, OVNIS, un Desafío a la Ciencia (I: 2), July-Aug. 1974, pp. 16-21.

**976**
Crexells, Joan, & Père Redón
 "La Espectacular Aventura de Maxi Iglesias, en Salamanca." Barcelona, Stendek (5: 17), Sept. 1974, pp. 2-10.

**977**
Ballester-Olmos, Vicente J.
 "Datos Biométricos en 19 Casos de Ocupantes." Córdoba, Arg., OVNIS (I:3), Sept.-Oct. 1974, pp. 13-19.

**978**
Schwartz, Berthold
 "Ocupantes: Realidad o Fantasía? Un Estudio Siquiátrico." Córdoba, OVNIS (I: 3), Sept.-Oct. 1974, pp. 1012.

**979**
Scorneaux, Jacques
 "El Supuesto Contacto Hickson-Parker." Córdoba, OVNIS (I: 3), Sept.-Oct. 1974, pp. 7-9.

**980**
Scorneaux, Jacques
 "Humanoides en Canada...". Translated from French by Oscar A. Galíndez. Córdoba, Arg., OVNIS (I: 4), Nov.-Dec. 1974, inside back cover.

**981**
Pease, Harry S.
 "Did Their Road Map Cover Some 220 Trillion Miles?" Milwaukee Journal, Dec. 8, 1974 (pt.5), p. 2.

NOTES:

## INTELLIGENT LIFE ON OTHER PLANETS

## BOOKS

**982**
Copland, Alexander
  The Existence of Other Worlds, Peopled With Living and Intelligent Beings, Deduced From the Nature of the Universe. London, J. G. & F. Rivington, 1834, 210p.

**983**
Brewster, David
  More Worlds Than one. New York, Robert Carter & Bros., 1854, 265p.

**984**
Flammarion, Camille
  Stories of Infinity: Lumen--History of a Comet--in Infinity. Boston, Roberts Brothers, 1873, 287p.

**985**
Warder, George W.
  The Cities of the sun. New York, G. W. Dillingham, 1901, 320p.

**986**
Wells, Herbert G.
  The First men on the Moon. London, G. Newnes, Ltd., 1901, 342p.

**987**
Moreux, Theophile.
  Les Autres Mondes, Sont-ils Habités? Paris, Editions Scientifica, 1912, 134p.

**988**
Maunder, E. W.
  Are the Planets Inhabited? New York, Harper Pubs., c/o Harper & Row, 1913. ($1.25)

**989**
Housden, C. E.
  Is Venus Inhabited? London, Longmans, Green, 1915, 39p.

**990**
Shipley, M.
  Are the Planets Inhabited? Girard, Kansas, Haldeman-Julius Co., 1924, 64p.

**991**
Younghusband, F.
  Life in the Stars; an Exposition of the View That on Some Planets of Some Stars Exist Beings Higher Than Ourselves, and on one a World Leader, the Supreme Embodiment of the Eternal Spirit Which Animates the Whole. 2d ed. London, John Murray, 1928, 222p.

NOTES:

## INTELLIGENT LIFE ON OTHER PLANETS

992
Younghusband, F.
  *The Living Universe.*  New York, Dutton, 1933, 252p.

993
Fontenelle, Bernard le Bovier de
  *Entretiens sur la Pluralité des Mondes.*
Paris, Éditions de la Nouvelle France, 1945, 201p.

994
White, George S.
  *A Book of Revelations: a True Narrative of Life on Many Planets.*  Los Angeles, 1945, 198p.

995
Bailey, James O.
  *Pilgrims Through Space and Time.*
New York, Argus, 1947, 341p.

996
Lönnqvist, C.
  *Människan på Världsteatern.*  Stockholm, Ljus, AB, 1947, 218p.

997
Heard, Gerald
  *The Riddle of the Flying Saucers. Is Another World Watching?*  London, Carroll & Nicholson, 1950, 157p.

998
Moreux, T.
  *Les Autres Mondes Sont-ils Habités?*  New Edition. Paris, G. Doin et Cie., 1950, 141p. (210 francs)

999
Hartlaub, G. F.
  *Bewusstsein auf Anderen Sternen? Ein Kleiner Leitfaden Durch die Menschheitsraume von den Planetenbewohnern.*  Munich, Ernst Reinhardt Verlag, 1951, 65p.

1000
Heuer, Kenneth
  *Men of Other Planets.*  London, Victor Gallancz, Ltd., 1951, 160p.

1001
  *The Mystery of Other Worlds Revealed.*  Greenwich, Conn., Fawcett Pubs., 1952, 144p.

1002
Mugler, Charles
  *Deux Thèmes de la Cosmologie Grecque: Devenir Cyclique et Pluralité des Mondes.*
Paris, Librairie C. Klincksieck, 1953, 92p.

1003
  *The Mystery of Other Worlds Revealed.*  New York, Sterling Pub. Co., 1953, 144p.

1004
Ashtar.  *In Days to Come.*  Los Angeles, New Age Pub. Co., 1955, 91p.

1005
Cove, Gordon
  *Who Pilots the Flying Saucers.*  London, 1955, 80p.

1006
Ferguson, William
  *A Message From Outer Space.*  Oak Park, Ill., Golden Age Press, 1955, 54p.

1007
Norman, Ernest L.
  *The Truth About Mars.*  Los Angeles, New Age Pub. Co., 1956, 61p.

1008
Bujanda, Jesús
  *Astronomía y Astros Habitados.*  Madrid, Ed. Razón y Fé, 1957, 284p.

NOTES:

# INTELLIGENT LIFE ON OTHER PLANETS

**1009**
Ferber, Adolph C.
  The Secret of Human Life on Other Worlds.
New York, Pageant Press, 1957, 105p.

**1010**
Gatland, Kenneth W., & Derek D. Dempster
  The Inhabited Universe. London, Alan
Wingate, 1957, 182p.

**1011**
McCoy, John
  They Shall Be Gathered Together.
Corpus Christi, 1957, 74p.

**1012**
Muller, Wolfgang D.
  Man Among the Stars. New York, Criterion
Books, 1957, 307p.

**1013**
Girvin, Calvin C.
  The Night Has a Thousand Saucers. El Monte,
Calif., Understanding Pub. Co., 1958, 168p.

**1014**
James, Trevor
  They Live in the sky. Los Angeles, New Age
Pub., 1958, 270p.

**1015**
Rocha, H.
  Outros Mundos, Outras Humanidades. Pôrto,
Editôra Educação Nacional, 1958, 371p.

**1016**
Rocha, Hugo
  Outros Mundos Outras Humanidades. Pôrto,
Editôra Educação Nacional, 1958, 371p.

**1017**
Kolosimo, Peter
  Il Pianeta Sconosciuto. Turin, S.E.I.,
1959, 320p. (1400 lira)

**1018**
Lugo, Francisco A.
  Los Visitantes del Espacio. Buenos Aires,
Ed. Continente, 1959.

**1019**
Menger, Howard
  From Outer Space to you. Clarksburg, W. Va.,
Saucerian Books, 1959, 256p.

**1020**
Williamson, George H.
  Road in the sky. London, Neville Spearman,
1959, 248p.

**1021**
Drake, Walter R.
  Spacemen in Antiquity. Sunderland, England,
196-

**1022**
Le Poer-Trench, Brinsley
  The Sky People. London, Neville, Spearman,
Ltd., 1960, 224p.

**1023**
Li, Ch'u
  Pieh ti Hsing Hsing Shang yu mei yu Sheng
Ming. Hong Kong, Yo-Ying Book Store, 1960,
37p.

**1024**
Stranges, Frank E.
  Danger From the Stars. Venice, Calif.,
International Evangelism Crusades, Inc.,
1960, 14p.

**1025**
Vries, Tjomme E.
  A la Découverte de L'Univers. Tr. from
Dutch by Andre M. Hendrickx. Paris, Editions
Sequoia, 1960, 155p.

**1026**
Lapp, Ralph E.
  Man and Space. New York, Harper, 1961,
183p. (see Chapters 9, 10).

NOTES:

## INTELLIGENT LIFE ON OTHER PLANETS

**1027**
Margaria, R.
 On the Possible Existence of Intelligent Living Beings on Other Planets. (In Baker, R. M. L., & M. W. Makemson, eds. International Astronautical Congress. 12th, Washington, Oct. 1-7, 1961. Proceedings. v.12. Vienna, Springer-Verlag, 1963, pp. 556-563).

**1028**
Pickering, J. S.
 Captives of the sun. New York, Dodd, Mead, 1961, 319p. ($4.95)

**1029**
Tikhov, G. A.
 Leben im Weltall. Tr. by Inge Rawald. Leipsig, Urania Verlag, 1961, 72p.

**1030**
Asimov, Isaac
 Fact and Fancy. Garden City, N.Y., Doubleday, 1962, 264p. (see chapter 11).

**1031**
Briggs, Michael H.
 The Distribution of Life in the Solar System: an Evaluation of the Present Evidence. Preprint. Wellington, N.Z., Victoria University, Sept. 1962, 25p.

**1032**
Drake, F. D.
 Intelligent Life in Space. New York, Macmillan, 1962, 128p. ($3.50)

**1033**
Jackson, C. D., & R. E. Hohmann
 An Historic Report on Life in Space. New York, American Rocket Society, 1962, 7p. (ARS Paper #2730-62).

**1034**
MacGowan, R. A.
 On the Possiblities of Extraterrestrial Intelligence. (In Ordway, F. I., ed. Advances in Space Science and Technology 4. New York, Academic Press, 1962, pp. 39-111).

**1035**
Misraki, Paul
 Les Extraterrestres, par Paul Thomas (pseud.). Paris, Plon, 1962, 224p.

**1036**
Ovenden, Michael W.
 Leben im Weltall? Eine Wissenschaftliche Diskussion. Tr. by Eberhard Böhringer. Basle, Desch, 1962, 186p. (4.35 Swiss francs)

**1037**
Rublowsky, J.
 Is Anybody out There? New York, Walker, 1962, 118p. ($3.95)

**1038**
Tikhov, G. A.
 Reaching for the Stars. Moscow, Foreign Languages Publishing House, 1962, 152p.

**1039**
Ahmad, I. I.
 Sukan al-Kwakib. Cairo, Al-Mussasa al-Misriyah al-Ahmmah Litahlif wa-Tarjama wa-Tibah wa-Nashir, 1963, 150p.

**1040**
Anderson, Poul
 Is There Life on Other Worlds? New York, Crowell-Collier Press, 1963, 223p.

**1041**
Briggs, Michael H.
 New Evidence of Martian Life. (In Gatland, K. W., ed. Spaceflight Today. London, Iliffe, 1963, pp. 223-227).

**1042**
Carrouges, Michel
 Les Apparitions de Martiens. Paris, Fayard, 1963, 275p.

NOTES:

**1043**
Firsoff, V. A.
  Life Beyond the Earth; a Study in Exobiology.
New York, Basic Books, 1963, 320p.

**1044**
Huang, Su-Shu, & R. H. Wilson, Jr.
  Astronomical Aspects of the Emergence of Intelligence. New York, Institute of the Aerospace Sciences, 1963, 15p. (IAS Paper # 63-48).

**1045**
Macvey, J. W.
  Alone in the Universe? New York, Macmillan, 1963, 273p.

**1046**
Motz, Lloyd
  Extra-terrestrial Intelligence and Stellar Evolution. New York, Institute of the Aerospace Sciences, 1963, 16p. (IAS Paper #63-49).

**1047**
Pearman, J. P. T.
  Extraterrestrial Intelligent Life and Interstellar Communication: an Informal Discussion. (In Cameron, A. G. W., ed. Interstellar Communication. New York, Benjamin, 1963, pp. 287-293).

**1048**
Schatzman, Evry
  La vie Existe-t-elle sur les Autres Planètes? Paris, Université de Paris, Palais de la Découverte, 1963, 15p.

**1049**
Tsung, Thomas
  Is There Life Beyond the Earth? A Scientific Approach in the Light of Past and Current Theories. New York, Exposition Press, 1963, 71p.

**1050**
Drake, Walter
  Gods or Spacemen? Amherst, Wis., Amherst Press, 1964, 176p.

**1051**
Fesenkov, V. G.
  Zhizn' vo Vselennoy. Moscow, "Znanie," 1964, 53p.

**1052**
Mundo, Laura
  Pied Piper From Outer Space. Los Angeles, The Planetary Space Center Working Committee, 1964, 152p.

**1053**
Richards, H. M. S.
  Look to the Stars. Washington, Review & Herald Publishing Association, 1964, 156p.

**1054**
Alessandri, Michelangelo, & Roberto Masi
  Altri Mondi Abitati? Assisi, Edizioni Pro Civitate Christiana, 1965, 226p.

**1055**
Allen, T. B.
  The Quest. A Report on Extraterrestrial Life. Philadelphia, Chilton, 1965, 323p.

**1056**
Mamikunian, Gregg, & Michael H. Briggs, eds.
  Current Aspects of Exobiology. New York, Pergamon, 1965, 420p.

**1057**
Moffat, Samuel
  Life Beyond the Earth. New York, Scholastic Book Services, 1965, 156p.

NOTES:

## INTELLIGENT LIFE ON OTHER PLANETS

**1058**
Ordway, F. I.
  Life in Other Solar Systems. Illustrated with Paintings by Harry H. K. Lange. New York, E. P. Dutton, 1965, 96p. ($3.75)

**1059**
Sagan, Carl
  The Quest for Life Beyond the Earth. (In Smithsonian Institution. Annual report 1964. Washington, 1965, pp. 297-306. Its Publication 4613)

**1060**
Shklovskiy, I. S.
  Vselennya, Zhizn', Razum. 2d ed. rev. & enl. Moscow, "Nauka," 1965, 283p.

**1061**
Vsesoyuznove Soveshchaniye No Probleme Vnezemnykh-Tsivilizatsiy, Byurakan, 20-23 May, 1965. Proceedings. Yerevan, Izd-vo-An Armyanskoy SSR, 1965, 152p.

**1062**
Berrill, Norman J.
  Worlds Apart: a Reflection on Planets, Life, and Time. London, Sidgwick & Jackson, 1966, 240p. (25 shillings)

**1063**
Cade, C. M.
  Other Worlds Than Ours. London, Museum Press, 1966, 248p.

**1064**
Doebel, Gunter
  Der Mensch Lebt Nicht Allein im All. Cologne, Schauberg, 1966, 212p. (16.80 marks)

**1065**
Knaggs, Oliver
  Let the People Know. Cape Town, South Africa, Howard Timmins, 1966, 113p.

**1066**
Lindsay, Gordon
  The Riddle of the Flying Saucers. Dallas, The Voice of Healing Pub. Co., 1966, 31p.

**1067**
MacGowan, R. A., & F. I. Ordway
  Intelligence in the Universe. Englewood Cliffs, Prentice-Hall, 1966, 402p. ($13.50)

**1068**
Shklovskiy, I. S., & Carl Sagan
  Intelligent Life in the Universe. San Francisco, Holden-Day, 1966, 509p. ($8.95)

**1069**
Silva, R. I. da.
  No Espaço não Estamos sós. São Paulo, Edart, 1966, 214p.

**1070**
Sullivan, Walter
  We are not Alone; the Search for Intelligent Life on Other Worlds. Rev. ed. New York, McGraw-Hill, 1966, 325p.

**1071**
  The Book of Spaceships in Their Relationship With the Earth. Los Angeles, DeVorss & Co., 1967, 47p.

**1072**
Firsoff, Valdemar A.
  Life, Mind and Galaxies. Edinburgh, Oliver & Boyd, 1967, 111p. (7 shillings, 6 pence)

**1073**
Golowin, Sergius
  Götter der Atom-Zeit. Moderne Sagenbildung um Raumschiffe und Sternenmenschen. Bern, Francke, 1967, 128p. (13.80 Swiss francs)

**1074**
Greenbank, Anthony
  Creatures From Outer Space. (In The Book of Survival. New York, Harper & Row, 1967, p. 34).

NOTES:

# INTELLIGENT LIFE ON OTHER PLANETS

**1075**
Greenfield, Irving A.
   *Why are They Watching us?*   by Allen L. Erskine (pseud). New York, Tower Pubs., 1967, 124p.

**1076**
Sykes, Egerton
   *The Extra Terrestrials.*  London, Markham House Press, 1967, 20p.

**1077**
Alvarez López, José, et al
   *La Vida Extraterrestre.* Buenos Aires, Kier, 1968, 173p.

**1078**
Drake, Walter R.
   *Spacemen in the Ancient East.* London, Neville Spearman, 1968, 240p. (30 shillings)

**1079**
Larson, Kenneth
   *The Discovery of the Graphic Message of Goodhue.* Los Angeles, 1968, 30p.

**1080**
Tyler, Steven
   *Are the Invaders Coming?*  New York, Tower Pubs., 1968, 139p.

**1081**
Varsavsky, Carlos M.
   *Vida en el Universo.* Buenos Aires, C. Perez, 1968, 123p. ($2.50)

**1082**
Cathie, Bruce L.
   *Harmonic 33.* San Francisco, Tri-Ocean Books, 1969, 208p. ($7.00)

**1083**
Dolezol, Theodor
   *Aufbruch zu den Sternen.* (Text-illustr.: Kurt Roschl.) Vienna, Ueberreuter, 1969, 255p. (129 marks)

**1084**
Douglas, Ulysee
   *The Phenomena of Flying Saucers and Spatial People.* New York, Exposition, 1969, 129p.

**1085**
Eugster, Jack
   *Die Forschung Nach Ausserirdischem Leben. Wissenschaftliche Grundlagen zu Einer Kosmobiologie.* Zürich, Orell Fussli, 1969, 303p.

**1086**
Friedmann, Eugenio
   *La Humanidad Dentro del Complejo de la Exploración Spacial.* Asunción, Ed. la Voz, 1969, 196p.

**1087**
Gindilis, L. M., et al
   *Vnezemnye Tsivilizatsii: Problemy Mezhvezdnoi Sviazi.* Moscow, Izd. Nauka, 1969, 439p.

**1088**
Have, P. J. ten
   *Buitenaards Leven?* Amsterdam, Stichting IVIO, 1969, 16p.  (.75 florins)

**1089**
Holmes, David C.
   *Cent Milliards de Mondes Habités?* Tr. by Albert Burnet. Neuilly, Dargaud, 1969, 192p.

**1090**
Monti, Adriano
   *Vita Nello Spazio. Mito o Realtà?* Rome, Edizioni Mediterranee, 1969, 160p. (2500 lira)

**1091**
Schwartz, Alan W.
   *Other Life in the Universe.* Nijmegen, Dekker & Van de Vegt, 1969, 18p. (2.40 kroner)

---

NOTES:

## INTELLIGENT LIFE ON OTHER PLANETS

**1092**
Bergier, Jacques
 Les Extra-terrestres dans l'Histoire.
Paris, Ed. J'ai Lu, 1970, 187p.

**1093**
Biraud, Francois, & Jean C. Ribes
 Le Dossier des Civilisations Extra-terrestres. Paris, Fayard, 1970, 238p.

**1094**
Däniken, Erich von
 Chariots of the Gods? Unsolved Mysteries of the Past. Tr. by Michael Heron. New York, Putnam, 1970, 189p. ($5.95)

**1095**
Elsasser, Hans, et al
 Sind wir Allein im Kosmos? Munich, Piper, 1970, 179p. (9.80 marks)

**1096**
Firsoff, Valdemar A.
 Vie, Intelligence et Galaxies. Paris, Dunod, 1970, 144p. (8.70 francs)

**1097**
Haber, Heinz
 Bruder im All; die Möglichkeit des Lebens auf Fremden Welten. Stuttgart, Deutsche Verlags-Anstalt, 1970, 135p.

**1098**
Imshenetskii, A. A., ed.
 Zhizn vne Zemli i Metody ee Obnaruzheniia. Moscow, Izd. Nauka, 1970, 207p.

**1099**
Sneath, P.
 Les Planètes et la Vie. Paris, Eds. du Groupe Express, 1970. (34 francs)

**1100**
Wegner, Willy
 Ufonauter. Copenhagen F., Fufos (Frederiskberg UFO Studiekreds); Distributed by Valentinersvej 15/5, 1970, 28p. (3.50 krones)

**1101**
Däniken, Erich von
 Gods from Outer Space... New York, Putnam, 1971, 190p. ($5.95)

**1102**
Däniken, Erich von
 Gods From Outer Space; Return to the Stars, or Evidence for the Impossible. Tr. by Michael Heron. New York, Putnam, 1971, 190p. ($5.95)

**1103**
Kaplan, S. A., et al
 Extraterrestrial Civilizations. Problems of Interstellar Communication. Jerusalem, Israel Program for Scientific Translations, 1971, 270p.

**1104**
Berlitz, Charles
 Mysteries From Forgotten Worlds. New York, Dell Pub., 1972, 225p.

**1105**
Biraud, François, & Jean C. Ribes
 Le Dossier des Civilisations Extra-terrestres. Paris, Fayard, 1972, 305p.

**1106**
Kolosimo, Peter
 Astronautes de la Préhistoire. Tr. from Italian by Simone de Vergennes. Paris, Michel, 1972, 393p.

**1107**
Mukhin, L. M.
 Sovremmenoe Polozhenie CETTs Tochki Zreniia Biologii. Paper Presented at International Astronautical Congress, Oct 8-15, 1972. Vienna, International Astronautical Federation, 1972, 10p.

**1108**
Von Krueger, Frederick
 The Transparent People. New York, Vantage Press, 1972, 33p. ($3.50)

NOTES:

**1109**
Wilson, Clifford
  Crash go the Chariots. Burnt Hills, N.Y., Word of Truth, 1972, 128p.

**1110**
Coarer-Kalondan, Edmond, & Gwezenn-Dana
  Les Celtes et les Extra-terrestres. Verviers, Belgium, Gérard, 1973, 186p.

**1111**
Charroux, Robert
  Le Livre du Passé Mystérieux. Paris, R. Laffont, 1973, 475p.

**1112**
  In Search of Ancient Astronauts. NBC Television Network Program, Jan. 5, 1973, 8-9 P.M., Eastern Standard Time.

**1113**
Keyhoe, Donald E.
  Aliens From Space; the Real Story of Unidentified Flying Objects. Garden City, Doubleday, 1973, 322p. ($7.95)

**1114**
Roulet, Alfred
  A la Recherche des Extra-terrestres. Paris, R. Julliard, 1973, 221p.

**1115**
Tomas, Andrew
  We are not the First. New York, Bantam Books, 1973, 180p.

**1116**
Collyns, Robin
  Did Spacemen Colonise the Earth? London, Pelham Books, 1974, 260p.

## PERIODICAL ARTICLES

**1117**
"Race of Flying men." Southern Review 10, 1832, p. 272.

**1118**
"Whewell on Plurality of Worlds." London, Fraser's Magazine 49, 1853, p. 245.

**1119**
Wills, J.
  "Other Worlds." Dublin, Dublin University Magazine 53, 1858, p. 330.

**1120**
Patterson, R. H.
  "Plurality of Worlds." Knickerbocker Magazine 61, 1862, p. 395.

**1121**
"Habitable Worlds." London, All the Year Round, 32, 1874, p. 127.

**1122**
"Worlds in the Sky." London, All the Year Round 38, 1877, p. 400.

**1123**
Wills, J.
  "Plurality of Worlds." Dublin, Dublin University Magazine 92, 1878, p. 14.

**1124**
Proctor, R. A.
  "Life in Other Worlds." London, Knowledge 11, 1887-88, p. 230.

NOTES:

# INTELLIGENT LIFE ON OTHER PLANETS

1125
Flammarion, Camille
"New Discoveries in the Heavens." *Arena*, Dec. 1891, pp. 1-11.

1126
Ball, R. S.
"The Possibility of Life in Other Worlds." London, *Fortnightly Review* 62, Nov. 1894, pp. 718-729.

1127
Delboeuf, J. R. L.
"In a World Half as Large; the law of Universal Attraction." *Popular Science* 52, March 1898, pp. 678-687.

1128
Macdougal, D. T.
"Life on Other Worlds." *Forum* 27, March 1899, pp. 71-77.

1129
Clerke, A. M.
"Life in the Universe." Edinburgh, *Edinburgh Review* 200, July 1904, pp. 59-74.

1130
Newcomb, S.
"Life in the Universe." *Harper's* Magazine 111, 1905, p. 404.

1131
"Habitability and Uninhabitability of Earth's Sister Planets." *Current Literature* 39, Dec. 1905, p. 655.

1132
Robinson, L.
"Are There men in Other Worlds?" *Current Literature* 44, June 1908, pp. 672-677.

1133
Mumford, N. W.
"Intelligence on Mars or Venus." *Popular Astronomy* 17, Oct. 1909, pp. 497-504.

1134
"Dr. P. Lowell Holds...Organized Life Exists on Mars." *New York Times*, Feb. 3, 1911, p. 6, col. 1.

1135
"Arrhenius Again Cites Doubt That Living Beings Inhabit Mars." *New York Times*, May 1, 1911, p. 3, col. 6.

1136
Wilson, H. C.
"Life in Other Worlds." *Popular Astronomy* 19, Oct. 1911, pp. 483-493.

1137
Wilson, H. C.
"Which of the Celestial Spheres are Inhabitable?" *Scientific American, Supplement* 72, Nov. 4, 1911, pp. 290-291.

1138
"Can Astronomy Ever say Positively That Other Planets are Inhabited?" *Current Literature* 52, Jan. 1912, pp. 64-66.

1139
"Other Worlds Than Ours." *Literary Digest* 46, June 14, 1913, pp. 1327-1328.

1140
Williams, H. S.
"Are the Planets Inhabited?" *Hearst's Magazine* 24, Aug. 1913, pp. 284-286.

1141
Howells, W.
"Are the Planets Inhabited?" *Harper's Monthly Magazine* 128, Dec. 1913, pp. 149-151.

1142
Macpherson, H.
"Man's Place in the Universe." *Popular Astronomy* 22, Oct. 1914, pp. 463-470.

NOTES:

# INTELLIGENT LIFE ON OTHER PLANETS

**1143**
Salmon, C. F.
"Are Martians People?" *Scientific American* (122:12), March 20, 1920, p. 301.

**1144**
"The Type of Mind That Believes in Life on Other Worlds." *Current Opinion* 71, Nov. 1921, pp. 630-631.

**1145**
"Editorial on Probability That the Earth is the Only Inhabited Planet in the Solar System." *New York Times*, Jan. 22, 1922, sec. II, p. 6, col. 4.

**1146**
"Prof. S. Arrhenius Predicts That Venus Will Carry Culture When our World Dies." *New York Times*, May 22, 1922, p. 15, col. 5.

**1147**
Becquerel, P.
"La vie Terrestre Provient-elle d'un Autre Monde?" Paris, *Astronomie* 38, 1924, pp. 393-417.

**1148**
Russell, H. N.
"Are There Other Habitable Worlds?" *Scientific American* 132, May 1925, p. 315.

**1149**
Younghusband, F.
"Life in the Stars." *Ninteenth Century and After* 102, Dec. 1927, pp. 851-861.

**1150**
Moseley, E. L.
"Are There Creatures Like Outselves in Other Worlds?" *Scientific American* 145, Nov. 1931, pp. 308-310.

**1151**
Chamberlin, R. V.
"Life in Other Worlds: a Study in the History of Opinion." *Bulletin of the University of Utah* (22:3), Feb. 1932.

**1152**
"Dean Inge Thinks They are Peopled." *New York Times*, Jan. 23, 1933, p. 7, col. 2.

**1153**
Leonard, F. C.
"Life on Other Worlds." *Popular Astronomy* 41, May 1933, pp. 260-263.

**1154**
"Can Life Exist on any Planet but the Earth?" *Literary Digest* 118, July 28, 1934, p. 17.

**1155**
Cleator, P. E.
"Extraterrestrial Life." London, *Journal of the British Interplanetary Society* 1, Oct. 1934, pp. 3-4.

**1156**
Cleator, P. E.
"Extraterrestrial Life." London, *Journal of the British Interplanetary Society* (2: 1), May 1935, p. 3.

**1157**
"McColley, Grant, & H. W. Miller Saint Bonaventure, Francis Mayron, William Vorilong and the Doctrine of a Plurality of Worlds." *Speculum* 12, July 1937, pp. 386-389.

**1158**
Bennett, D. A.
"Men From Mars." *Sky, Magazine of Cosmic News* 3, Dec. 1938, pp. 8-9.

**1159**
Underwood, R. S.
"Are we Alone in the Universe?" *Scientific Monthly* 49, Aug. 1939, pp. 155-159.

**1160**
Langer, R. M., & T. E. Stimson
"Is There Life Among the Stars?" *Popular Mechanics* 78, July 1942, pp. 82-85.

NOTES:

## INTELLIGENT LIFE ON OTHER PLANETS

**1161** Russell, H. N.
"Anthropocentrism's Demise: new Discoveries Lead to Probability That There are Thousands of Inhabited Planets in our Galaxy." Scientific American 169, July 1943, pp. 18-19.

**1162** "Many Habitable Worlds." Science Newsletter 47, June 30, 1945, p. 402.

**1163** "The sky is Haunted." Saturday Evening Post, March 6, 1948, p. 12.

**1164** Carlisle, N., & M. Carlisle
"Is There Life on Other Worlds?" Coronet 26, Sept. 1949, pp. 30-33.

**1165** "Comment on F. Hoyle Theory That Life Exists Outside our Solar System." New York Times, Sept. 25, 1949, sec. IV, p. 9, col. 6.

**1166** "Are There Other Living Worlds?" Science Digest 26, Dec. 1949, pp. 68-69.

**1167** Gregory, J. C.
"Plurality of Worlds." London, Fortnightly Review 174, July 1950, pp. 28-33.

**1168** Cooper, J.
"Cities in the sky." Science Digest 28, Nov. 1950, pp. 78-80.

**1169** Strughold, Hubertus
"Physiological Considerations on the Possibility of Life Under Extraterrestrial Conditions." (In Marbarger, John P., ed. Space Medicine. Urbana, Ill., University of Illinois Press, 1951, pp. 31-48).

**1170** Fishbein, J.
"Life on 100,000 Planets?" Science Digest 29, March 1951, Inside Cover.

**1171** Wilkins, Harold T.
"1,000 Years of Flying Saucers." Fate 4, April 1951, pp. 23-30.

**1172** Heuer, K.
"Men of Other Planets." Science Digest 30, July 1951, pp. 9-13.

**1173** England, J.
"On the Possible Nature of Extraterrestrial Life." Journal of Spaceflight (4:2), 1952, pp. 1-3.

**1174** Kind, S. S.
"Energy Fixation and Intelligent Life." London, Journal of the British Interplanetary Society 11, 1952, pp. 168-172.

**1175** Lambert, Richard S.
"Flying Saucers--Their Lurid Past." Toronto, Saturday Night 67, May 10, 1952, pp. 9, 18.

**1176** "The Angels." E. Hartford, Bee-Hive 28, Jan. 1953, p. 22.

**1177** "J. V. Miller Letter on Nov. 16 on Prof. Urey's Theory of Possibility of Life on Other Planets." New York Times, Jan. 10, 1953, sec. IV, p. 12, col. 6.

**1178** Eiseley, Loren C.
"Is man Alone in Space?" Scientific American 189, July 1953, pp. 80-82.

NOTES:

**1179**
Marais, D.
"Are we Being Watched?" Bloemfontein, South Africa, Personality (53:1401), Jan. 1, 1954, pp. 26-28.

**1180**
"J. Hillaby Series on Fifth International Astronautical Federation Congress, Innsbruck...Sees Life Impossible..." New York Times, Aug. 4, 1954, p. 19, col. 1.

**1181**
Oberth, Hermann
"Wir Werden Beobachtet." Berlin, Deutsche Illustrierte, Sept. 11, 1954, pp. 10-11, 24-28.

**1182**
Edwards, Frank
"The Spies From Outer Space." Real 5, Nov. 1954, pp. 20-21, 58-60.

**1183**
Faust, H.
"Zur Frage Ausserirdischen Lebens." Frankfurt-am-main, Weltraumfarht 6-9, 1955-58, pp. 23-24.

**1184**
Tihkov, G. A.
"Is Life Possible on Other Planets?" London, Journal of the British Astronomical Association (65:5), 1955, pp. 193-204.

**1185**
Salisbury, F. B.
"The Inhabitants of Mars." Engineering and Science 18, April 1955, pp. 23-32.

**1186**
"Soviet Scientist E. Krinov Allays Fears of Invasion of USSR by Mars, Radio Broadcast." New York Times, May 1, 1955, p. 20, col. 7.

**1187**
Moseley, James W.
"Peruvian Desert, map for Saucers?" Fate 8, Oct. 1955, pp. 28-33.

**1188**
Roberts, A. W.
"Are the Stars Inhabited?" Scientific American 195, Sept. 1956, p. 73.

**1189**
"Man not Alone in Cosmos." Science Newsletter 70, Sept. 15, 1956, p. 163.

**1190**
"Space Visitors: From Which Planets do They Originate, and is There Life on Them?" London, Practical Mechanics, Jan. 1957, pp. 203-207.

**1191**
"Is the U.S. Government Expecting Invaders From Space?" Flying Saucers From Other Worlds, Aug. 1957, pp. 16-21.

**1192**
Edwards, Frank
"Frank Edwards' Report, What do Flying Saucers Want?" Fate 10, Sept. 1957, pp. 19-27.

**1193**
Clarke, A. C.
"On the Morality of Space." Saturday Review 40, Oct. 5, 1957, pp. 8-10.

**1194**
Clarke, A. C.
"Where's Everybody?" Harper's Magazine 215, Nov. 1957, pp. 73-77.

**1195**
Slater, Alan E.
"The Probability of Intelligent Life Evolving on a Planet." (In VIIth International Astronautical Congress, Barcelona, 1957. Proceedings. Vienna, Springer-Verlag, 1958, pp. 395-402).

NOTES:

## INTELLIGENT LIFE ON OTHER PLANETS

1196
"Other Beings on Other Planets?" Life 44, Jan. 6, 1958, p. 66.

1197
Ley, W.
"What Will Space People Look Like?" Science Digest 43, Feb. 1958, pp. 61-64.

1198
Helland, Albert E.
"They Caught a Spacemen." Fate 11, March 1958, p. 62.

1199
Ley, Willy
"Mighty Invaders From Outer Space." Catholic Digest 22, April 1958, pp. 25-29.

1200
Gatland, K. W., & D. D. Dempter
"Resident of Fahrenheit; Excerpt From Inhabited Universe." Saturday Review 41, April 5, 1958, pp. 56-57.

1201
Evans, S.
"New Thought About Life on Other Planets." American Mercury 86, June 1958, pp. 83-88.

1202
Kind, S. S.
"Speculation on Extraterrestrial Life." London, Spaceflight (1:8), July 1958, pp. 288-290.

1203
"Suspect Human Life on Millions of Planets." Science Newsletter 74, Nov. 1958, p. 328.

1204
"Do Other Humans Live?" Newsweek 52, Nov. 17, 1958, p. 56.

1205
"Suspect Human Life on Millions of Planets." Science Newsletter 74, Nov. 22, 1958, p. 328.

1206
DuShane, G.
"Next Question." Science 130, 1959, p. 1733.

1207
Firsoff, Valdemar A.
"Life?" (In his Strange World of the Moon. New York, Basic Books, 1959, pp. 170-180).

1208
Howells, W.
"Would Other Humans Look Like us?" (In Mankind in the Making. New York, Doubleday & Co., 1959, pp. 53-58).

1209
"Other Beings on Other Planets?" Space Journal 1, Winter 1959, p. 4.

1210
Breig, Joseph A.
"Are There Rational Beings on Planets far Away?" Ave Maria 89, March 1959, p. 19.

1211
Breig, J. A.
"Are There Rational Beings on Planets far Away?" Ave Maria 89, March 7, 1959, p. 19.

1212
Ross, John C.
"A Scientist Looks at Life on Other Worlds." Fate 12, April 1959, pp. 86-89.

1213
"Little Inhabited Stars." Time 73, April 27, 1959, p. 65.

1214
"There's Intelligent Life on the Moon." Flying Saucers, May 1959, pp. 67-74.

NOTES:

1215
"Dr. Schlovsky (USSR) Holds 2 Moons are Probably Artificial Satellites put into Orbit Millions of Years ago by Then Existing Martians." New York Times, May 3, 1959, p. 42, col. 5.

1216
Kind, S. S.
"Speculations on Extraterrestrial Life." London, Spaceflight (1: 8), July 1959, p. 288.

1217
Randolph, James R.
"Is There Life on Mars?" Ordnance 44, July-Aug. 1959, pp. 72-73.

1218
"Dr. Huang Holds 2 Stars Within 16 Light Years of Earth may Have Planets With Intelligent Life." New York Times, July 12, 1959, sec. IV, p. 9, col. 5.

1219
Perego, A
"Rational Life Beyond the Earth?" Theology Digest 7, Fall 1959, pp. 177-178.

1220
Briggs, Michael H.
"Terrestrial and Extraterrestrial Life." London, Spaceflight 2, Oct. 1959, pp. 120-121.

1221
Flanders, C. M.
"Superman--Does he Really Exist?" Flying Saucers, Oct. 1959, pp. 16-19, 45.

1222
Palmer, Ray
"Mars' Moons Artificial." Flying Saucers, Oct. 1959, pp. 13-15.

1223
Fernández, Juan M.
"La Pluralidad de los Mundos Habitados." Santander, Spain, Salterrae 11, Nov. 1959, pp. 606-622.

1224
"W. L. Laurence on Theories That Intelligent Life Exists in Distant Solar Systems and may be Signalling Earth; Discusses Plan of National Radio Astronomy Observatory, Green Bank, W. Va., to Tune in on Possible Signals; Plan, Known as Project Ozma, under Direction of Dr. F. D. Drake." New York Times, Nov. 22, 1959, sec. IV, p. 14, col. 6.

1225
"Anybody out There?" Time 74, Nov. 23, 1959, pp. 84-85.

1226
"Somebody out There?" Newsweek 54, Nov. 23, 1959, p. 60.

1227
Tombaugh, Clyde W.
"Could the Satellites of Mars be Artificial?" Astronautics 4, Dec. 1959, pp. 38-39.

1228
"Dr. Struve's Letter Expands on Laurence Comment on Possibility of Life in Distant Systems." New York Times, Dec. 6, 1959, sec. IV, p. 10, col. 6.

1229
"Dr. Calvin's Discovery of pre-Biological Chemical Compounds in Meteorite Molecules Seen as First Clue to Existence of Spatial Conditions Favorable to Development of Living Forms." New York Times, Dec. 20, 1959, sec. IV, p. 7, col. 6.

---

NOTES:

# INTELLIGENT LIFE ON OTHER PLANETS

**1230**
Briggs, Michael H.
"Other Astronomers in the Universe?" Wellington, <u>Southern Stars</u> 18, 1960, pp. 147-151.

**1231**
Lear, J.
"The Search for Intelligent Life on Other Planets." <u>Saturday Review</u> 43, Jan. 2, 1960, pp. 39-40.

**1232**
Howells, W.
"Would Other Humans Look Like us? Excerpt From <u>Mankind in the Making</u>." <u>Science Digest</u> 47, Feb. 1960, pp. 53-58.

**1233**
"USSR Prof. Vorontsov-Velyaminov Sees Infinite Number of Worlds Inhabited by Intelligent Beings." <u>New York Times</u>, Feb. 21, 1960, p. 7, col. 1.

**1234**
Diamond, E.
"Life out There?" <u>Newsweek</u> 55, Feb. 22, 1960, pp. 68-71.

**1235**
"Enoch and Other Cosmonauts; Soviet Theories." <u>Time</u> 75, Feb. 22, 1960, p. 26.

**1236**
Esnaola Auzmendi, Francisco M.
"Viajes Interplanetarios y Mundos Habitados." Vitoria, Spain, <u>Lumen</u> (9: 2), March-April 1960, pp. 115-171.

**1237**
Michel, Aimé
"Do Flying Saucers Originate From Mars?" London, <u>Flying Saucer Review</u> 6, March-April 1960, pp. 13-15.

**1238**
"Russian Report: Is There Life on Mars?" <u>Fate</u> 13, April 1960, pp. 27-33.

**1239**
"Dr. Von Braun Holds Existence of Life Elsewhere in Space a 'Logical Asssumption'." <u>New York Times</u>, April 29, 1960, p. 20, col. 1.

**1240**
"United States Report: Is There Life on Mars?" <u>Fate</u> 13, May 1960, pp. 27-34.

**1241**
Rich, Valentin, & Mikhail Chernenko
"Tracks That Lead to Space." London, <u>Flying Saucer Review</u> 6, May-June 1960, pp. 3-6.

**1242**
Bracewell, R. N.
"Communications From Superior Galactic Communities." London, <u>Nature</u> (186: 4726), May 28, 1960, p. 670.

**1243**
Fuhs, Allen E.
"Visual Sensitivity of Residents of Other Planets." <u>American Rocket Society Journal</u> 30, June 1960, p. 577.

**1244**
"Dr. F. J. Dyson Suggests Astronomers Seek Signs of Stars Surrounded by Gigantic Habitable Shells Built by Sentient Beings (Article in <u>Science</u>)." <u>New York Times</u>, June 3, 1960, p. 11, col. 3.

**1245**
"Shells Around Suns may Have Been Built." <u>Science Newsletter</u> 77, June 18, 1960, p. 389.

**1246**
"Dr. Shapley Doubts Other Worlds Seek Contact With Earth." <u>New York Times</u>, June 24, 1960, p. 31, col. 4.

**1247**
Binder, Otto
"Is Anybody out There?" <u>Space World</u> 1, July 1960, pp. 22-25.

NOTES:

**1248**
"Rev. D. C. Raible Sees Possibility of Existence of Rational Beings With Cultures and Civilizations Superior to Those of Earth, Living in 'State of Innocence' (Article in Jesuit Publication, America)." New York Times, Aug. 7, 1960, p. 14, col. 1.

**1249**
Raible, D. C.
"Rational Life in Outer Space?" America 103, Aug. 13, 1960, pp. 532-535.

**1250**
Briggs, Michael H.
"New Evidence of Martian Life." London, Spaceflight 2, Oct. 1960, pp. 237-238.

**1251**
Eisnau, M.
"Is There Intelligent Life on the Planets?" American Mercury 91, Oct. 1960, pp. 32-45.

**1252**
McHugh, L.
"Others out Yonder." America 104, Nov. 1960, pp. 295-297.

**1253**
Breig, J. A.
"Man Stands Alone." America 104, Nov. 26, 1960, pp. 294-295.

**1254**
McHugh, L.
"Others out Yonder." America 104, Nov. 26, 1960, pp. 295-297.

**1255**
Raible, D. C.
"Men From Other Planets?" Catholic Digest 25, Dec. 1960, pp. 104-108.

**1256**
"Brookings Institute Study for NASA Warns on Implications of Possible Discovery of Intelligent Life in Space." New York Times, Dec. 15, 1960, p. 16, col. 3.

**1257**
"Editorial on Brookings Institute Study of Intelligent Life in Space." New York Times, Dec. 16, 1960, p. 32, col. 2.

**1258**
Koch, Howard
"I Marziani Invaderanno la Terra?" (In Pieroni, P., ed. Destionazione Universo. Florence, 1961, pp. 315-345).

**1259**
Horowitz, Norman
"Is There Life on Other Planets?" Engineering and Science 24, March 1961, pp. 11-15.

**1260**
Abelson, Philip H.
"Extraterrestrial Life." (In U.S. National Academy of Sciences. Proceedings 47, April 1961, pp. 575-581).

**1261**
Edwards, Frank
"Frank Edwards' Report: Soviet Scientists Claim Life on Mars." Fate 14, April 1961, pp. 40-46.

**1262**
Asimov, Isaac
"We, the In-betweens." Mademoiselle 53, May 1961, pp. 136-137.

**1263**
Briggs, Michael H.
"Superior Galactic Communities." London, Spaceflight (3:3), May 1961, pp. 109-110.

**1264**
Howells, W.
"The Evolution of Humans on Other Planets." London, Discovery (22:6), June 1961, pp. 237-241.

NOTES:

## INTELLIGENT LIFE ON OTHER PLANETS

1265
"Space men Predicted." Science Newsletter 80, July 15, 1961, p. 35.

1266
Zarco de Egea, J.
"En que Situación se Hallarían Frente a la Redención de los Posibles Seres de Otros Mundos?" Rosario, Argentina, Didascalia, Aug. 1961, pp. 331-333.

1267
Zigel, F.
"There is Intelligent Life on Mars." Space World 1, Sept. 1961, pp. 20-21.

1268
Krasovskii, Y. I.
"Astronautics and Extraterrestrial Civilizations." London, Flying Saucer Review 7, Sept.-Oct. 1961, pp. 3-5.

1269
Mann, M.
"Man-horse From Outer Space." Popular Science 179, Oct. 1961, p. 19.

1270
"Man not in Space." Science Newsletter 80, Oct. 14, 1961, p. 251.

1271
"Astronomer, Dr. Gadomski, of Poland, in Recent Paper, Says High Developed Civilizations on Planets of Suns in our Galaxy Should be the Rule Rather Than Exception; Biologist, Dr. Margaria, of Italy, Disagrees...Sees all Conditions of Life Having to be Duplicated to Produce Life Known to man...Life, if it Exists, Will be Different...Unlikely man Will Find Intelligent Beings...to Communicate With." New York Times, Oct. 15, 1961, sec. IV, p. 7, col. 7.

1272
Barr, R.
"Who's Taking up Space?" Today 17, Nov. 1961, pp. 23-25.

1273
Rublowsky, John
"Is There Life on the Moon?" Space World 1, Nov. 1961, pp. 12-15.

1274
"Danger From Space?" Time 78, Nov. 17, 1961, p. 76.

1275
"Prof. Lovell's Article on Probability of Sentient Life Existing Elsewhere in Cosmos and on Radio-Astronomical Studies for Clues." New York Times, Dec. 24, 1961, sec. VI, p. 10.

1276
Margaria, R.
"On the Possible Existence of Intelligent Living Beings on Other Planets." Rome, Rivista Medica Aeronautica Spaziale (25:1), 1962, pp. 24-25.

1277
Miller, Stanley L.
"Extraterrestrial Life." (In U.S. Air Force. School of Aerospace Medicine. Lectures in Aerospace Medicine. Brooks Air Force Base, Texas, 1962, 277p.).

1278
Tombaugh, C.
"Life on Mars." Space World (2:2), Jan. 1962, pp. 36-37, 58, 61-62.

1279
"Adamski's Hieroglyphics." London, Flying Saucer Review 8, Jan.-Feb. 1962, pp. 7-8.

1280
Creighton, Gordon W.
"What do the Russians Know?" London, Flying Saucer Review 8, Jan.-Feb. 1962, pp. 20-21.

---

NOTES:

**1281**
Rublowsky, John
"Life on Venus." *Space World* 2, March 1962, pp. 34-35.

**1282**
Harford, J.
"Rational Beings in Other Worlds." *Jubilee*, May 1962, pp. 17-21.

**1283**
Slater, A. E.
"Life in the Universe." London, *Spaceflight* 4, May 1962, p. 88.

**1284**
"Ancient Records of UFO in Japan." Moorabin, Australia, *Australian Saucer Record* 2, June 1962, pp. 15-17.

**1285**
Macvey, John W.
"Alone in the Universe?" London, *Spaceflight* (4:4), July 1962, pp. 125-127.

**1286**
Bernal, J. D.
"Is There Life Elsewhere in the Universe?" Calcutta, *Science and Culture* (28:8), Aug. 1962, pp. 356-357.

**1287**
Austin, R. R.
"Extraterrestrial Life." London, *Spaceflight* 4, Sept. 1962, p. 176.

**1288**
Ledger, Joseph R.
"Saucers or Ghosts?" London, *Flying Saucer Review* 8, Sept.-Oct. 1962, pp. 19-20.

**1289**
Briggs, Michael H.
"The Distribution of Life in the Solar System: an Evaluation of the Present Evidence." London, *Journal of the British Interplanetary Society* (18: 11-12), Sept.-Dec. 1962, p. 431.

**1290**
"W. L. Laurence on Possibility of Sentient Life on Other Worlds as Indicated by Prof. Calvin and Other Studies Showing Life can Evolve in 'Primitive Atmosphere' of Organic Molecules by Physio-Chemical Reactions." *New York Times*, Sept. 9, 1962, sec. IV, p. 11, col. 7.

**1291**
Imshenetskii, A. A.
"Ekzobiologia: Novaya Oblast Nauchnykh Issledovanii." Moscow, Akademiya Nauk SSR, *Vestnik* 32, Nov. 1962, pp. 58-63.

**1292**
"Dr. Shlovsky Says Most Logical Place to Look for Signs of Intelligent Life Outside Earth is in Spiral Nebula of Andromeda... Holds U.S. Project Ozma Used Wrong Approach in Scanning Comparatively Close Stars for Radio Signals." *New York Times*, Dec. 3, 1962, p. 36, col. 3.

**1293**
Imshenetskii, A. A.
"Outer Space and Life." (Tr. into English from *Vestnik Akad. Nauk. SSR*. Moscow, #9, Sept. 1963, pp. 23-29). (JPRS-22015; OTS-63-41224). Washington, NASA, 1963, 12p. (Available from NTIS: 50¢).

**1294**
Imshenetskii, A. A.
"Perspectives for the Development of Exobiology." (In *Space Research III. Proceedings of the Third International Space Science Symposium, Washington, D. C., May 2-8, 1962.* Ed. by R. B. Livingston, et al. New York, Interscience, 1963, pp. 19-32).

NOTES:

# INTELLIGENT LIFE ON OTHER PLANETS

**1295**
Imshenetskii, A. A.
"The Possibility of Existence and Methods of Detecting Extraterrestrial Life." (In Problems of Space Biology, by N. M. Sisakyan, et al. Tr. into English. (NASA-TT-F-174). Washington, NASA, 1963, pp. 153-160).

**1296**
Underwood, R. S.
"Are we Alone in the Universe?" Navigator 9, Winter 1963, pp. 16-22.

**1297**
Kleczek, Josip
"Possibilities for Life in the Universe." (Associazione Internazionale Uomo nello Spazio. Proceedings of the International Congress-Man and Technology in the Nuclear and Space Age, Milan, April 18-21, 1962. Rome, 1963, pp. 135-138).

**1298**
Margaria, Rodolfo
"On the Possible Existence of Intelligent Living Beings on Other Planets." (In 12th International Astronautical Congress. Proceedings, Vol. 2. Washington, D. C., Oct. 1-7, 1961. New York, Academic Press, 1963, pp. 556-563).

**1299**
Drake, Walter R.
"UFO's Over Ancient Rome." London, Flying Saucer Review 9, Jan.-Feb. 1963, pp. 11-13.

**1300**
Golomb, S. W.
"When is Extraterrestrial Life Interesting?" Engineering and Science (26:5), Feb. 1963, pp. 15-17.

**1301**
Cade, C. M.
"Are we Alone in Space?" London, Discovery 24, April 1963, pp. 27-29.

**1302**
Oparin, A.I.
"Extraterrestrial Life" (Tr. from Russian. In Soviet Studies in Space Biology and Medicine. Washington, Joint Publications Research Service, April 4, 1963, pp. 1-7).

**1303**
Sclanders, I.
"Space Race: What-or who-is Waiting at the Finish Line?" Toronto, Macleans Magazine 76, April 20, 1963, p. 48.

**1304**
Muller, H. J.
"Life Forms to be Expected Elsewhere Than on Earth." London, Spaceflight (5:3), May 1963, pp. 74-85.

**1305**
Sagan, Carl
"Direct Contact Among Galactic Civilizations by Relativistic Interstellar Spaceflight." Planetary and Space Science 11, May 1963, pp. 485-498.

**1306**
Murdin, Paul
"The Plurality of Worlds." London, Axle Quarterly 4, Summer 1963, pp. 29-33.

**1307**
Barnes, Sam
"Life on Other Planets." Machine Design 35, June 6, 1963, pp. 118-123.

**1308**
Bongers, Leonard H.
"Is There Life on Mars?" Space-Aeronautics 40, Aug. 1963, pp. 86-88.

**1309**
Blaher, Damian J.
"Is Anybody There?" Friar 20, Sept. 1963, pp. 15-17.

NOTES:

**1310**
Rea, D. G.
"Evidence for Life on Mars." London, *Nature* 200, Oct. 1963, pp. 114-116.

**1311**
Slater, A. E.
"Life in the Universe." London, *Spaceflight* (5:6), Nov. 1963, pp. 198-201.

**1312**
O'Brien, R.
"Somebody up There Like us." *Esquire* 60, Dec. 1963, pp. 185-187.

**1313**
Simpson, G. G.
"Nonprevalence of Humanoids: Excerpts From This View of Life." *Science* 143, Feb. 21, 1964, pp. 769-775.

**1314**
Voronin, M.A.
"In Quest of Signs of Civilization in Other Worlds." Tr. into English from *Priroda* (Moscow), #11, 1962, pp. 78-83. (NASA-TT-F-8590). Alexandria, Va., NTIS, March 1964, 9p. ($1.00)

**1315**
Goddard, J.
"New Light on Ancient Tracks." London, *Flying Saucer Review* 10, March-April 1964, pp. 15-16.

**1316**
Kardashev, N. S.
"Peredacha Informatsii Vnezemnymi Tsivilzatsiiami." Moscow, *Astronomicheskii Zhurnal* 41, March-April 1964, pp. 282-287.

**1317**
Drake, Walter R.
"Spacemen in the Middle Ages." London, *Flying Saucer Review* 10, May-June 1964, pp. 11-13.

**1318**
Friedman, Bruno
"Millions of Inhabited Planets." London, *Flying Saucer Review* 10, May-June 1964, pp. 7-10.

**1319**
Evans, Gordon H.
"Three Unsolved Martian Mysteries." *Fate* 17, June 1964, pp. 27-33.

**1320**
"Dr. F. J. Dyson Says Civilizations Which may Exist on Planets in Orbit Around Pairs of dim Stars Called 'White Dwarfs' Could use Gravitational Energy of These Stars to Power Space Vehicles (Anthology of Articles on Search for Intelligent Life Outside Solar System)". *New York Times*, June 9, 1964, p. 20, col. 3.

**1321**
Pendergast, R.
"Terrestrial and Cosmic Polygenism." Oxford, *Downside Review* 82 (Downside Newman Bookshop), July 1964, pp. 189-198.

**1322**
Ross, H. E.
"A Contribution to Astrosociology." London, *Spaceflight* 6, July 1964, pp. 120-124.

**1323**
Drake, Walter R.
"Spacemen in Saxon Times." London, *Flying Saucer Review* 10, Sept.-Oct. 1964, pp. 10-12.

**1324**
Krasovskii, V. I.
"Astronautics and Extraterrestrial Civilizations." (In U.S. Air Force Systems Command. Foreign Technology Div. *Space: Collection of Articles*. Tr. into English of *Kosmos*, Moscow, #1, 1963, pp. 60-81. Dayton, Wright-Patterson Air Force Base, Sept. 3, 1964).

NOTES:

1325
Posin, D. Q.
"Other Suns, Other Planets, but is There Other Life?" Today's Health 42, Nov. 1964, pp. 56-59.

1326
"N. J. Berrill Book, Worlds Without End, on Life in Space and Possibly on Other Planets, Reviewed." New York Times, Nov. 22, 1964, sec. VII, p. 5.

1327
"I. Asimov Article on Possible Life on Other Planets and Ways of Communicating With Such Planets." New York Times, Nov. 29, 1964, sec. VI, p. 52.

1328
Bieri, R.
"Humanoids on Other Planets?" American Scientist (52:4), Dec. 1964, pp. 452-458.

1329
Krasovskii, V.I.
"Astronautics and Extraterrestrial Civilizations." Tr. from Kosmos (USSR), May 27, 1965, pp. 32-54. (FTD-MT-63-259; AD-619559). Wright-Patterson Air Force Base, Ohio, Air Force Systems Command, Foreign Technology Div., 1965, 22p.

1330
Siforov, V. I.
"Nekotor ye Voprosy Poiska i Analiza Radioizluchenii ot Drugikh Tsivilizatsii." (In Vnezemnye Tsivilizatsii. Vsesoiuznoe Soveshshanie. 1st, Yerevan, Armenian SSR, May 20-23, 1964. Trudy. Ed. by G. M. Tovmasian. Yerevan, Izd. Akad. Nauk. Armianskoi SSR, 1965, pp. 121-128).

1331
Swart, Hans
"Über den Nachweis Extraterrestrischen Lebens." Berlin, Astronomie und Raumfahrt 6, 1965, pp. 161-171.

1332
"How to Tell a Martian." Christian Science Monitor, Jan. 27, 1965, p. 2, col. 1.

1333
"L'Etrange Histoire des Satellites de Mars." Paris, Phénomènes Spatiaux, Feb. 1965, pp. 9-15.

1334
"New Evidence of Intelligent Life on Other Worlds." Science Digest 57, Feb. 1965, pp. 8-10.

1335
Ruzic, Neil P.
"The Case for Going to the Moon. VI-the Case for Life Beyond the Earth." Industrial Research 7, Feb. 1965, pp. 79-88, 90.

1336
Aharon, Y. N. ibn
"The Recent Literature of Extraterrestrial-ism." Saucer News 12, March 1965, pp. 11-12.

1337
Creighton, Gordon
"A Russian Wallpainting and Other 'Spacemen'." London, Flying Saucer Review 11, July-Aug. 1965, pp. 11-14.

1338
"W. Sullivan Discusses Dr. Lederberg's Views, Expressed in Interview...That Life and Water may Have Existed on Planet at one Time but That Most Races Have Since Been Eradicated by Meteorite Bombardment." New York Times, Aug. 8, 1965, p. 59, col. 4.

1339
"End of the Myths About men on Mars." U.S. News and World Report 59, Aug. 9, 1965, p. 4.

NOTES:

1340
Asimov, Isaac
"Anatomy of a man From Mars." Esquire 64, Sept. 1965, pp. 113-117.

1341
Good, I. J.
"The Human Preserve." London, Spaceflight (7: 5), Sept. 1965, p. 167.

1342
Leiber, F.
"Homes for men in the Stars." Science Digest 58, Sept. 1965, pp. 53-57.

1343
Farish, Lucius
"Myths, Legends, and UFO's." London, Flying Saucer Review 11, Nov.-Dec. 1965, pp. 19-21.

1344
Siguret, B.
"La vie sur Mars." Paris, Phénomènes Spatiaux, Dec. 1965, pp. 3-7.

1345
Imshenetskii, A. A., et al
"The Possibility of Life in Outer Space." (In International Space Science Symposium. 6th, Mar del Plata, Argentina, May 11-19, 1965. Papers. Life Sciences and Space Research, v. 4. Ed. by A. H. Brown and M. Florkin. Washington, Spartan Books, 1966, pp. 121-130).

1346
Kaufmann, Richard
"Ist Irgend Jemand da Oben? Fliegende Untertassen: Märchen, Massenwahn, Sensation Oder Wahrheit." Stuttgart, Christ und Welt (19:36), 1966, p. 28.

1347
Ordway, Frederick I.
"Some Implications of Extrasolar Intelligence." (In Planetology and Space Mission Planning Monograph. New York, New York Academy of Sciences, 1966).

1348
"Eight Eyes on Strange new Worlds; Opinions of Science-fiction Writers." Esquire 65, Jan. 1966, pp. 56-59.

1349
Keyhoe, Donald E.
"I Know the Secret of the Flying Saucers." True, January 1966.

1350
Sullivan, W.
"Who is out There? Excerpt From We are not Alone." Catholic Digest 30, Jan. 1966, pp. 96-100.

1351
"USSR Astronomers Call for 5-Year Cooperative Effort by Radio-Astronomy Centers Around the World to Attempt to Make Contact With Possible Extraterrestrial Civilizations." New York Times, Feb. 1966, p. 2, col. 3.

1352
Smith, S. L.
"The Problems of Exobiology. Part II: Life in the Solar System." London, BUFORA Journal and Bulletin 1, Spring 1966, pp. 13-18.

1353
Fouéré, René
"Surhumains ou Sous-humains, Anges ou Démons, que Sont les Extraterrestres?" Paris, Phénomènes Spatiaux, March 1966, pp. 5-11.

1354
Drake, Walter R.
"Spacemen in Norman Times." London, Flying Saucer Review 12, March-April 1966, pp. 17-19.

1355
"Prof. Wald Sees Intelligent Life as Culmination of Cosmic Evolution, Lecture, Museum of Natural History." New York Times, March 8, 1966, p. 23, col. 8.

NOTES:

# INTELLIGENT LIFE ON OTHER PLANETS

**1356**
"Rabbi N. Lamm Says Existence of Rational Beings on Other Planets is Compatible With Jewish Theology (Article in Tradition)." New York Times, April 10, 1966, p. 36, col. 1.

**1357**
Pinotti, Roberto
"Space Visitors in Ancient Egypt." London, Flying Saucer Review 12, May-June 1966, pp. 16-18.

**1358**
Bernstein, J.
"Books." New Yorker 42, May 28, 1966, pp. 117-120.

**1359**
Smith, S. L.
"The Problems of Exobiology. Part III: Life Beyond the Solar System." London, BUFORA Journal and Bulletin 1, Summer 1966, pp. 10-14.

**1360**
"Voorname Kronologie." Amsterdam, Het Interplanetair Nieuwsbulletin, June 1966, pp. 1-6.

**1361**
Willems, Louis
"Wordt de Aarde Door Andere Planeten Bespied?" Amsterdam, A.B.C., June 1966, pp. 6-11.

**1362**
Davy, J.
"Is Anybody There? New Speculation About Life Beyond the Earth." London, Observer, July 31, 1966, pp. 17-18.

**1363**
Pinotti, Roberto
"Space Visitors in Ancient Egypt." Flying Saucers From Other Worlds, Aug. 1966, pp. 18-19.

**1364**
"Me Earthman, you Gasbag: Theories of Josef Shklovskii and Carl Sagan." Newsweek 68, August 22, 1966, pp. 88-89.

**1365**
Oparin, Alexander I.
"Is There Life on the Moon?" Space World 3-7-33, Sept. 1966, p. 44.

**1366**
Larson, Kenneth
"The Great Pyramid a UFO Beacon?" Flying Saucers From Other Worlds, Oct. 1966, pp. 31-33.

**1367**
"ABC-TV Presents Special Show, We are not Alone, Based on Walter Sullivan's Book on Possibility of Extraterrestrial Life." New York Times, Oct. 22, 1966, p. 63, col. 3.

**1368**
Tomas, Andrew
"Their Purpose of Coming?" Moorabin, Australia, Australian Flying Saucer Review 9, Nov. 1966, pp. 5-7.

**1369**
Ordway, Frederick I.
"Some Implications of Extrasolar Intelligence." New York Academy of Science, Annals 140, Dec. 16, 1966, pp. 653-658.

**1370**
Kazantsev, Aleksandr
"Vizitnye Kartochi s Stru." Moscow, Tekhnika Molodiozhi 1, 1967, pp. 22-25.

**1371**
Zaitsev, Vyacheslav
"Visitors From Outer Space." Moscow, Sputnik 1, Jan. 1967, pp. 162-179.

NOTES:

**1372**
Evans, Gordon H.
 "UFO's in History and Myth." London, Science and Mechanics 38, Feb. 1967, pp. 52-55, 86, 88, 90.

**1373**
Lovitch, A.
 "UFO's--Science or Sorcery?" Engineering Digest 61, Feb. 1967, pp. 29-34.

**1374**
Cade, C. M.
 "A Long, Cool Look at Alien Intelligences: Part I--the Non-uniqueness of man." London, Flying Saucer Review 13, March-April 1967, pp. 24-25.

**1375**
Cade, C. M.
 "A Long, Cool Look at Alien Intelligences: Part II--the Forms of Intelligent Organisms." London, Flying Saucer Review 13, May-June 1967, pp. 13-15, 19.

**1376**
Cyr, Guy J.
 "Are There Humanoids in Space?" Flying Saucers, Aug. 1967, pp. 12-14.

**1377**
Gaddis, Vincent
 "Are Flying Saucers Really Creatures That Live in Outer Space?" True (48:363), Aug. 1967, pp. 26-27, 59-62.

**1378**
"Little Men From Mars." Science Digest 62, Nov. 1967, p. 37.

**1379**
"Belgische UFO-Waarnemingen." Amsterdam, Het Interplanetair Nieuwsbulletin 2, Oct.-Nov.-Dec. 1967, pp. 3-7.

**1380**
Hatcher, Judi Anne
 "Towers on the Moon?" Fate 20, Dec. 1967, pp. 32-37.

**1381**
Öpik, E. J.
 "Life and Intelligence in the Universe: Bottomless Speculations." Armagh, Irish Astronomical Journal 8, Dec. 1967, pp. 128-139.

**1382**
Berger, Rainer
 "The Solar System and Extraterrestrial Life." Advances in the Astronautical Sciences 13, 1968, pp. 649-665.

**1383**
Bobrovnikoff, N. T.
 "Soviet Attitudes Concerning the Existence of Life in Space." (In Handbook of Soviet Space Science Research. Ed. by G. E. Wukelic. New York, Gordon and Breach, 1968, pp. 453-472).

**1384**
Imshenetskii, A. A.
 "A Comparative Evaluation of Different Methods for Detection of Extraterrestrial Life." (In Life Sciences and Space Research VI; COSPAR Plenary Meeting. 10th, Open Meeting of Working Group V, London, England, July 27-28, 1967. Proceedings. Ed. by A. H. Brown and F. G. Favorite. Amsterdam, North-Holland Pub., 1968, pp. 170-182).

**1385**
Boncompagni, Solas
 "Attualita' del Mito di Osiris." Turin, Clypeus 5, Feb. 1968, pp. 9-12.

**1386**
Tobin, Michael
 "Our Earth in Space: Material Evolution and the Emergence of a Planetary Civilization." Midwest Quarterly 9, April 1968, pp. 297-318.

NOTES:

**1387**
Sullivan, Walter
"The Universe is not Ours Alone." New York Times Magazine, Sept. 29, 1968, pp. 40-41.

**1388**
Walosin, Frank
"Danger From Space Civilization: Look to Security UN Warned." Flying Saucers 60, Oct. 1968, pp. 6-7.

**1389**
Drake, Walter R.
"Spacemen in History." Flying Saucers 61, Dec. 1968, pp. 12-14.

**1390**
Wagner, Bernard M.
"Sociological Aspects of Exobiology." (In Exobiology: the Search for Extraterrestrial Life. American Astronautical Society and American Association for the Advancement of Science Symposium, New York, N.Y., Dec. 30, 1967. Proceedings. Ed. by M. M. Freundlich & Bernard M. Wagner. Tarzana, Calif., American Astronautical Society, AAS Science and Technology Series, Vol. 19, 1969, pp. 117-130).

**1391**
Zaitsev, Vyacheslav
"Visitors From Outer Space." Flying Saucers 63, April 1969, pp. 16-21.

**1392**
"Other Earths." London, The Economist 232, July 26, 1969, p. 13.

**1393**
Drake, Walter R.
"Space Chronicles of Ancient Rome." Flying Saucers 67, Dec. 1969, pp. 9-12.

**1394**
King, David W.
"Tell it Like it is." Navigator (18: 1), 1970, p. 13.

**1395**
Wiley, John P., Jr.
"Space: a Barrier to the Species." Natural History 79, Jan. 1970, pp. 70-73.

**1396**
"Persons, by R. Pucetti (Review, by Bernard Murchland)." Commonweal 91, Feb. 27, 1970, pp. 592-594.

**1397**
Drake, Walter R.
"Spacemen in Ancient Greece." Flying Saucers 68, March 1970, pp. 2-5.

**1398**
Kuttner, Robert E.
"Traces of Alien Influences From UFO's." Fate 247, Oct. 1970, pp. 75-79.

**1399**
Mallan, Lloyd
"Air Force Academy Textbook Warns Cadets That UFO's may be Spacecraft Operated by Aliens From Other Worlds." National Enquirer, Oct. 11, 1970, back page.

**1400**
"Life Elsewhere." London, Spaceflight 12, Nov. 1970, pp. 438-439.

**1401**
Panovkin, B. N.
"Extraterrestrial Civilizations: Problem and Considerations." (Tr. into English from Priroda, Moscow, #7, 1971, pp. 56-61, by Joint Publications Research Service, Washington, D.C.) Available on microfilm from NTIS.

**1402**
Ribes, J. C., & F. Biraud
"Extraterrestrial Civilizations." (In Astronomical Society of Australia. Annual General Meeting. 5th, University of Sydney, Sydney, Australia, May 19-21, 1971. Proceedings 2. Sydney, Australia, Sydney University Press, 1971, pp. 11-13).

NOTES:

**1403**
Drake, Walter R.
"Spacemen in the Ancient West." Flying Saucers 72, March 1971, pp. 20-21.

**1404**
Moll, Horacio M.
"Bioastronautics and Extraterrestrial Life." (Tr. into English from Revista de Aeronautica y Astronautica, Madrid, v. 30, #358, Sept. 1970, pp. 663-673). NASA-TT-F-13467. March 1971, 20p. (Available on microfilm from NTIS).

**1405**
Krebs, Columba
"How we Look to Spacemen." Search 97, May 1971, pp. 51-52.

**1406**
Kreifeldt, J. G.
"A Formulation for the Number of Communicative Civilizations in the Galaxy." Icarus 14, June 1971, pp. 419-430.

**1407**
Rapp, Daniel J.
"The Astroufologist's Resumé of the Planet Mars in Anthrochronology, 1608-1971." Flying Saucers 73, June 1971, pp. 4-6.

**1408**
"Gallup Poll Survey of World Leaders in Various Fields Shows 53% Believe in Existence of Life on Other Planets." New York Times, June 13, 1971, p. 20, col. 3.

**1409**
"Seeking Extraterrestrial Civilizations (Symposium at Byurakan Astrophysical Observatory, Armenia, Sept. 6-11, 1971)." London, Spaceflight 13, Aug. 1971, pp. 278-281.

**1410**
Aggen, Erich
"Life Beyond the Asteroid Belt." Flying Saucers 74, Sept. 1971, p. 28.

**1411**
Prytz, John M.
"Applied Exobiology." Space World 11-11-95, Nov. 1971, pp. 34-35.

**1412**
"Panel of Four Leading Scientists and one Theologian Agrees on Nov. 19, That Highly Advanced Civilizations Flourish Elsewhere in Universe, but Differ Over Whether it is Prudent to Establish Contact With Them..." New York Times, Nov. 20, 1971, p. 29, col. 1.

**1413**
Wiley, John P., Jr.
"Sky Reporter: Intelligent Life in the Universe." Natural History 81, Feb. 1972, pp. 50-53.

**1414**
Agnew, Irene
"Life on Other Planets?" Science Digest 72, Sept. 1972, p. 86.

**1415**
Renshaw, Jr., Charles C., & Frank D. Drake, eds.
"Is There Life Out There?" National Wildlife 10, Oct. 1972, pp. 50-53.

**1416**
Aguilar, Xavier F.
"Origin of the Saucers: Terrestrial, Extraterrestrial, Intratelluric?" Flying Saucers 79, Dec. 1972, pp. 8-11.

**1417**
Balagezyan, Jivan
"Extraterrestrial Life: Quests Continue." Flying Saucers 79, Dec. 1972, pp. 49-50.

**1418**
Gallup, Jr., George, & John O. Davies
"Does Human Life Exist on Other Planets?" Fate 273, Dec. 1972, pp. 73-76.

NOTES:

## INTELLIGENT LIFE ON OTHER PLANETS

**1419**
Powell, Conley
 "Interstellar Flight and Intelligence in the Universe." London, Spaceflight 14, Dec. 1972, pp. 442-447.

**1420**
Rorvik, David M.
 "Present Shock." Esquire 78, Dec. 1972, p. 134.

**1421**
Molton, P.
 "Limits of Terrestrial Life." London, Spaceflight 15, Jan. 1973, pp. 27-30.

**1422**
Ozick, Cynthia
 "If you can Read This, You are too far out." Esquire 79, Jan. 1973, p. 59.

**1423**
Sagan, Carl
 "Ancient and Legendary Gods of Old; Excerpt from The Cosmic Connection, an Extraterrestrial Perspective." Natural History 82, June 1973, pp. 92-93.

**1424**
Ball, J. A.
 "The Zoo Hypothesis." Icarus 19, July 1973, pp. 347-349.

**1425**
Gould, Bernard H.
 "Astronaut Gordon Cooper: Intelligent Life From Outer Space has Visited Earth." National Enquirer, July 1, 1973, p.3.

**1426**
Downing, Barry H.
 "Crash go the Chariots, by Clifford Wilson (Review)." Christianity Today 17, Aug. 31, 1973, pp. 30-31.

**1427**
Stanley, R. R.
 "Did Planetary Beings Visit Earth?" Two Worlds 3957, Oct. 1973, pp. 273-275.

**1428**
Oberbeck, S. K.
 "Deus ex Machina." Newsweek 82, Oct. 8, 1973, p. 104.

**1429**
Gastonguay, P. R.
 "Ancient Astronauts and UFO's." America 129, Nov. 17, 1973, p. 391.

**1430**
Ostriker, Alicia
 "What if We're Still Scared, Bored and Broke? Theories of Erich von Däniken." Esquire 80, Dec. 1973, pp. 238-240.

**1431**
Bersadski, Rudolf
 "Keskustelu Marsista." Turku, Finland, Argos 1974, pp. 26-29.

**1432**
Mitton, Simon, & Roger Lewin
 "Is Anyone out There?" Saturday Evening Post 246, Jan. 1974, pp. 52-53.

**1433**
"Robert Jastrow." Tomorrow Show, NBC-TV Network, 1 A.M., Eastern Daylight Time, June 6, 1974.

**1434**
Valentine, Tom
 "Spacemen Lived Here 40,000 Years Ago, Bringing Culture and Technology to Earth." National Tattler, Dec. 8, 1974 (Special Pullout Section), pp. 17-32.

NOTES:

# INTELLIGENT LIFE WITHIN EARTH

## BOOKS

**1435**
Bernard, Raymond
  The Hollow Earth...Mokelumne Hill, California, Health Research, 1963, 105p.

**1436**
Bernard, Raymond W.
  The Hollow Earth; the Greatest Geographical Discovery in History, made by Admiral Richard E. Byrd in the Mysterious Land Beyond the Poles; the true Origin of the Flying Saucers. New York, Universe Books, Inc., 1969, 254p.

**1437**
Sanderson, Ivan T.
  Invisible Residents; a Disquisition from Certain Matters Maritime, and the Possibility of Intelligent Life Under the Waters of This Earth. Cleveland, The World Pub. Co., 1970, 248p.

**1438**
Spencer, John W.
  Limbo of the Lost. Actual Stories of Sea Mysteries. Rev. New York, Bantam Books, 1973, 136p.

## PERIODICAL ARTICLES

**1439**
Goble, H. C.
  "Atlantis Sank Later Than you Think." Fate 10, Aug. 1957, pp. 19-24.

**1440**
Ormond, Ron
  "Is Siam a Secret Base for Flying Saucers?" Flying Saucers From Other Worlds, May 1959, pp. 51-56.

**1441**
Murray, Jacqueline
  "Flying Saucers and Atlantis." London, Flying Saucer Review 5, May-June 1959, pp. 18-19, 25.

**1442**
Palmer, Ray
  "Saucers From Earth: a Challenge to Secrecy." Flying Saucers, Dec. 1959, pp. 8-21.

**1443**
Ribera, Antonio
  "UFO's and the sea." London, Flying Saucer Review 10, Nov.-Dec. 1964, pp. 8-10.

NOTES:

# INTELLIGENT LIFE WITHIN EARTH

**1444**
Fouéré, René
 "Existe-t-il des Bases Sous-Marines de
Soucoupes Volantes?"  Paris, <u>Phénomènes
Spatiaux</u>, Feb. 1965, pp. 16-25.

**1445**
Ribera, Antonio
 "More About UFO's and the sea."  London,
<u>Flying Saucer Review</u> 11, Nov.-Dec. 1965, pp.
17-18.

**1446**
Fouéré, René, & Francine Fouéré
 "Le Plateau de Valensole Serait-il un Haut
Lieu du Tourisme Insolite?"  Paris, <u>Phénomènes
Spatiaux</u>, Sept. 1966, pp. 10-20.

**1447**
Keel, John A.
 "Secret UFO Bases Across the U.S."  <u>Saga</u> 36,
April 1968, pp. 30-33, 86, 89-90, 92-94, 96.

**1448**
Steiger, Brad, & Joan Whritenour
 "Unidentified Underwater Saucers."  <u>Saga</u> 36,
June 1968, pp. 34-37, 54-57.

**1449**
Earley, George W.
 "10,000 Says UFO's Aren't Extraterrestrial."
<u>Fate</u> 247, Oct. 1970, p. 83.

**1450**
Stewart-Gordon, James
 "What's the Truth About the Bermuda
Triangle?"  <u>Reader's Digest</u>, July 1975, pp.
75-79.

NOTES:

# INTERSTELLAR COMMUNICATION

## BOOKS

**1451**
Pelley, William D.
   Earth Comes...Design for Materialization.
Indianapolis, Fellowship Press, 1941, 303p.

**1452**
Williamson, George H., & A. C. Bailey
   The Saucers Speak! A Documentary Report of Interstellar Communication by Radiotelegraphy.
Los Angeles, New Age Pub. Co., 1954, 127p.

**1453**
Rowe, Kelvin
   A Call at Dawn. El Monte, Calif., Understanding Pub. Co., 1958, 198p.

**1454**
Handelsman, M.
   Considerations on Communication With Intelligent Life in Outer Space; Paper Presented at the Western Electronic Show & Convention, Los Angeles, Aug. 21-24, 1962; Wescon Paper #4.
North Hollywood, Western Periodicals Co., 1962, 16p.

**1455**
Jackson, C. D., & R. E. Hohmann
   An Historic Report of Life in Space: Tesla, Marconi, Todd. Paper #2730-62. New York, American Rocket Society, 1962.

**1456**
Bergier, Jacques
   A L'écoute des Planètes. Paris, Fayard, 1963, 182p.

**1457**
Cameron, A. G. W., editor
   Interstellar Communication, a Collection of Reprints and Original Contributions (Item 28: Extraterrestrial Intelligent Life and Interstellar Communication: an Informal Discussion).
New York, Benjamin, 1963, 320p. ($8.50)

**1458**
Hoerner, S. von
   Search for Signals From Other Civilizations. (In Cameron, A. G. W., ed. Interstellar Communication. New York, Benjamin, 1963, pp. 272-286). ($8.50)

**1459**
Morrison, P.
   Interstellar Communication. (In Cameron, A. G. W., ed. Interstellar Communication. New York, Benjamin, 1963, pp. 249-271). ($8.50)

NOTES:

# INTERSTELLAR COMMUNICATION

**1460**
Shlovskiy, I. S.
  Is Communication Possible With Intelligent Beings on Other Planets? (In Cameron, A. G. W., ed. Interstellar Communication. New York, Benjamin, 1963, pp. 5-16). ($8.50)

**1461**
Webb, J. W., & W. H. Webb
  Detection of Intelligent Signals From Space. (In Cameron, A. G. W., ed. Interstellar Communications. New York, Benjamin, 1963, pp. 178-191). ($8.50)

**1462**
Williamson, George H.
  The Saucers Speak; a Documentary Report of Interstellar Communication and Radiotelegraphy. London, N. Spearman, 1963, 160p.

**1463**
Melpar, Inc.
  Research and Development of an Instrument for Detection of Extraterrestrial Life by Optical Rotatory Dispersion. Springfield, Va., For sale by the Clearinghouse for Federal Scientific and Technical Information, 1966, 96p. (NASA contractor report, NASA CR-423)

**1464**
Dole, Stephen H.
  The Search for a Rationale for Interstellar Communications. Presented at the 132d Annual Meeting of the American Association for the Advancement of Science, Berkeley, Calif., Dec. 30, 1965. (P-3296; AD-626675). Alexandria, Va., NTIS, Jan. 1966, 8p.

**1465**
MERINT Radiotelegraph Procedure. Authorized by the Secretary of the Navy. No. JANAP 146-E. Washington, G.P.O., March 1966, 6p.

**1466**
Reeve, Bryant
  The Advent of the Cosmic Viewpoint. Amherst, Wis., Amherst Press, 1967, 256p.

**1467**
North Atlantic Treaty Organization. Advisory Group for Aerospace Research and Development
  Propagation Factors in Space Communications; AGARD Conference Proceedings, #3. Ed. by W.T. Blackband. Pelham, N.Y., Circa Pubs., 1968, 569p. ($37.50)

**1468**
Vnezemnye Tsivilizatsii. Problemy Mezhzvezdnoi Sviazi. Moscow, "Nauka" 1969, v.5, 438p.

**1469**
Mills, David M.
  Extragalactic Radio Sources. Final Report. AD-727145; SU-IPR-408. 1971, 84p. (Available on microfilm from NTIS).

**1470**
Aerospace Corp. Lab Operations, El Segundo, Calif.
  Radio Sources: 3.3mm Flux and Variability Measurements. Report for April 1965-March 1971, by William G. Fogarty, et al. (AD-731-552). Report #TR-0172 (2230-20)-5 SAMSO-TR-71-225, Aug. 15, 1971, 32p. (Available from NTIS: $3.00).

**1471**
California. University at Berkeley. Space Sciences Lab.
  The 30-36 GHz Venus Microwave Emission, Final Report, by Michael Allen Janssen, et al. Berkeley, 1972, 138p. (AD-743982; SSL-Series-13-Issue-51). Available from NTIS.

---

NOTES:

**1472**
Oliver, B. M.
State of the art in the Detection of Intelligent Extraterrestrial Signals. Paper Presented at 23rd International Astronautical Congress, Vienna, Austria, Oct. 8-15, 1972. Vienna, International Astronautical Federation, 1972, 30p.

**1473**
Sukhotin, B. V.
Problems of Decoding. Paper Presented at 23rd International Astronautical Congress, Vienna, Austria, Oct. 8-15, 1972. Vienna, International Astronautical Federation, 1972, 24p.

**1474**
Macvey, John
Whispers From Space. New York, Abelard-Schuman, 1974, 272p.

**1475**
Sagan, Carl, ed.
Communication With Extraterrestrial Intelligence (CETI). Cambridge, MIT Press, 1974, 428p. ($10.00)

## PERIODICAL ARTICLES

**1476**
Guillemin, A. V.
"Communication With the Planets." *Popular Science* 40, Jan. 1892, pp. 361-363.

**1477**
Flammarion, Camille
"Inter-astral Communication." *American Review of Reviews* 5, Feb. 1892, p. 90.

**1478**
Galton, F.
"Intelligible Signals Between Neighbouring Stars." London, *Fortnightly Review* 66, Nov. 1896, pp. 657-664.

**1479**
"That Prospective Communication With Another Planet." *Current Opinion* 66, March 1919, pp. 170-171.

**1480**
"W. Marconi Says Queer Signals Occur in London and New York Simultaneously." *New York Times*, Jan. 27, 1920, p. 7, col. 3.

**1481**
"Comment by Prof. Einstein on Signals." *New York Times*, Feb. 2, 1920, p. 24, col. 2.

**1482**
Nieman, H. W., & C. W. Nieman
"What Shall we say to Mars?" *Scientific American* (122:12), March 20, 1920, pp. 298-312.

**1483**
"J. H. C. Macbeth Says W. Marconi is Sure That Mars Flashes Messages and That it Would be Simple Matter to Arrange Code." *New York Times*, Sept. 2, 1921, p. 1, col. 4.

**1484**
"Marconi Reports That he has no Message From Mars Yet." *New York Times*, June 16, 1922, p. 19, col. 4.

NOTES:

# INTERSTELLAR COMMUNICATION

1485
"French War Ministry Investigates Radio Signals Which Experts Think may Have Been Sent From Mars." New York Times, March 21, 1923, p. 18, col. 2.

1486
"Dr. Steinmetz Predicts Communication With Mars." New York Times, Aug. 20, 1923, p. 1, col. 7.

1487
"C. Flammarion Predicts That we Will Talk With Mars." New York Times, Dec. 12, 1923, p. 3, col. 2.

1488
"Special Article by C. Flammarion on Mars, Possibility of Life and of Communications Discussed." New York Times, March 2, 1924, sec. IX, p. 3, col. 1.

1489
"Government to Cooperate With Prof. Todd in his Effort to Obtain air Silence for Attempted Communication." New York Times, Aug. 21, 1924, p. 11, col. 3.

1490
"Editorial on Probable Illusion of Wireless Communication, Wisdom of Possible Martian Inhabitants." New York Times, Aug. 23, 1924, p. 8, col. 4.

1491
"C. F. Jankins Takes 30 Feet of Radio Film of Mysterious Signals During Closest Approach to Earth With Device he Invented." New York Times, Aug. 28, 1924, p. 6, col. 2.

1492
Todd, D.
"Radio Messages From Mars?" Literary Digest (82:10), Sept. 6, 1924, p. 28.

1493
Fitch, C. J.
"Interplanetary Communication." Astronautics 4 (ARS Journal), Oct. 1930, pp. 1-2, 5-7.

1494
"N. Tesla Predicts Intercommunication." New York Times, July 10, 1932, p. 19, col. 1.

1495
"Signalisation Interplanétaire." Paris, La Nature 2944, Jan. 1, 1935, p. 1.

1496
"Comment on Attempt to Signal to Them." New York Times, Feb. 10, 1935, sec. VIII, p. 15, col. 2.

1497
"Mars: Drs. B. J. Bok and F. Watson on Possibility of Life and Contact With Earth." New York Times, July 23, 1939, p. 3, col. 1.

1498
"Study of Radio Noises Generated in Space Outside the Earth Discussed." New York Times, Dec. 21, 1947, sec. IV, p. 9, col. 6.

1499
"Radar Detection of Cosmic Radio Waves Revealed, Cornell University Conference... Dr. D. M. Menzel Reports Sounds Discordant." New York Times, Oct. 6, 1948, p. 1, col. 2.

1500
"Buck Rogers Baedeker." Newsweek 33, Feb. 28, 1949, pp. 48-49.

1501
Balta Elias, J.
"Radio-Ondas Extraterrestres." Madrid, Arbor 105-106, Sept.-Oct. 1954, pp. 1-27.

NOTES:

# INTERSTELLAR COMMUNICATION

1502
Dye, Clarkson
"Radio to Other Worlds." Fate 8, March 1955, pp. 88-96.

1503
Sanford, Ray
"Contact With a Flying Saucer." Fate 9, May 1956, pp. 12-18.

1504
"Scientists Detect Radio Waves From Planet Venus." Army-Navy-Air Force Register 77, June 9, 1956, p. 7.

1505
"Kraus Reports new Radio Signals From Venus." New York Times, June 23, 1956, p. 9, col. 1.

1506
Satterthwaite, Gilbert E.
"Radio Noise From Jupiter." London, Spaceflight 1, Oct. 1956, p. 14.

1507
Smith, Bernard
"People From Outer Space Contact Earth man by Radio." London, Flying Saucer Review 4, July-Aug. 1958, pp. 28-29.

1508
Rodgers, Philip
"Spacemen Speaking." London, Flying Saucer Review 4, Sept.-Oct. 1958, pp. 20-21.

1509
"Communication to Mars (and Interplanetary Communication)." Signal 13, Nov. 1958, p. 64.

1510
Drake, Frank D.
"How can we Detect Radio Transmissions From Distant Planetary Systems?" Sky & Telescope 19, 1959, p. 140.

1511
"Mysterious Broadcast From Space Ship?" Flying Saucers, Feb. 1959, pp. 35-37, 46.

1512
Miller, Max B.
"Scientists Track Space Radio Signals." Fate 12, June 1959, pp. 57-58.

1513
Miller, Max B.
"Scientists Track Space Radio Signals." Fate 12, June 1959, pp. 57-58.

1514
Dreyer, H. R.
"The man who Knows the Languages of the Stars." Fate 12, Aug. 1959, pp. 78-79.

1515
Cocconi, G., & P. Morrison
"Searching for Interstellar Communications." London, Nature 184, #4690, Sept. 19, 1959, pp. 844-846.

1516
McClain, Edward F.
"The 600-foot Radio Telescope." Scientific American 202, Jan. 1960, pp. 45-51.

1517
"Spacious Talk." Scientific American (202:1), Jan. 1960, pp. 74, 76, 79.

1518
DuShane, G.
"Hello out There." New Republic 142, Jan. 25, 1960, p. 9.

1519
Drake, Frank D.
"How we can Detect Radio Transmissions From Distant Planetary Systems." Sky and Telescope (19:3), Feb. 1960, pp. 140-143.

NOTES:

## INTERSTELLAR COMMUNICATION

**1520**
Kneitel, Thomas S.
"The Radio Voices of Space." *Space Age* 2, March 1960, pp. 11-14.

**1521**
Edson, J. B.
"Tuning in on Other Worlds." *New York Times Magazine*, March 13, 1960, p. 31.

**1522**
Calder, N.
"Stowaways and Starfish." *New Statesman* 59, March 19, 1960, p. 388.

**1523**
Ball, G.
"Listen in on Other Suns." *Science Newsletter* 77, April 30, 1960, pp. 283-283.

**1524**
"U.S. National Bureau of Standards Reports Radio Signals Emitted With Energy of 2000 Tons of TNT About Once a Second Indicate Disturbances on or in Planet With Energies of 100 Million Tons of TNT...Speculation on Source of Radio Waves." *New York Times*, April 30, 1960, p. 2, col. 5.

**1525**
Edson, James B.
"Tuning in on Other Worlds." *Air Force & Space Digest* 43, May 1960, pp. 55-56.

**1526**
Bracewell, R. N.
"Communications From Superior Galactic Communities." London, *Nature* (186:4726), May 28, 1960, pp. 670-671.

**1527**
Koestler, A.
"Trying to Talk With Other Planets." *Science Digest* 47, June 1960, pp. 69-72.

**1528**
"Walter Sullivan, on Dr. Bracewell's Suggestion, Published in *Nature*, That Some Civilized Galactic Community may be Trying to Communicate With our Solar System by Orbiting Satellite With Message Awaiting our Radio Recognition,Unexplained Radio 'Echoes' in 1927, and 1928 and 1934 Cited; Bracewell...Believes Sending out Satellites With Recorded Messages More Practical Than Continuous Direct Radio Signals." *New York Times*, June 20, 1960, p. 1, col. 3.

**1529**
Beller, William
"How to Contact 'People' in Space?" *Missiles & Rockets* 7, July 25, 1960, pp. 42-44.

**1530**
Briggs, Michael H.
"Other Astronomers in the Universe?" Wellington, New Zealand, *Southern Stars* 18, Sept. 1960, pp. 147-151.

**1531**
Hicks, Clifford B.
"We're Listening for Other Worlds." *Popular Mechanics* 114, Sept. 1960, pp. 81-85.

**1532**
"U.S. Navy Reports First Conclusive Radio Signals Picked up by Michigan University Radio Telescope From Saturn." *New York Times*, Sept. 9, 1960, p. 3, col. 6.

**1533**
"Universal Decoding Plan for Interstellar Messages: Project Ozma." *Science Newsletter* 78, Oct. 22, 1960, p. 265.

**1534**
Golay, M. J. E.
"Coherence in Interstellar Signals." *IRE Proceedings* 49, 1961, pp. 958-959.

NOTES:

# INTERSTELLAR COMMUNICATION

**1535**
Webb, J. A.
"Detection of Intelligent Signals From Space." (In *Communications: Bridge or Barrier*. Institute of Radio Engineers, Seventh National Communication Record. New York, 1961, p. 10).

**1536**
Golomb, S. W.
"Extraterrestrial Linguistics." *Astronautics (ARS Journal)* 6, Jan.-June 1961, pp. 46-47, 96.

**1537**
Boehm, A. W.
"Are we Being Hailed From Interstellar Space?" *Fortune* 63, March 1961, pp. 144-149.

**1538**
Lapp, R. E.
"How to Talk to People, if any, on Other Planets. Excerpt From *Man and Space: the Next Decade*." *Harper's Magazine* 222, March 1961, pp. 58-63.

**1539**
Drake, Frank D.
"Project Ozma." *Physics Today* (14: 4), April 1961, p. 40.

**1540**
Drake, Frank D.
"Radio Emission From the Planets." *Physics Today* (14:4), April 1961, pp. 30-34.

**1541**
Schwartz, R. N., & C. H. Townes
"Interstellar and Interplanetary Communication by Optical Masers." London, *Nature* (190: 4772), April 15, 1961, p. 205.

**1542**
Briggs, Michael H.
"Superior Galactic Communities." London, *Spaceflight* 3, May 1961, pp. 109-110.

**1543**
Golay, M. J. E.
"Notes on the Probable Character of Intelligent Radio Signals From Other Planetary Systems." *Proceedings of the Institute of Radio Engineers* (49: 1, part 1), May 1961, p. 959.

**1544**
Golomb, Solomon W.
"Extraterrestrial Linguistics." *Astronautics* 6, May 1961, pp. 46-47, 96.

**1545**
"Signals From Space?" *Science Newsletter* 79, May 31, 1961, p. 295.

**1546**
Oakley, C. O.
"Math, our Link With Space People." *Science Digest* 49, June 1961, pp. 7-13.

**1547**
Von Hoerner, Sebastian
"The Search for Signals From Other Civilizations." *Science* 134, Dec. 1961, pp. 1839-1843.

**1548**
Von Hoerner, S.
"The Search for Signals From Other Civilizations." *Science* (134: 3493), Dec. 8, 1961, p. 1839.

**1549**
Lovell, A. C. B.
"Search for Voices From Other Worlds." *New York Times Magazine*, Dec. 24, 1961, p. 10.

**1550**
Handelsman, M.
"Considerations on Communication With Intelligent Life in Outer Space." (In *Wescon Convention Record*. Proceedings 6, Pt. 5; 1962. Los Angeles, Wescon, 1962, 15p. Paper #4.4).

NOTES:

## INTERSTELLAR COMMUNICATION

1551
"Dr. P. Morrison Proposes Devising Easily-Broken Codes for Contacts With Intelligent Beings on Other Planets." New York Times, Jan. 24, 1962, p. 12, col. 5.

1552
Bracewell, R. N.
"Radio Signals From Other Planets." Proceedings of the Institute of Radio Engineers (50: 2), Feb. 1962, p. 214.

1553
"Secret Conference Sponsored by National Sciences Academy at Green Bank, W. Va., in November 1961 on Possibilities and Problems of Communicating With Other Worlds Revealed and Discussed." New York Times, Feb. 4, 1962, p. 1, col. 4.

1554
"Sir Bernard Lovell, Dr. H. S. Brown, Others Discuss Possibility of Life on Other Worlds and of Radio Contact With Them..." New York Times, March 2, 1962, p. 1, col. 5.

1555
Ascher, Robert, & Marcia Ascher
"Interstellar Communication and Human Evolution." London, Nature (193: 8419), March 10, 1962, p. 940.

1556
"Star Search: new Instruments Search the Heavens for Intelligent Signals From Other Worlds." Space World 3, July 1962, p. 27.

1557
Rosenberg, Paul
"Communication With Extraterrestrial Intelligence." Aerospace Engineering 21, Aug. 1962, pp. 68-69, 111.

1558
Calvin, Melvin
"Communication: From Molecules to Mars." Bulletin of the American Institute of Biological Sciences (12:5), Oct. 1962, pp. 29-44.

1559
Oliver, Bernard M.
"Radio Search for Distant Races." International Science & Technology 10, Oct. 1962, p. 55.

1560
Cade, C. M.
"Communicating With Life in Space." London, Discovery (24:5), 1963, pp. 36-41.

1561
Webb, J. A.
"Detection of Intelligent Signals From Space." (In Cameron, A. G. W., ed. Interstellar Communication. New York, Benjamin, 1963, pp. 178-191).

1562
Calvin, Melvin
"Talking to Life on Other Worlds. Excerpt From Address, With Deciphering of a Coded Message." Science Digest 53, Jan. 1963, pp. 14-19.

1563
Mowbray, Lionel
"Communication Avec le Cosmos et Langage Universel." Paris, Science, Progrès et la Nature 3334, Feb. 1963, p. 79.

1564
"Mars Radar Contact Made." Los Angeles, Western Aerospace 43, March 1963, p. 20.

1565
Cade, C. M.
"Communicating With Life in Space." London, Discovery 24, May 1963, pp. 36-41.

1566
Motz, Lloyd
"Intelligent Life Beyond the Solar System." Signal 17, May 1963, pp. 75-77.

NOTES:

**1567**
Smith, R. F. W.
"Communication With Extraterrestrial Beings Called Improbable Unless man can Signal in two Systems of Thought." London, Science Fortnightly, Oct. 30, 1963, p. 8.

**1568**
Cameron, A. G. W.
"Communicating With Intelligent Life on Other Worlds." Sky & Telescope 26, Nov. 1963, pp. 258-261.

**1569**
Miller, E. C., & J. L. Smith
"Some Considerations Regarding the Possibility of Contact With Intelligent Extraterrestrial Beings." London, BUFORA Journal and Bulletin 1, Winter 1964, pp. 4-7.

**1570**
Newman, J. R.
"Interstellar Communication, edited by A. G. W. Cameron, a Review." Scientific American 210, Feb. 1964, pp. 141-146.

**1571**
Mannel, Cliff
"Interstellar Communications." Challenge (3: 1), Spring 1964, p. 24.

**1572**
Kardeshev, N. S.
"Transmission of Information by Extraterrestrial Civilizations." (Tr. into English from Astron. Zhurnal, Moscow (41: 2), March-April 1964, pp. 282-287). (JPRS-25307; TT-64-31588). Washington, Joint Publications Research Service, 1964, 10p. (Available from NTIS: 50¢).

**1573**
"Two USSR Writers Hypothesize That Light Signals Have Been Received on Earth 3 Times From Inhabitants of Planet 61 Cygni." New York Times, March 21, 1964, p. 4, col. 3.

**1574**
Creighton, Gordon W.
"What the Soviets are Saying." London, Flying Saucer Review 10, Sept.-Oct. 1964, p. 26.

**1575**
"N. S. Kardashev Suggests CTA-21 and CTA-102, Powerful Sources of Radio Emission in Constellations Aries and Pegasus, May be Sites of Super-civilizations Attempting to Communicate (Article in Astronomical Journal of the Soviet Academy of Sciences)". New York Times, Oct. 26, 1964, p. 1, col. 2.

**1576**
Asimov, Isaac
"Hello, CTA-21, is Anyone There?" New York Times Magazine, Nov. 29, 1964, p. 52.

**1577**
"Message From Space." America 111, Dec. 12, 1964, pp. 770-771.

**1578**
Gambling, W. A.
"Possibilities of Optical Communications." London, Engineering 198, Dec. 18, 1964, p. 776.

**1579**
"Tass Reports Astronomer G. Sholomitsky Finds CTA-102 Radio Source Emits 'Flickering' Radio Waves Every 100 Days; Quotes N. Kardashev, Other Scientists as Saying Another Civilization has Been Discovered." New York Times. April 13, 1965, p. 1, col. 6.

**1580**
Asimov, Isaac
"Microwave Radio Beams From Space Might be Signals From Other Intelligences." Catholic Digest 29, June 1965, pp. 37-40.

NOTES:

## INTERSTELLAR COMMUNICATION

**1581**
Vertregt, M.
"Cosmic Squeak and Gibber." London, Spaceflight 7, July 1965, pp. 122-128.

**1582**
Gardner, M.
"Mathematical Games: Communications With Intelligent Organisms on Other Worlds." Scientific American 213, Aug. 1965, pp. 96-100.

**1583**
Martynov, Dmitri
"Russian Scientists Discover Mysterious Radio Transmissions From the Milky Way." Space World B-8-22, Aug. 1965, pp. 41-42.

**1584**
Mutschall, Vladimir
"Interstellar Communications." Library of Congress, Aerospace Technology Div., Foreign Science Bulletin (I: 9), Sept. 1965, pp. 56-63.

**1585**
Lewis, R. S.
"Message From Mariner IV." Bulletin of the Atomic Scientist 21, Nov. 1965, pp. 38-40.

**1586**
Wooster, H., et al
"Communication With Extraterrestrial Intelligence." IEEE Spectrum 153, 1966, p. 3.

**1587**
Mercer, D. M. A.
"Messages to the Stars." Courier, Jan. 1966, p. 4.

**1588**
"Dr. B. M. Oliver Reports he has Devised Antenna System That he Believes can Dectect TV Emissions of Civilizations up to 200 Light Years Away, AAAS Symposium (W. Sullivan Report)." New York Times, Jan. 2, 1966, p. 14, col. 3.

**1589**
"USSR Academy of Sciences Astronomical Council Pres. Mustel Urges International Astronomical Union 1967 General Assembly Discuss International Effort to Locate Signals from Other Worlds; Poll of 8 Leading Soviet Scientists Shows all Believe Existence of Extraterrestrial Life is Reasonable Hypothesis." New York Times, May 10, 1966, p. 19, col. 1.

**1590**
"Space Radio Noise Picked up." Air Force Times 26, June 8, 1966, sec. M-10.

**1591**
Saunders, Alex
"Alien Life Contemplation." Flying Saucers, Oct. 1966, pp. 21-22.

**1592**
Crenshaw, James
"What can we Expect of 'Civilization' in Outer Space?" Fate 20, Jan. 1967, pp. 79-89.

**1593**
Pascalis, Bernardino
"Werkelijkheid of Fantasie?" Haarlem, Netherlands, Panorama, Feb. 1967, pp. 31-38.

**1594**
Burns, J. A.
"Jupiter's Decametric Radio Emission and the Radiation Belts of its Galilean Satellites." Science 159, March 1, 1967, pp. 971-972.

**1595**
Cade, C. M.
"A Long, Cool Look at Alien Intelligences: Part II--Modes of Communication." London, Flying Saucer Review 13, July-Aug. 1967, pp. 15-17.

NOTES:

1596
"Observations of 'Mysterium Phenomen' of Radio Emissions From Milky Way Show Emissions Meet Almost all Criteria for Artificial Communications, but Scientists see Natural Causes Rather Than Signals by Intelligent Beings." New York Times, Aug. 26, 1967, p. 1, col. 3.

1597
"Maser Here. Hello There! Radiation Waves From Hydroxyl Clouds." Newsweek 70, Sept. 11, 1967, p. 84.

1598
Cade, C. M.
"A Long, Cool Look at Alien Intelligence. Part IV--Possible Significance of Parapsychology." London, Flying Saucer Review 13, Nov.-Dec. 1967, pp. 13-15.

1599
Kaplan, S. A.
"A Possible Interpretation of the Radio Emission From Small Cosmic Sources." Soviet Astronomy 11, Nov.-Dec. 1967, pp. 416-418.

1600
Cade, C. M.
"A Long, Cool Look at Alien Intelligence: Part V--It's all in the Mind." London, Flying Saucer Review 14, March-April 1968, pp. 7-9.

1601
Kublin, D.
"How man Might Communicate With Other Planets." Tr. by Joint Publications Research Service, from Nauk i Zhizn #3 (Moscow), 1968, pp. 110-111. New York, Sept. 1968, 23p. (Available on microfilm from NTIS: 65¢).

1602
"W. Sullivan on Pulsar Sources: Views That They are Manifestations of Intelligent Life in Other Civilizations..." New York Times, April 14, 1968, sec. IV, p. 18, col. 1; Sept. 29, 1968, sec. VI, p. 40.

1603
"Signals From Outer Space." Flying Saucers 60, Oct. 1968, pp. 22-23.

1604
Whalley, Paul
"UFO's and Radio Waves." Flying Saucers 60, Oct. 1968, p. 26.

1605
Elliott, G.
"Mysterious Radio Signals." London, Flying Saucer Review (15: 2), March-April 1969, pp. 29-30.

1606
"American Astronomers at Arecibo Ionospheric Observatory Detect Radio Signals From Space That Might be From Other Civilizations." New York Times, March 10, 1969, p. 10, col. 2.

1607
Kaplan, S. A., & V. I Eidman
"Ob Odnom Vozmozhnom Antennon Mekhanizme Kosmicheskogo Radioiz-lucheniia." Moscow, Astronomicheskii Zhurnal (48: 1), 1971, pp. 440-443.

1608
Rees, M. J.
"New Interpretation of Extragalactic Radio Sources." London, Nature 229, Jan. 29, 1971, pp. 312-317.

1609
"U.S. Unmanned Pioneer 10 Spacecraft...Will Contain Message in Scientific Symbols Stating who Sent it and Where They Live... Represents First Direct Attempt by man to Communicate With Intelligent Beings Elsewhere in Universe." New York Times, Feb. 25, 1971, p. 7, col. 1.

NOTES:

## INTERSTELLAR COMMUNICATION

**1610**
Sullivan, Walter
  "Comments on Scheduled U.S. Pioneer 10 Flight...Drawing Shows Message, Including Sketch of Earth Inhabitants That Will be Carried by Spacecraft in Effort to Communicate With Intelligent Beings Elsewhere in Universe." New York Times, Feb. 27, 1971, p. 60, col. 1.

**1611**
Lawton, A. T.
  "Infrared Interstellar Communication." London, Spaceflight 13, March 1971, pp. 83-85.

**1612**
Prytz, John
  "Pulsars." Flying Saucers 74, Sept. 1971, pp. 6-7.

**1613**
Lawrence, L. G.
  "Interstellar Communications: What are the Prospects?" Electronics World 86, Oct. 1971, pp. 34-35.

**1614**
  "Is Anybody out There Sending?" Science News 100, Oct. 2, 1971, pp. 223-224.

**1615**
Belitsky, Boris
  "Signals From Other Worlds (First Soviet-U.S. Conference on Communications With Extra-Terrestrial Intelligence (CETI)." London, Spaceflight 14, Jan. 1972, pp. 17-19.

**1616**
Walker, J. C. G.
  "The Search for Signals From Extraterrestrial Civilizations." London, Nature 241, Feb. 9, 1973, pp. 379-381.

**1617**
Balfour, Malcolm
  "Scientists Attempting Contact With Alien Space Probe Believed to be Orbiting Earth." National Enquirer, March 18, 1973, p.26.

**1618**
Lawton, A. T.
  "The Interpretation of Signals From Space." London, Spaceflight 15, April 1973, pp. 132-137.

**1619**
Robertson, J. G.
  "An All-sky Catalog of Strong Radio Sources at 408 MHz." Melbourne, Australian Journal of Physics 26, June 1973, pp. 403-416.

**1620**
Oliver, B. M.
  "Project Cyclops Study. Conclusions and Recommendations." Icarus 19, July 1973, pp. 425-428.

**1621**
Sagan, Carl
  "On the Detectivity of Advanced Galactic Civilizations." Icarus 19, July 1973, pp. 350-352.

**1622**
  "...USSR on October 16 Reports Detection of Radio Signals that may have Originated with Another Civilization... in Form of Pulses Detected Several Times per day..." New York Times, Oct. 17, 1973, p. 94, col. 1.

**1623**
  "Is Anybody out There Sending?" Spaceview (2:6), Nov.-Dec. 1973, pp. 12-13, 21.

NOTES:

**1624**
Gris, Henry
 "Two Top Soviet Astrophysicists Reveal Russians Receive Radio Signals From Civilization in Outer Space." <u>National Enquirer</u>, Nov. 25, 1973, p.17.

**1625**
Wilhelm, John
 "Communication With Extraterrestrial Intelligence, CETI, Edited by Carl Saga (Review)." <u>Time</u> 103, Jan. 21, 1974, p. 74.

**1626**
Keyhoe, Donald E.
 "U. S. Scientists Have Been in Contact With a Civilization in Outer Space." <u>National Enquirer</u>, March 24, 1974, p.20.

**1627**
Kotulak, Ronald
 "Is Anyone Listening? Man Finally Speaks to the Unknown." <u>Chicago Tribune</u>, Dec. 22, 1974, sec. 2, p. 18.

NOTES:

## LIFE ON OTHER PLANETS IN GENERAL AND IN SPACE (EXCLUDING INTELLIGENT LIFE)

**BOOKS**

**1628**
Serviss, Garrett P.
  Curiosities of the sky. New York, Harper & Bros., 1909, 268p.

**1629**
Jeans, J.
  Life in the Universe. (In The Universe Around us. New York, Macmillan, 1929, 351p.)

**1630**
von Cues, N.
  Cusanus Studien...(Sitzungsberichte der Heidelberger Akademie der Wissenschaften. Philosophisch-historische Klasse. v.20. 1929-30, Proceedings, #3). Heidelberg, C. Winter, 1930-

**1631**
Muller, Hermann J.
  Out of the Night: a Biologist's View of the Future. New York, Vanguard Press, 1935, 127p.

**1632**
Jones, H. S.
  Life on Other Worlds. New York, Macmillan, 1940, 299p. ($3.00)

**1633**
Jeans, J.
  Is There Life on the Other Worlds? Smithsonian Institution Annual Report for 1942. Washington, 1943, pp. 145-150.

**1634**
Krafft, Carl F.
  Ether and Matter. Richmond, Va., Dietz Printing Co., 1945, 117p.

**1635**
Rocquet, R.
  La vie Dans la Matière et Dans le Cosmos. Paris, Omnium Littéraire, 1950, 221p. (370 francs)

**1636**
Sytinskaia, N. N.
  Est' li Zhizn' na Drugikh Planetakh. Moscow, Gosudarstvennoe Izdatel'Stvo Kul'turnoprosvetitel'noi literatury, 1952, 64p.

**1637**
Nelson, A. F. J. H. N.
  Life and the Universe. London, Staples Press, 1953, 223p. (16 shillings)

**1638**
Shapley, H.
  On Climate and Life. (In Climatic Change: Evidence, Causes and Effects). Cambridge, Harvard University Press, 1954, pp. 3-7. ($7.00)

NOTES:

# LIFE ON OTHER PLANETS IN GENERAL AND IN SPACE (Excluding Intelligent Life)

**1639**
Eddington, A. S.
  The Nature of the Physical World. London, Dent, 1955, 341p. (7 shillings)

**1640**
Nelson, Albert F.
  There is Life on Mars. New York, Citadel Press, 1955, 140p.

**1641**
Moore, P.
  The World Around us. New York, Abelard-Schuman, 1956, 157p. ($2.50)

**1642**
Nelson, A. F. J. H. N.
  There is Life on Mars. American Edition. New York, Citadel Press, 1956, 152p. ($3.00)

**1643**
Dufay, J.
  Nebulae and Interstellar Matter. Trans. From the French by A. J. Pomerans. New York, Philosophical Library, 1957, 352p. ($15.00)

**1644**
Menzel, D. H.
  The Universe in Action. Birmingham, Alabama, Rushton Lectures Foundation, 1957, 71p.

**1645**
Rush, J. H.
  Life in the Universe. (In his The Dawn of Life, chapter 8. New York, New American Library, 1957, 218p.). ($.75)

**1646**
De Marcus, W. C.
  Planetary Interiors. (In Encyclopedia of Physics. 52nd Edition. Berlin, Springer, 1958).

**1647**
Pereira, Flavio A.
  Introdução a Astrobiologia. São Paulo, Sociedade Interplanetária Brasileira, 1958, 126p.

**1648**
Shapley, H.
  Of Stars and men: the Human Response to an Expanding Universe. New York, Washington Square Press, 1958, 157p. ($.50)

**1649**
Anfinsen, C. B.
  The Molecular Basis of Evolution. New York, Wiley, 1959, 228p. ($7.00)

**1650**
Calvin, M., & S. K. Vaughn
  Extraterrestrial Life: Some Organic Constituents of Meteorites and Their Significance for Possible Extraterrestrial Biological Evolution. (In Space Research: Proceedings of the First International Space Science Symposium, ed. by H. Kallman. New York, Wiley, 1960, pp. 1171-1191). ($26.00)

**1651**
Dollfus, A.
  Recherches Concernant la vie sur les Planètes. (In Bijl, H. K., ed. Space Research; Proceedings of the First International Space Science Symposium, Nice, France, Jan. 11-16, 1960. New York, Interscience, 1960, pp. 1146-1152). ($26.00)

**1652**
Gauroy, P.
  Les Mondes du Ciel: Terres Vivantes ou Cimétières? Paris, Librairie Arthème Fayard, 1960, 298p. (12 N.F.)

**1653**
Prince, A. E., et al
  Space age Microbiology. (In Developments in Industrial Microbiology 1. New York, Plenum Publishing Corp., 1960, pp. 13-20).

**1654**
Urey, H. C.
  Science in Space; Report for 1960. Washington, Space Science Board, National Academy of Sciences, National Research Council, 1960, Chapters 4, 5.

Notes:

# LIFE ON OTHER PLANETS IN GENERAL AND IN SPACE (EXCLUDING INTELLIGENT LIFE)

**1655**
Lederberg, J.
Experimental Approaches to Life Beyond the Earth. (In Berkner, L. V., & H. Odishaw, eds. Science in Space. New York, McGraw-Hill, 1961, pp. 407-425).

**1656**
Oparin, Aleksandr I., & V. Fesenkov
Life in the Universe. New York, Twayne Pubs., Inc., 1961, 245p.

**1657**
Pannekoek, A.
A History of Astronomy. New York, Barnes & Noble, 1961, 521p.

**1658**
Sagan, Carl, & W. W. Kellogg
The Atmospheres of Mars and Venus. Washington, National Academy of Sciences-National Research Council, 1961, 151p. (National Research Council. Publication 944).

**1659**
U.S. National Aeronautics & Space Administration
First Planning Conference on Biomedical Experiments in Extraterrestrial Environments, Held Under Auspices of NASA, Washington, June 1960; Proceedings. NASA TN D-781; N62-71355. Washington, 1961, 85p.

**1660**
Young, R. S.
Basic Research in Astrobiology. Proceedings of the 6th Annual Meeting, A. A. S., New York, Jan. 18-21, 1960. (In Advances in Astronautical Sciences, 6. New York, Macmillan, 1961, 898p).

**1661**
Advances in Space Science and Technology, Vol. 4. Ed. by Frederick I. Ordway. New York, Academic Press, 1962, 431p.

**1662**
Fox, Sidney W., et al
Emergent Organic Chemistry Under Various Planetary Conditions and Extraterrestrial Matrices and Environments. First Annual Report, 1 Oct. 1961-30 Sept. 1962. (NASA-CR-56526). Alexandria, Va., NTIS, 1962, 52p.

**1663**
Jackson, F. L., & P. Moore
Life in the Universe. New York, Norton, 1962, 140p.

**1664**
Menzel, Donald
The Universe in Action. Birmingham, The Rushton Lecture Foundation, 1962, 71p.

**1665**
NASA-University Conference on the Science and Technology of Space Exploration, Chicago, Nov. 1-3, 1962
Bioastronautics, NASA SP-18. Washington, Office of Scientific and Technical Information, National Aeronautics and Space Administration; for sale by the Superintendent of Documents, U.S. Government Printing Office, Washington, 1962, 35p.

**1666**
Ordway, F. I., et al
Astrobiology. (In Basic Astronautics. Englewood Cliffs, Prentice-Hall, 1962, pp. 244-305).

**1667**
Ovendon, M. W.
Life in the Universe; a Scientific Discussion. New York, Doubleday, 1962, 160p. ($.95)

**1668**
Posin, D. Q.
Life Beyond our Planet; a Scientific Look at Other Worlds in Space. New York, McGraw-Hill, 1962, 128p. ($3.25)

NOTES:

## LIFE ON OTHER PLANETS IN GENERAL AND IN SPACE (Excluding Intelligent Life)

**1669**
Rand Corporation
 The Environment of the Planets, by William W. Kellogg. Presented at the Space Exploration Lecture Series, University of California, and at Moffett Field, Anaheim, Los Angeles, and San Diego, Oct. 1-4, 1962. P-2640; N63-19338. Santa Monica, 1962, 13p.

**1670**
Sagan, Carl
 Life Beyond the Earth; USIA Voice of America Forum Series on Space Science, Jan. 8-May 21, 1962; Lecture #20. Washington, U.S. Information Agency, 1962, pp. 297-310.

**1671**
U.S. Congress. House of Representatives. Committee on Science & Astronautics
 Panel on Science and Technology, Washington, 87th Congress, Second Session, Hearings at Fourth Meeting, March 22, 1962, pp. 73-74.

**1672**
Victoria University, Wellington, New Zealand.
 The Distribution of Life in the Solar System: An Evaluation of the Present Evidence, by Michael H. Briggs. N63-10771. Wellington, 1962, 25p.

**1673**
Young, R. S.
 Exobiology. (In NASA--University Conference on the Science & Technology of Space Exploration. 1st, Chicago, Nov. 1-3, 1962. Proceedings. v.1. Washington, U.S. National Aeronautics & Space Administration, 1962, pp. 423-429).

**1674**
Berger, R.
 The Solar System and Extraterrestrial Life; Paper Presented at the A.A.S. 9th Annual Meeting, Interplanetary Missions Conference, Los Angeles, Jan. 15-17, 1963. (In Advances in the Astronautical Sciences 13, North Hollywood, Western Periodicals Co., 1963, pp. 649-655).

**1675**
California. University at San Diego
 Remarks Concerning the Chemical Composition of the Atmosphere of Venus, by Hans E. Suess. NASA-CR-55367; N64-14152. San Diego, 1963, 12p.

**1676**
Cameron, A. G. W.
 Stellar Life Zones. (In his Interstellar Communication. New York, Benjamin, 1963, pp. 107-110). ($8.50)

**1677**
Hagen, Charles A.
 Life in Extraterrestrial Environments. Quarterly Status Report, Feb. 15-May 15, 1963. (NASA-Cr-50883; ARF 3194-9). Alexandria, Va., NTIS, 15p. ($1.60)

**1678**
Hermann, J.
 Leben auf Anderen Sternen? Guutersloh, C. Bertelsmann, 1963, 190p.

**1679**
Kleczek, J.
 Possibilities for Life in the Universe. (In International Congress on the man and Technology in the Nuclear and Space age, Milan, April, 1962. Proceedings. Rome, Associazione Internationale Uomo Nello Spazio, 1963, pp. 135-138).

**1680**
Kocherhans, Ernst
 Kosmisches Leben. Basle, E. Reinhardt, 1963, 100p. (8.50 Swiss francs)

**1681**
Liubarskii, K. A.
 History, Nature and Theory of Astrobiology. (In Ocherki po Astrobiologii. Moscow, 1962, pp. 5-11, 104-111). English Trans. by Joint Publications Research Service, Washington, April 16, 1963: JPRS-18724; OTS 63-21584; N64-10643.

NOTES:

# LIFE ON OTHER PLANETS IN GENERAL AND IN SPACE (Excluding Intelligent Life)

**1682**
Pikelner, Solomon B.
  Soviet Science of Interstellar Space.
New York, Philosophical Library, 1963,
230p.

**1683**
Rand Corporation
  A Survey of Exobiology, by P. G. Seybold.
Project Rand; RM-3178-PR; N63-13966. Santa
Monica, 1963, 48p.

**1684**
Sagan, Carl, & W. W. Kellogg
  The Terrestrial Planets. (In Goldberg, L.,
ed. Annual Review of Astronomy and Astrophysics
1, Palo Alto, Annual Review, Inc., 1963, pp.
235-266).

**1685**
Seybold, Paul G.
  A Survey of Exobiology. (RM-3178-PR).
Santa Monica, Rand Corp., March 1963, 48p.

**1686**
U.S. Joint Publications Research Service
  Extraterrestrial Life. (Tr. of 2 articles
into English from Priroda, Moscow, #2, 1963;
JPRS-21026; OTS-63-3173b). Washington,
NTIS, 1963, 20p. (50¢).

**1687**
U.S. National Aeronautics & Space Administration
  After Now, What Then in Space? News Release.
Presented at the Polytechnic Institute on
Artificial Satellites, Aug. 12, 1963. Washington, 1963, 12p.

**1688**
U.S. National Aeronautics & Space Administration
  The Search for Extraterrestrial Life.
(NASA-EP-100). Washington, U.S. Government Printing Office, 1963, 23p. (20¢).

**1689**
Huang, Su-shu
  Life in Space and Humanity on the Earth.
A Joint Review of 5 Books. (NASA-TM-X-
56136). Alexandria, Va., NTIS, 1964, 19p.
($1.00)

**1690**
International Space Sciences Symposium. 4th,
Warsaw, Poland, 1963
  Life Sciences and Space Research 2, edited
by M. Florkin & A. Dollfus. New York, Wiley,
1964, 439p. ($16.50)

**1691**
Oparin, Alexander I.
  The Chemical Origin of Life. Springfield,
Ill., Thomas, 1964, 124p.

**1692**
Rocky Mountain Bioengineering Symposium.
  IEEE Conference Record. 1st- New York,
1964- (annual).

**1693**
Sagan, Carl
  Exobiology: a Critical Review. (In Florkin,
M., & A. Dollfus, eds. Life Sciences and Space
Research II; International Space Science Symposium. 4th, Warsaw, June 3-12, 1963. New
York, Interscience, 1964, pp. 35-53).

**1694**
Allen, T. B.
  The Quest; a Report on Extraterrestrial Life.
Philadelphia, Chilton Co., 1965, 323p. ($4.95)

**1695**
Mamikunian, G., & M. H. Briggs, editors
  Current Aspects of Exobiology. Jet Propulsion Laboratory Technical Report #32-428.
Pasadena, Jet Propulsion Laboratory, California
Institute of Technology, 1965, 420p.

**1696**
Menzel, Donald H.
  Life in the Universe. (n.p., 1965), 219p.

Notes:

# LIFE ON OTHER PLANETS IN GENERAL AND IN SPACE (Excluding Intelligent Life)

**1697**
Moffat, Samuel, & Elie A. Shneour
 *Life Beyond the Earth*. Epilogue by Joshua Lederberg. New York, Scholastic Book Services, 1965, 156p.

**1698**
Ordway, Frederick I.
 *Life in Other Solar Systems*. New York, Dutton, 1965, 96p.

**1699**
Schiaparelli, G. V.
 *Mars den Gådefulde Planet*. Oversat af E. Slej. Randers, Denmark, UFO-NYTS Forlag, 1965, 73p.

**1700**
Sullivan, Walter, et al
 *The Search for Extraterrestrial Life*. Report #356. New Haven, Yale Reports, Woodbridge Hall, Yale University, 1965, 8p.

**1701**
Pittendrigh, Colin S., Wolf Vishniac, & J. P. T. Pearman, eds.
 *Biology and the Exploration of Mars*. Washington, National Academy of Sciences--National Research Council, 1966, 516p. (NAS-NRC Publication 1296).

**1702**
Young, R. S.
 *Extraterrestrial Biology*. New York, Holt, Rinehart & Winston, 1966, 119p. ($2.95)

**1703**
U.S. National Aeronautics & Space Administration
 *Significant Achievements in Space Science 1966*. (NASA-SP-155). Washington, U.S. Government Printing Office, 1967, 224p. ($1.50).

**1704**
Bellcomm, Inc., Washington, D.C.
 *On the Possibility of Exotic Biochemistries*, by S.G. Schulman. (NASA-CR-104095; TR-68-710-9). Washington, NTIS, Dec. 17, 1968, 30p. (Available from NTIS).

**1705**
Dauvillier, Alexandre
 *The Primitive Terrestrial Atmosphere and the Atmospheres of Venus and Mars*. (Tr. into English from *Comptes Rendus de l'Académie des Sciences*, Paris, series D, v. 267, Aug. 12, 1968, pp. 697-700). Defense Scientific Information Service, Ottawa, Canada: Report 70-444, 5pp. Available from NTIS.

**1706**
Johnson, Richard D., & Harold P. Klein
 *Extraterrestrial Biology; Prospects and Problems in the Early 1970's*. Paper 68-1122. American Institute of Aeronautics and Astronautics, Annual Meeting and Technical Display, 5th, Philadelphia, Pa., Oct. 21-24, 1968. New York, 1968, 7p.

**1707**
Klein, Harold P.
 *Extraterrestrial Biology: Prospects and Problems*. Presented at the Fourth International Symposium on Bioastronautics and Exploration of Space, San Antonio, Texas, 24-27 June, 1968. Moffett Field, Calif., NASA-Ames Research Center, 1968, 42p. ($3.00) Available from NTIS.

**1708**
Rubenchik, L. I.
 *Mikroorganizmy i Kosmos*. Kiev, Izd. Naukova Dumka, 1968, 116p.

**1709**
Serdobolskii, V.I.
 "Problema Rasprostraneiia Razumnoi Zhizn vo Vselennoi" (In *Fizika, Matematika, Mekhanika*. Moskovskaia Konferentsiia Molodykh Uchenykh, 1st, Moscow, USSR, Dec. 1964. Trudy. Ed. by A. Iu. Ishlinskii and V.A. Petukhov. Moscow, Izd. Nauka, 1968, pp. 145-154).

**1710**
Techtran Corp., Glen Burnie, Md.
 *Life in Space; New Advances in Space Biology*. (NASA-TT-F-11540). Tr. from German. Glen Burnie, Md, Feb. 1968, 7p. ($3.00) Available on microfilm from NTIS.

Notes:

# LIFE ON OTHER PLANETS IN GENERAL AND IN SPACE (EXCLUDING INTELLIGENT LIFE)

**1711**
U.S. National Aeronautics & Space Administration
  Planetary Atmospheres. Continuing Bibliography With Indexes, June 1966-Dec. 1967. (NASA-SP-7017, 02). Washington, July 1968, 126p. ($3.00) Available from NTIS.

**1712**
Tel Aviv University. Dept. of Environmental Sciences
  Papers of Staff Members of the Institute of Planetary and Space Science and the Dept. of Environmental Sciences, Tel Aviv University since 1960. Tel Aviv, 1969, 36p. (Available from NTIS).

**1713**
Young, R. S.
  Life Beyond Earth. Morristown, N. J., Silver Burdett Co., 1969, 63p.

**1714**
U.S. National Aeronautics & Space Administration
  Planetary Program Review, July 1969. (NASA-TM-X-62569). Washington, July 1969, 75p. Available from NTIS.

**1715**
Krupenio, N. N.
  New Information on Planetary Atmospheres (Symposium in the United States). Washington, NASA, 1970, 9p. (Tr. into English from Vestn. Akad. Nauk. SSR, Moscow, #3, 1970, pp. 67-70 of Conference Held at Marfa, Texas, 26-31 Oct. 1969. NASA-TT-F-13813. Available from NTIS).

**1716**
Bova, Benjamin
  Planets, Life, & LGM. Reading, Mass., Addison-Wesley, 1970, 109p. ($4.25)

**1717**
Dole, Stephen H.
  Habitable Planets for Man. 2d ed. New York, American Elsevier Pub. Co., 1970, 158p.

**1718**
European Space Research Organization
  Missions Spatiales Vers les Planètes 4: Physique des Surfaces Planétaires. Paris, 1970, 47p. (Available from NTIS).

**1719**
Martin, C.
  Le Cosmos et la vie. Paris, Encyclopédie Planète; dist. by Hachette, 1970, 346p. (4.85 francs)

**1720**
Sneath, Peter H. A.
  Planets and Life. London, Thames & Hudson, 1970, 216p. (42 shillings)

**1721**
California. University at Los Angeles. Brain Research Institute
  UCLA 1969 Summer Institute in Space Biology. Final Report. Los Angeles, Jan. 1970, 216p. (NASA-Cr-108120). Available on microfilm from NTIS.

**1722**
Lavrentyev, G.A.
  The Method of Separation and Identification of Amino Acids to Detect Extraterrestrial Life. Washington, NASA, 1971, 13p. (NASA-TT-F-13765). (Tr. into English of USSR Academy of Sciences, Institute for Space Research Report PR-59). Available on microfilm from NTIS.

**1723**
Schwartz, A. W., ed.
  Theory and Experiment in Exobiology. v. 1. Groningen, Wolters-Noordhoff, 1971, 160p.

**1724**
Derpgolts, V.
  Water on Planets. Tr. into English from Nauka i Zhizn #4, Moscow, 1972, 28p. (JPRS-57709). Available from NTIS: $3.25.

NOTES:

## LIFE ON OTHER PLANETS IN GENERAL AND IN SPACE (Excluding Intelligent Life)

**1725**
Ponnamperuma, Cyril, ed.
  Exobiology. Amsterdam, North-Holland Pub., 1972, 504p.

**1726**
Schwartz, Alan W., ed.
  Theory and Experiment in Exobiology. 2v. Groningen, Netherlands, Wolters-Noordhoff, 1972.

**1727**
Teyfel, V.G., ed.
  Physical Characteristics of the Giant Planets. Washington, NASA, 1972, 227p. ($3.00) (NASA-TT-F-717) (Available from NTIS).

**1728**
Vishniac, Wolf, ed.
  Life Sciences and Space Research X. Proceedings of the 14th Plenary Meeting, Seattle, Wash., June 21-July 2, 1971. Berlin, Akademie-Verlag, 1972, 238p.

**1729**
American Chemical Society
  In Search of Life, Chemical Evolution (Script #640, Man and Molecules Radio Program). Washington, 1973, 5p.

**1730**
  Life Sciences and Space Research XI. Proceedings of the 15th Plenary Meeting, Madrid, Spain, May 10-24, 1972. Ed. by P. H. A. Sneath. Berlin, Akademie-Verlag, 1973, 288p.

**1731**
Sagan, Carl
  The Cosmic Connection. An Extraterrestrial Perspective. Garden City, Anchor Press, 1973, 274p.

## PERIODICAL ARTICLES

**1732**
Flight, W.
  "Meteorites and the Origin of Life." London, Popular Science Review 16, 1877, p. 390.

**1733**
  "M. Faye on Interplanetary air." London, Knowledge 2, 1882, p. 399.

**1734**
Serviss, G. P.
  "Are There Planets Among the Stars?" Popular Science 52, Dec. 1897, pp. 171-177.

**1735**
Mason, E. C.
  "Life in Other Worlds." Popular Astronomy 6, 1899, p. 520.

**1736**
Newcomb, S.
  "Life in the Universe." Harper's Monthly Magazine 111, Aug. 1905, pp. 404-408.

**1737**
  "Life on Other Worlds." Scientific American 93, Oct. 28, 1905, pp. 334-335.

**1738**
  "The Weight of man on the Planets." Scientific American, Supplement 63, Feb. 23, 1907, p. 26042.

**1739**
Henkel, F. W.
  "Life in Other Worlds: the Interpretation of Planetary Markings." Scientific American, Supplement 67, May 15, 1909, pp. 318-319.

Notes:

## LIFE ON OTHER PLANETS IN GENERAL AND IN SPACE (EXCLUDING INTELLIGENT LIFE)

**1740**
See, T. J. J.
"Rotation of Venus and Life on Planets Other Than the Earth." Popular Astronomy 18, Jan. 1910, pp. 1-3.

**1741**
"Limits of Organic Life in our Solar System." American Review of Reviews 43, Feb. 1911, pp. 242-243.

**1742**
"The Struggle for Existence on the Planet Mars." Current Opinion 54, Jan. 1913, pp. 40-42.

**1743**
"Life on the Planets." Harper's Weekly 57, Feb. 1, 1913, p. 26.

**1744**
"The Transmission of Life to a Dead World Despite the Ultra-violet ray." Current Opinion 56, April 1914, p. 286.

**1745**
Francis, V.
"Is our Earth the Only Life Supporting Body in the Universe?" Popular Astronomy 25, Aug. 1917, pp. 441-453.

**1746**
Francis, V.
"The Problem of the Universe: is our Earth the Only Life Supporting Body?" Scientific American Supplement 34, Sept. 29, 1917, pp. 206-208.

**1747**
Mathew, W. D.
"Life in Other Worlds." Science 54, Sept. 16, 1921, pp. 239-241.

**1748**
Mathew, W. D.
"Life in Other Worlds." Science (54:1394), 1921, pp. 241-289.

**1749**
"Improbability of Life on the Planets." Literary Digest 71, Nov. 5, 1921, p. 18.

**1750**
"C. G. Abbott Discusses Habitability of Other Worlds." New York Times, June 18, 1922, sec. VII, p. 9, col. 2.

**1751**
Becquerel, P.
"La vie Terrestre Provient-elle d'un Autre Monde?" Paris, Astronomie 38, 1924, pp. 393-417.

**1752**
"Prof. E. Campbell on Chance of Vegetation and Life." New York Times, June 16, 1925, p. 5, col. 3.

**1753**
"Why are we Keen About Mars?" Literary Digest 86, Aug. 1925, p. 19.

**1754**
Coblentz, W. W.
"Is There Life on Other Planets?" Forum 74, Nov. 1925, pp. 688-696.

**1755**
"Dr. W. Coblentz Will Make Further Investigation of Existence of Life on Mars." New York Times, Oct. 3, 1926, sec. II, p. 19, col. 6.

**1756**
"L. Campbell Believes Life on Planets of This and Other Solar Systems may Exist." New York Times, Jan. 26, 1931, p. 5, col. 1.

**1757**
Chamberlin, Ralph V.
"Life on Other Worlds." Bulletin of the University of Utah 22, Feb. 1932, pp. 1-52.

NOTES:

# LIFE ON OTHER PLANETS IN GENERAL AND IN SPACE (Excluding Intelligent Life)

**1758**
Maxim, H. P.
  "What is it all About?" *Scientific American* 146, April 1932, pp. 199-203.

**1759**
"Feature Article on Possibility of Life on (Planets)." *New York Times*, June 26, 1932, sec. VI, p. 10.

**1760**
Leonard, Frederick C.
  "Life on Other Worlds." *Popular Astronomy* 41, May 1933, pp. 260-263.

**1761**
"E. W. Barnes Believes There is Life on Others Besides Earth." *New York Times*, Oct. 20, 1933, p. 13, col. 6.

**1762**
"Life on the Planets." *Science, Supplement* 6, Dec. 15, 1933, p. 78.

**1763**
Russell, H. N.
  "Fading Belief in Life on Other Planets." *Scientific American* 150, June 1934, pp. 296-297.

**1764**
"Planets." *Newsweek* 3, June 30, 1934, p. 18.

**1765**
"The Atmosphere of the Planets." *Science, Supplement* 8, July 6, 1934, p. 80.

**1766**
"Editorial on Lowell Observatory Report on Possibilities of Life." *New York Times*, July 7, 1934, p. 12, col. 3.

**1767**
Russell, H. N.
  "The Atmosphere of the Planets." *Science* (81:2088), 1935, pp. 1-9.

**1768**
Uphof, J. C. T.
  "The Dynamics of the Distribution of Life in the Universe." Milan, *Scientia* 58, July 1935, pp. 30-38.

**1769**
Sternfeld, A. J
  "La vie Dans l'Univers." Paris, *La Nature* 63, part 2, July 1, 1935, pp. 1-12.

**1770**
Demoulin, Maurice
  "La vie Dans L'univers." Paris, *La Nature* 64, Part 1, April 1936, pp. 369-371.

**1771**
Lipman, C. B.
  "Bacteria in Meteorites." *Popular Astronomy* 44, Oct. 1936, pp. 442-446.

**1772**
Jones, H. S.
  "Is There Life in Other Worlds?" London, *Discovery* 2, Jan. 1939, pp. 36-47.

**1773**
"Life Beyond Earth?" *Time* 36, Oct. 7, 1940, p. 62.

**1774**
"Plenty of Lebensraum in Other Parts of Universe." *Science Newsletter* 39, Feb. 1, 1941, p. 71.

**1775**
Gamow, G.
  "Many More Worlds Like Ours?" *Science Digest* 11, May 1942, pp. 85-87.

---

Notes:

# LIFE ON OTHER PLANETS IN GENERAL AND IN SPACE (EXCLUDING INTELLIGENT LIFE)

1776
Jeans, J.
"Is There Life on Other Worlds?" Science 95, June 12, 1942, pp. 589-592.

1777
Lafleur, L. J.
"Surface Gravity and Behavior." Popular Astronomy 51, April 1943, pp. 197-202.

1778
"Prof. H. N. Russell Holds Many (Planets) May be Inhabited." New York Times, June 25, 1945, p. 19, col. 7.

1779
"Evergreens Will Greet Visitors in Mars." Science Digest 18, Nov. 1945, p. 88.

1780
"USSR Scientists Study Mars Plant Life." New York Times, Nov. 28, 1949, p. 27, col. 3.

1781
Fears, Francis R.
"Silicon and Extraterrestrial Life Forms." London, Journal of Spaceflight 2, Nov. 1950, pp. 5-6.

1782
Chaikin, G. L.
"A Transitional Hypothesis Concerning Life on Interstellar Bodies." Popular Astronomy (59:1), Jan. 1951, pp. 50-51.

1783
Bowman, Norman J.
"Silicon as a Base for Life Forms." London, Journal of Space Flight 3, March 1951, pp. 1-6.

1784
"What Goes on up There?" America 87, August 23, 1952, p. 489.

1785
"Dr. Urey Estimates 1 Quadrillion Worlds in Known Universe Might Originate and Support Living Organisms; Cites Conducive Chemical Conditions." New York Times, Nov. 14, 1952, p. 25, col. 2.

1786
Shapley, Harlow
"Life on Other Planets; Excerpt From Climatic Change." Atlantic Monthly 192, Nov. 1953, pp. 29-32.

1787
Bliven, B.
"Is There Life on Other Planets?" Reader's Digest 66, Feb. 1955, pp. 103-107.

1788
Moore, P.
"Life on Other Planets? Excerpt From Guide to the Planets." Science Digest 37, Feb. 1955, pp. 53-57.

1789
Struve, O.
"Life on Other Worlds." Sky and Telescope (14:4), Feb. 1955, pp. 4, 137-140, 146.

1790
Moore, Patrick
"Life on the Moon?" Armagh, Irish Astronomical Journal 3, March 1955, pp. 133-137.

1791
Salisbury, Frank
"The Inhabitants of Mars." Engineering and Science 18, April 1955, pp. 23-32.

1792
Randolph, J. R.
"Are Planets Habitable?" Ordnance 40, Jan.-Feb. 1956, pp. 608-610.

NOTES:

## LIFE ON OTHER PLANETS IN GENERAL AND IN SPACE (Excluding Intelligent Life)

**1793**
Dingle, H.
"Cosmology and Science." *Scientific American* 195, Sept. 1956, p. 228.

**1794**
"Fanciful Preview to new Facts." *Life* 41, Sept. 24, 1956, pp. 40-41.

**1795**
Urey, H C.
"Diamond, Meteorites and the Origin of the Solar System." *Astrophysical Journal* 124, Nov. 1956, pp. 623-627.

**1796**
Webster, G.
"Is There Life on Other Worlds?" *Natural History* 65, Dec. 1956, pp. 526-531.

**1797**
"Problems Common to the Fields of Astronomy and Biology: a Symposium." *Publications of the Astronomical Society of the Pacific* 70, 1957, p. 41.

**1798**
Briggs, Michael H.
"Life on Other Planets." London, *The Humanist* (72:8), Aug. 1957, pp. 18-20.

**1799**
Abel, R. C.
"Archipelago." London, *The Dublin Review* (232:475), 1958, pp. 13-18.

**1800**
Beischer, D. E.
"Potentialities and Ramifications of Life Under Extreme Environmental Conditions." *Journal of Aviation Medicine* (29:7), 1958, pp. 500-503.

**1801**
Bun, Thomas P., & Flavio A. Pereira
"Biospheric Index, a Contribution to the Problem of Determination of the Existence of Extra-solar Planetary Biospheres." (In *VIIth International Astronautical Congress, Barcelona, 1957. Proceedings.* Vienna, Springer-Verlag, 1958, p. 63).

**1802**
"Developments of International Effects to Avoid Contamination of Extraterrestrial Bodies." *Science* 128, 1958, pp. 887-889.

**1803**
Fenn, W. O.
"The Challenge of Space Biology." *Bulletin of the American Institute of Biological Sciences* (8:2), 1958, p. 15.

**1804**
Strughold, Hubertus
"Life on Mars in View of Physiological Principles." (In *Epitome of Space Medicine. Article 3.* Brooks AFB, Texas, School of Aviation Medicine, 1958, pp. 1-8).

**1805**
Lamb, I. M.
"Remarkable Lichens." *Natural History* 67, Feb. 1958, pp. 86-93.

**1806**
"Life on a Billion Planets?" *Time* 71, March 31, 1958, p. 42.

**1807**
Hald, T. S.
"Of Stars and men: Human Response to an Expanding Universe; Review, by Harlow Shapley." *Saturday Review* 41, May 17, 1958, pp. 22-23.

**1808**
"Life-supporting Planets." *Science Newsletter* 73, May 24, 1958, p. 322.

NOTES:

## LIFE ON OTHER PLANETS IN GENERAL AND IN SPACE (EXCLUDING INTELLIGENT LIFE)

**1809**
Stuhlinger, Ernst
"Life on Other Stars." Space Journal, part 1. Spring 1958, pp. 10-16; Summer 1958, pp. 21-30.

**1810**
Stuhlinger, Ernst
"Life on Other Stars." Space Journal, part 1 (I: 2), Spring 1958, p. 10; part 2 (I: 3), Summer 1958, p. 21.

**1811**
"Dr. Shapley Sees at Least 100,000,000 Planets in Universe Suitable for Human Life." New York Times, June 1, 1958, sec. IV, p. 9, col. 5.

**1812**
Kind, S. S.
"Speculations on Extraterrestrial Life." London, Spaceflight 1, July 1958, pp. 288-289.

**1813**
Hess, S. L.
"Atmospheres of Other Planets." Science 128, Oct. 10, 1958, pp. 809-814.

**1814**
"Dr. Calvin on Probability That Plants, Animals and Other Forms of Life Exist on Millions of Earthlike Planets...' New York Times, Nov. 9, 1958, p. 55, col. 1.

**1815**
Rhodes, George
"100 Million Habitable Planets in Universe." Missiles & Rockets 4, Dec. 15, 1958, p. 44.

**1816**
Hulley, John
"Dynamics of Life in the Universe." Space Journal 1, Winter 1959, pp. 33-41.

**1817**
Jones, H. S.
"Life on Other Worlds." London, New Scientist (5:112), Jan. 8, 1959, pp. 65-67.

**1818**
"Martian Vegetation." Sky and Telescope 18, March 1959, p. 252.

**1819**
Keilin, D.
"The Problem of Anabiosis on Latent Life: History and Current Concept." London, Proceedings of the Royal Society of London 150B, March 17, 1959, p. 149.

**1820**
Loftin, H.
"Rambling far Afield." Science Newsletter 75, March 28, 1959, p. 208.

**1821**
Strughold, Hubertus
"Advances in Astrobiology." (In Proceedings of Lunar and Planetary Exploration Colloquium (I: 6), April 25, 1959, p. 1).

**1822**
Parkes, A. S., & A. U. Smith
"Transport of Life in the Frozen or Dried State." London, British Medical Journal 1, May 1959, pp. 1295-1297.

**1823**
"Prof. P. Morrison Sees Jupiter, Mars and Venus Capable of Supporting 'Something Like Life'." New York Times, June 9, 1959, p. 76, col. 1.

**1824**
"Planetary Biology Soon." Science Newsletter 75, June 20, 1959, p. 391.

NOTES:

## LIFE ON OTHER PLANETS IN GENERAL AND IN SPACE (Excluding Intelligent Life)

1825
Gordon, Samuel
  "There's Something in the Moon." Fate 12, July 1959, pp. 47-53.

1826
Tikhov, G. A.
  "What is Astrobotany?" London, Spaceflight 2, July 1959, pp. 74-77.

1827
Huang, Su-shu
  "Occurrence of Life in the Universe." American Scientist (47: 3), Sept. 1959, p. 397.

1828
  "Life on Other Planets?" Science Digest 46, Sept. 1959, Inside Back Cover.

1829
Wald, G.
  "Life and Light." Scientific American (201:4), Oct. 1959, pp. 92-108.

1830
Sinton, W. M.
  "Evidence of the Existence of Life on Mars." Science 130, Nov. 6, 1959, pp. 1234-1237.

1831
  "Astrobotany." London, Research 13, 1960, p. 374.

1832
Sloane, E.
  "Artificial Biosphere." Science (132:3421), 1960, p. 252.

1833
Steiner, G.
  "Questionnaire." The Reporter 22, Jan. 7, 1960, pp. 37-38.

1834
  "Space and Bugs." Time 75, Jan. 25, 1960, p. 50.

1835
Lederberg, J.
  "The Search for Life Beyond the Earth." London, New Scientist (7:170), Feb. 18, 1960, pp. 386-388.

1836
Calvin, Melvin, & S. K. Vaughn
  "Meteorites Yield Clue to Life in Space." Chemical & Engineering News 38, March 14, 1960, p. 38.

1837
Huang, Su-shu
  "Life Outside the Solar System." Scientific American (202: 4), April 1960, p. 55.

1838
Shapley, Harlow
  "Extraterrestrial Life." Astronautics (ARS Journal) (5:4), March 1960, pp. 32-33, 50-52; April 1960, pp. 321-33, 350.

1839
Tsvetikov, A. N.
  "Next Question and K. E. Tsiolkovsky." Science 131, April 22, 1960, pp. 1262-1263.

1840
Gold, Thomas
  "Cosmic Garbage (may Have Brought Life to Other Planets)." Air Force & Space Digest 43, May 1960, p. 65.

1841
Shapley, Harlow
  "The Human Response to an Expanding Universe." London, The Hibbert Journal 58, July 1960, pp. 319-328.

1842
Sloan, Eugene A.
  "Artificial Biosphere." Science 132, July 1960, p. 252.

1843
  "Are There Spores in Space?" Science Digest 48, Aug. 1960, Inside Back Cover.

Notes:

# LIFE ON OTHER PLANETS IN GENERAL AND IN SPACE (Excluding Intelligent Life)

**1844**
Vishniac, Wolf
"Extraterrestrial Microbiology." <u>Aerospace Medicine</u> 31, Aug. 1960, pp. 678-680.

**1845**
Lederberg, J.
"Exobiology: Approaches to Life Beyond the Earth." <u>Science</u> 132, Aug. 12, 1960, pp. 393-400.

**1846**
Ewing, A.
"Space Biology Experiments." <u>Science Newsletter</u> 78, Oct. 1, 1960, pp. 218-219.

**1847**
"Animal, Vegetable, or..." <u>Newsweek</u> 56, Nov. 28, 1960, p. 78.

**1848**
Huang, Su-shu
"The Limiting Sizes of Habitable Planets." <u>Publications of the Astronomical Society of the Pacific</u> 72, Dec. 1960, p. 489.

**1849**
Strughold, Hubertus
"An Introduction to Astrobiology." <u>Astronautics</u> 5, Dec. 1960, pp. 20-21.

**1850**
Mukerji, B.
"Advancing Frontiers of Life Sciences." <u>Bulletin of the American Institute of Biological Sciences (11:2)</u>, 1961, pp. 21-24, 32.

**1851**
Shapley, Harlow
"The Probable Environment on Other Planets." (In Pirie, N.W., ed. The Biology of Space Travel. London, the Institute of Biology, 1961, pp. 107-116).

**1852**
Wynne, E. S.
"Sterilization of Space Vehicles: the Problem of Mutual Contamination." (In <u>Lectures in Aerospace Medicine</u>. Brooks Air Force Base, Texas, U.S. Air Force Aerospace Medical Center, 1961).

**1853**
Tello, Luis R.
"Introducción Breve a la Exobiología." Buenos Aires, <u>Revista Nacional Aeronáutica y Espacial</u> (21:14), Jan. 1961, p. 15.

**1854**
Ovendon, M. W.
"Is There Life on the Planets? A Scientific Discussion." London, <u>Illustrated London News</u>, 12 parts, Jan. 21, 1961-April 8, 1961.

**1855**
Drawert, F.
"Estraterrestrisches Leben." Stuttgart, <u>Naturwissenschaftliche Rundschau</u> (14:2), Feb. 1961, pp. 68-70.

**1856**
Horowitz, Norman H.
"Is There Life on Other Planets?" <u>Engineering and Science</u> (24: 6), March 1961, p. 11.

**1857**
Anders, Edward
"The Moon as a Collector of Biological Material." <u>Science</u> 133, April 1961, pp. 1115-1116.

**1858**
"Holds Life on Other Planets Unlikely." <u>Science Digest</u> 49, April 1961, p. 82.

**1859**
"Extraterrestrial Life?" <u>America</u> 105, April 1, 1961, p. 4.

Notes:

## LIFE ON OTHER PLANETS IN GENERAL AND IN SPACE (Excluding Intelligent Life)

1860
Bergamini, David
 "Wax and Wigglers: Life in Space?" Life 50, May 1961, pp. 57, 59-60, 62.

1861
Novick, A., & J. Lederberg
 "Challenges to Biology." Bulletin of the Atomic Scientists (17:5-6), May-June 1961, pp. 203-206.

1862
McHugh, L.
 "Life in Meteorites?" America 105, May 27, 1961, pp. 379-382.

1863
 "Does Life Exist in Space?" Science Newsletter 79, May 20, 1961, pp. 314-315.

1864
Horowitz, N. H.
 "Is There Life on Other Planets?" California Institute of Technology Quarterly (2:3), Summer 1961, pp. 12-17.

1865
Huang, Su-Shu
 "Some Astronomical Aspects of Life in the Universe." Sky and Telescope 21, June 1961, pp. 312-316.

1866
Mann, M.
 "The Month in Science." Popular Science 178, June 1961, pp. 31-32.

1867
 "Report Finding Evidence of Outer-space Life in Meteorite." Science Digest 49, June 1961, p. 13.

1868
Heinlein, R., & H. Wooster
 "Xenobiology." Science (134:3473), July 21, 1961, pp. 223-225.

1869
Ashbrook, J.
 "Astronomical Scrapbook; Beginnings of the Space age." Sky and Telescope 22, Aug. 1961, p. 85.

1870
Berland, Theodore
 "Meteorites: Proof of Life on Other Planets?" Popular Mechanics 116, Aug. 1961, pp. 108-112, 208, 210.

1871
Litynski, Z.
 "Life on Other Planets; View of a Polish Scientist." Science Digest 50, Aug. 1961, pp. 71-75.

1872
Neville, T.
 "Life Inside a Meteorite." Science Digest 50, Aug. 1961, pp. 56-58.

1873
Novick, Aaron, & Joshua Lederberg
 "Biology's Stake in Astronautics." Air Force & Space Digest 44, Aug. 1961, pp. 65-66.

1874
 "Ecosphere may Shape Life on Distant Planets." Science Newsletter 80, Oct. 1961, p. 272.

1875
 "Life on Largest Planet?" Science Digest 50, Oct. 1961, p. 59.

1876
Bergamini, David
 "Fossils That say Life may Really be out of This World." Life 51, Dec. 1961, p. 45.

1877
 "Life in Time and Space." Time 78, Dec. 1, 1961, p. 73.

Notes:

# LIFE ON OTHER PLANETS IN GENERAL AND IN SPACE (Excluding Intelligent Life)

**1878**
Bergamini, D.
"Fossils That say Life may Really be out of This World." Life 51, Dec. 8, 1961, p. 45.

**1879**
Andrews, F. F.
"Bioastronomy." Wellington, New Zealand, Southern Stars 19, 1962, pp. 155-160.

**1880**
Horowitz, N. H.
"Lecture Before the Federation of American Societies for Experimental Biology." Bulletin of the American Institute of Biological Sciences (12:5), 1962, p. 74.

**1881**
Ordway, Frederick I., et al
"Astrobiology." (In Basic Aeronautics. Englewood Cliffs, Prentice-Hall 1962, pp. 244-305).

**1882**
Tobias, C., & J. Slater
"Our View of Space Biology Widens." Astronautics (ARS Journal) (7:1), 1962, pp. 20-22, 47-52.

**1883**
"Mystery of Life Nearly Solved." Science Newsletter 81, Feb. 24, 1962, p. 120.

**1884**
Bernal, J. D.
"Comments," London, Nature (193:4821), March 24, 1962, pp. 1127-1129.

**1885**
"Life's Spread Through the Universe." Sky and Telescope 23, April 1962, p. 183.

**1886**
Sagan, Carl, et al
"Planetary Studies: Part IV." The Spectrum (1:7), April 1962, pp. 8-9, 18, 20.

**1887**
Salisbury, F. B.
"Martian Biology: Accumulating Evidence Favors the Theory of Life on Mars, but we can Expect Surprises." Science (136:3510), April 1962, pp. 17-26.

**1888**
Ewing, A.
"Tiny Stars With Life." Science Newsletter 81, May 26, 1962, p. 323.

**1889**
"Icarus, International Journal of the Solar System, Edited by 35 Scientists From East and West, Starts Publication to Deal With Flood of Space Science Inquiries." New York Times, July 8, 1962, p. 31, col. 1.

**1890**
Ewing, A., & H. Shapley, editors
"Life on Tiny, Dark Stars." Science Newsletter 82, July 21, 1962, p. 39.

**1891**
Bernal, J. D.
"Is There Life Elsewhere in the Universe?" Science and Culture 28, Aug. 1962, pp. 356-357.

**1892**
Beller, W.
"Moon may now be Contaminated," Missiles and Rockets (11:9), Aug. 27, 1962, pp. 24-27.

**1893**
Steen, J.
"Why I Believe There is Life on Mars. Interview." Popular Science 181, Sept. 1962, pp. 100-104.

Notes:

## LIFE ON OTHER PLANETS IN GENERAL AND IN SPACE (Excluding Intelligent Life)

**1894**
Briggs, Michael H.
"The Distribution of Life in the Solar System: an Evaluation of the Present Evidence." London, *Journal of the British Interplanetary Society* 18, Sept.-Dec. 1962, pp. 431-437.

**1895**
Revelle, R.
"Sailing in new and old Oceans." *Bulletin of the American Institute of Biological Sciences,* Oct. 1962, pp. 45-47.

**1896**
Reynolds, O. E.
"Space Biosciences." *Bulletin of the American Institute of Biological Sciences* (12:5), Oct. 5, 1962, pp. 49-51.

**1897**
Bracewell, R. N.
"Life in the Galaxy." (In Cameron, A.G.W., ed. *Interstellar Communication*. New York, W. A. Benjamin, Inc., 1963, pp. 232-242).

**1898**
Cameron, A. G. W.
"Stellar Life Zones." (In Cameron, A. G. W., ed. *Interstellar Communication*. New York, W. A. Benjamin, Inc. 1963, pp. 107-110).

**1899**
Imshenetskii, A. A.
"Exobiology, new Area for Scientific Research." (Tr. into English by John E. Holman & Co. from *Vestnik Akad. Nauk. SSR*, Moscow, #11, 1962, pp. 58-63). Washington, NASA, 1963, 10p. (NASA-TT-F-166). (Available from NTIS: 50¢).

**1900**
Ley, Willy
"The History of the Concepts About Mars." (In Morgenthaler, George W., ed. *Exploration of Mars*. North Hollywood, Calif., Western Periodicals Co., 1963, pp. 435-455). (Advances in the Astronautical Science 15).

**1901**
Liubarskii, K. A.
"History, Nature and Theory of Astrobiology." (Tr. into English of 2 chapters from the book *Ocherki po Astrobiologii*, Moscow, 1962, pp. 5-11, 104-111). (JPRS-18724; OTS-63-21584). Washington, NASA, 1963, 12p. (Available from NTIS: 50¢).

**1902**
Sinton, William M.
"Evidence of the Existence of Life on Mars." *Advances in the Astronautical Sciences* 15, 1963, pp. 543-551.

**1903**
Strughold, Hubertus
"The Ecological Profile of Mars: Bioastronautical Prospect." *Advances in the Astronautical Sciences* 15, 1963, pp. 543-551.

**1904**
Young, Louis B., ed.
"Is There Other Life in the Universe?" (In his *Exploring the Universe*. New York, McGraw-Hill, 1963, part 9).

**1905**
Firsoff, V. A.
"An Ammonia-based Life." London, *Discovery* 23, Jan. 1963, pp. 36-42.

**1906**
"Space-life Study Lags." *Science Newsletter* 83, Jan. 19, 1963, p. 38.

**1907**
Shapley, Harlow
"Life on Unseen Planets." *Science Digest* (53:2), Feb. 1963, pp. 59-64.

**1908**
Mason, B.
"Organic Matter From Space." *Scientific American* 208, March 1963, pp. 43-49.

NOTES:

## LIFE ON OTHER PLANETS IN GENERAL AND IN SPACE (Excluding Intelligent Life)

**1909**
Muller, H. J.
"Life Forms to be Expected Elsewhere Than on Earth." London, Spaceflight 5, May 1963, pp. 74-85.

**1910**
Wellman, Wade
"Phobos and Deimos: an Inquiry." London, Flying Saucer Review 9, May-June 1963, pp. 26-27.

**1911**
Berrill, N. J.
"Search for Life." Atlantic Monthly 212, Aug. 1963, pp. 35-40.

**1912**
Von Braun, Werner
"Mars: are its Canals Full of Water; is There Really Life There?" Popular Science 183, Aug. 1963, 548p.

**1913**
Bauchau, A.
"Les Météorites Sont-elles Chargées d'un Message Biologique?" Paris, La Revue Nouvelle 38, Sept. 1963, pp. 240-243.

**1914**
"Experimental Exobiology." Sky and Telescope 26, Sept. 1963, p. 132.

**1915**
"Life in Meteorites?" Sky and Telescope 26, Sept. 1963, p. 132.

**1916**
"Extraterrestrial Life." Moscow, Priroda 2, 1963, 20p. English Trans. by Joint Publications Research Service, Washington, Sept. 11, 1963, JPRS-21026; OTS-63-3137B.

**1917**
Bok, B. J.
"Life on Other Worlds." Melbourne, Australian Science Teachers' Journal 9, Nov. 1963, p. 4.

**1918**
Lederberg, J.
"Exobiology." Science 142, Nov. 29, 1963, p. 1126.

**1919**
Imshenetskii, A. A.
"Life and Space." (In Life Sciences and Space Research II. International Space Science Symposium, 4th. Warsaw, June 3-12, 1963. Ed. by M. Florkin and A. Dollfus. New York, Interscience, 1964, pp. 25-34).

**1920**
Sagan, Carl
"Exobiology, a Critical Review." (In Life Sciences and Space Research II. International Space Science Symposium. 4th, Warsaw, June 3-12, 1963. Ed. by M. Florkin and A. Dollfus. New York, Interscience, 1964, pp. 35-53).

**1921**
Salisbury, Frank B.
"Exobiology" (In U.S. Air Force Academy. Proceedings of the First Annual Rocky Mountain Bioengineering Symposium, Held at the Academy, Colorado Springs, Colorado, 4-5 May, 1964. (AD-450818). Colorado Springs, U.S. Air Force Academy, 1964, pp. 75-83).

**1922**
Cohen, Gaston
"La Météorite d'Orgeuil, a-t-elle Apporté les Traces d'une vie Extraterrestre?" Paris, Science, Progrès et la Nature 3345, Jan. 1964, p. 18.

**1923**
Straiser, J. A.
"Is There Life on Mars?" Electronics 37, Jan. 3, 1964, pp. 71-73.

NOTES:

# LIFE ON OTHER PLANETS IN GENERAL AND IN SPACE (Excluding Intelligent Life)

**1924**
Hovis, W. A.
"Infrared Emission Spectra of Organic Solids From 5 to 6.6 Microns." *Science* 143, Feb. 7, 1964, pp. 587-588.

**1925**
Hibben, R. D.
"Space Sciences Challenge." *Aviation Week and Space Technology* 80, Feb. 17, 1964, pp. 91-93.

**1926**
Ewing, Ann
"Life in all Solar Systems?" *Science Newsletter* 86, Sept. 1964, p. 199.

**1927**
"Prevalence of Planets and the Probability of Life." *Time* 84, Sept. 25, 1964, pp. 49-50.

**1928**
Posin, Daniel Q.
"Other Suns, Other Planets, but is There Other Life?" *Today's Health* 42, Nov. 1964, pp. 56-69.

**1929**
Hoyle, F.
"Can we Learn From Other Planets? Excerpt From Of men and Galaxies." *Saturday Review* 47, Nov. 7, 1964, pp. 63-67.

**1930**
Hayatsu, Ryoichi
"Orgueil Meteorite--Organic Nitrogen Contents." *Science* 146, Dec. 1964, pp. 1291-1293.

**1931**
Gilvarry, J. J.
"The Possibilities of a Primordial Lunar Life." (In Mamikunian, Gregg, & M. H. Briggs, eds.) *Current Aspects of Exobiology*. Pasadena, Jet Propulsion Laboratory, 1965, pp. 179-241).

**1932**
Hawrylewicz, Ervin J., et al
"Response of Microorganisms to a Simulated Martian Environment." (In *Life Sciences and Space Research, Vol. 3*. International Space Science Symposium, 5th, Florence, Italy, May 12-16, 1964. Ed. by Marcel Florkin. New York, Wiley, 1965, pp. 64-73).

**1933**
Jackson, F., & P. Moore
"Possibilities of Life on Mars." (In Mamikunian, Gregg and M. H. Briggs. *Current Aspects of Exobiology*. Pasadena, Jet Propulsion Laboratory, 1965, pp. 243-259).

**1934**
Botan, E. A., et al
"Study of an Instrumented Analytical System for Extraterrestrial Study of Atmospheres, Possible Life Forms and Soils." *Aerospace Medicine* 36, Jan. 1965, pp. 21-25.

**1935**
Levin, G. V.
"Significance and Status of Exobiology; Presented at the A.I.B.S. Annual Meeting, Boulder, Colorado, Aug. 26, 1964." *Bioscience* 15, Jan. 1965, pp. 17-20.

**1936**
"Dr. S. M. Siegel Reports Union Carbide Research Institute Studies Show Many Life Forms Have Ability to Adapt to Conditions Similar to Those on Mars." *New York Times*, April 1, 1965, p. 11, col. 3.

**1937**
Smoluchowski, R.
"Is There Vegetation on Mars?" *Science* 148, May 1965, pp. 946-947.

**1938**
"Where There's Life..." *Newsweek* 65, May 10, 1965, p. 100.

Notes:

## LIFE ON OTHER PLANETS IN GENERAL AND IN SPACE (EXCLUDING INTELLIGENT LIFE)

**1939**
"Dr. S. M. Siegel and C. Giammarro Report Some Bacteria and Fungi Thrive in Atmosphere With High Concentrations of Methane and Ammonia Like That on Jupiter...Dr. Siegel Notes Similarity of Atmosphere to That Presumed for Earth Long Ago." New York Times, May 12, 1965, p. 10, col. 1.

**1940**
Asimov, Isaac
"Science in Search of a Subject: Exobiology." New York Times Magazine, May 23, 1965, pp. 52-53.

**1941**
"Quarantine for Space Travelers?" Time 85, June 4, 1965, p. 70.

**1942**
Smart, R.
"Science." Toronto, Saturday Night 80, July 1965, pp. 28-30.

**1943**
Mumford, G. S.
"Life in an Ammonia-rich Atmosphere." Sky and Telescope 30, Aug. 1965, pp. 84-85.

**1944**
Ponnamperuma, Cyril
"Life in the Universe." Astronautics & Aeronautics (3: 10), Oct. 1965, p. 66.

**1945**
"Meteoritic Life Studied." Science Newsletter 88, Oct. 23, 1965, p. 259.

**1946**
"Organic Life Debated." Christian Science Monitor, Jan. 3, 1966, p. 15, col. 1.

**1947**
Urey, H. C.
"Biological Material in Meteorites: a Review." Science 151, Jan. 14, 1966, pp. 157-166.

**1948**
Horowitz, N. H.
"Search for Extraterrestrial Life." Science 151, Feb. 18, 1966, pp. 789-792.

**1949**
"Biomedical Space Technology Lagging." Missiles and Rockets 18, Feb. 21, 1966, p. 18.

**1950**
Cohen, D.
"Life on Earth and Elsewhere." Science Digest 59, March 1966, pp. 92-93.

**1951**
"American Astronautical Society Sponsors Symposium on Search for Extraterrestrial Life, Anaheim, Calif." New York Times, May 25, 1966, p. 22, col. 1.

**1952**
"Life on Other Planets--What are the Possibilities?" Flying Saucers, Oct. 1966, pp. 8-11.

**1953**
"What's Stirring on Mars? Franco-U.S. Findings." Newsweek 68, Oct. 1966, p. 66.

**1954**
Reynolds, O. E., & F. H. Quimby
"The Search for Organic Material on the Moon." (In Malina, Frank J., ed. Life Sciences Research and Lunar Medicine. Proceedings. London, Pergamon Press, 1967, pp. 51-53).

**1955**
Chappelle, E. W., et al
"Prevention of Protein Denaturation During Exposure to Sterilization Temperatures." Science 155, March 10, 1967, pp. 1287-1288.

**1956**
Harrison, Philip
"The Case for the Missing Planet." Flying Saucers, Aug. 1967, pp. 16-18.

NOTES:

## LIFE ON OTHER PLANETS IN GENERAL AND IN SPACE (Excluding Intelligent Life)

**1957**
Burbidge, G. R., & E. M. Burbidge
"Absorption Lines in Quasistellar Objects."
London, Nature 216, Dec. 16, 1967, p. 1092.

**1958**
Ensanian, Minas
"Does Life Exist Elsewhere in the Universe? A Review of Scientific Theory and the Potential of Space Exploration." (In Canaveral Council of Technical Societies. Space Congress, 5th, Cocoa Beach, Florida, March 11-14, 1968. Proceedings, v. 3. Cape Canaveral, Fla., 1968, pp. 25. 5-1 to 25. 5-20).

**1959**
Klein, Harold P.
"Some Biological Problems in the Search for Extraterrestrial Life." (In Space Projections From the Rocky Mountain Region; Proceedings of the Symposium, Denver, Colorado, July 15-16, 1968. Vol. 3. Tarzana, Calif., American Astronautical Society, 1968, 21p.).

**1960**
"Chlorophyll and the red Spot." Time 91, Feb. 9, 1968, p. 48.

**1961**
Allison, Anthony
"Possible Forms of Life." London, Journal of the British Interplanetary Society 21, March 1968, pp. 48-51.

**1962**
Kremp, Gerhard O. W.
"Observations on Fossil-Like Objects in the Orgeuil Meteorite." London, Journal of the British Interplanetary Society, 21, March 1968, pp. 99-112.

**1963**
Libby, Willard F.
"Life in Space." Dordrecht, Space Life Sciences 1, March 1968, pp. 5-9.

**1964**
Meadows, A. J.
"Planetary and Space Environments."
London, Journal of the British Interplanetary Society 21, March 1968, pp. 2-11.

**1965**
Sagan, Carl
"Simulating Extraterrestrial Environments."
London, Science Journal 4, March 1968, pp. 75-79.

**1966**
Prytz, John
"Exobiology and the Universe." Flying Saucers 57, April 1968, pp. 12-13.

**1967**
Hartmann, William K.
"Craters, a Tale of 3 Planets." Natural History 77, May 1968, pp. 58-63.

**1968**
Frisch, B.
"Dyson's Cool Worlds." Science Digest 63, June 1968, pp. 35-41.

**1969**
Hotchin, John
"The Microbiology of Space." London, Journal of the British Interplanetary Society 21, June 1968, pp. 122-130.

**1970**
Liebl, V., & J. Lieblova
"Coacervate Systems and Life." London, Journal of the British Interplanetary Society 21, Sept. 1968, pp. 295-312.

**1971**
Klein, Harold P.
"Extraterrestrial Biology: Prospects and Problems." (In Bioastronautics and the Exploration of Space. Brooks Air Force Base, Texas, School of Aerospace Medicine, Nov. 1968, pp. 569-593. (AD-687893). (Available from NTIS).

NOTES:

## LIFE ON OTHER PLANETS IN GENERAL AND IN SPACE (Excluding Intelligent Life)

**1972**
Ebel, Hans F.
"Über Enstehung und Verbreitung des Lebens im Kosmos." Mannheim, Sterne und Weltraum 7, Dec. 1968, pp. 296-301.

**1973**
Lovelock, James E., & C. E. Giffin
"Planetary Atmospheres, Compositional and Other Changes Associated With the Presence of Life." (In Advanced Space Experiments. American Astronautical Society Conference, Ann Arbor, Mich., Sept. 16-18, 1968. Proceedings. Ed. by O. L. Tiffany and E. M. Zaitseff. Tarzana, Calif., American Astronautical Society, Advances in the Astronautical Sciences, Vol. 25, 1969, pp. 179-193).

**1974**
Saunders, Joseph F.
"A Cryobiologist's Conjecture of Planetary Life." Cryobiology (6: 3), 1969, pp. 151-159.

**1975**
"National Radio Astronomy Observatory Reports...Clouds of Formaldehyde Thought to be Chemical Precursor of Life, in many Parts of Milky Way...Reinforces Growing Suspicion That Life...as it Exists on Earth, Began in Outer Space." New York Times, March 21, 1969, p. 43, col. 1.

**1976**
"Life Search Turns up More Clues in Space." Christian Science Monitor, March 31, 1969, p. 15, col. 3.

**1977**
Duchene, Jules
"Meteorites and Extraterrestrial Life." London, Science Journal 5, April 1969, pp. 33-38.

**1978**
Eshleman, Von R.
"Similarities in Mars, Venus and Earth." Space World F-6-66, June 1969, pp. 34-35.

**1979**
"Hint of Martian Life Found." Christian Science Monitor, June 3, 1969, p. 14, col. 2.

**1980**
Barth, Charles A.
"Planetary Ultraviolet Spectroscopy." Applied Optics 8, July 1969, pp. 1295-1304.

**1981**
Hodgson, G. W., & B. L. Baker
"Porphyrins in Meteorites; Metal Complexes in Orgueil, Murray, Cold Bokkeveld, and Mokoia Carbonaceous Chondrites." Oxford, England, Geochimica et Cosmochimica Acta 33, Aug. 1969, pp. 943-958.

**1982**
Brooks, J. & G. Shaw
"Evidence for Extraterrestrial Life: Identity of Sporopollenin With the Insoluble Organic Matter Present in the Orgueil and Murrary Meteorites and Also in Some Terrestrial Microfossils." London, Nature 223, Aug. 16, 1969, pp. 754-756.

**1983**
Bylinsky, Gene
"Penetrating the Secrets of the Planets." Fortune 80, Nov. 1969, pp. 138-143.

**1984**
Fesenkov, V. G.
"Usloviia Zhiznivo Vselennoi." Moscow, Priroda 1, 1970, pp. 20-27.

**1985**
Schwartz, A. W.
"Exobiology." (In Recent Advances in Aerospace Medicine. Proceedings of the 18th International Congress of Aviation and Space Medicine, Amsterdam, Netherlands, Sept. 15-18, 1969. Ed. by D. E. Busby. Dordrecht, Reidel, 1970, pp. 48-52).

Notes:

## LIFE ON OTHER PLANETS IN GENERAL AND IN SPACE (Excluding Intelligent Life)

**1986**
Wolfgang, Richard
 "Carbon Monoxide as a Basis for Primitive Life on Other Planets." London, Nature 225, Feb. 28, 1970, p. 876.

**1987**
Siegel, S. M.
 "Experimental Biology of Extreme Environments and its Significance for Space Bioscience." London, Spaceflight 12, March 1970, pp. 128-130.

**1988**
Biemann, K.
 "Organic Analysis." Applied Optics 9, June 1970, pp. 1282-1288.

**1989**
Moll, Horacio M.
 "Bioastronáutica y Vida Extraterrestre." Madrid, Revista de Aeronáutica y Astronáutica 30, Sept. 1970, pp. 663-673.

**1990**
Michel, Aimé
 "Orthoténie: Réalités et Illusions." Paris, Phénomènes Spatiaux 26, Dec. 1970, pp. 11-15.

**1991**
Toulet, F.
 "L'Orthoténie n'est-elle Qu'une Hypothèse?" Paris, Phénomènes Spatiaux 26, Dec. 1970, pp. 3-11.

**1992**
"U.S. Scientists, Drs. C. Ponnamperuma, I. R. Kaplan, and C. Moore Report They Have Discovered 17 Amino Acids, Including 6 That are Precursors of Life in Meteorite That Fell Near Murchison, Australia in '69." New York Times, Dec. 2, 1970, p. 1, col. 4.

**1993**
Young, R. S.
 "The Planets and Life." (In Chemical Evolution and the Origin of Life. Proceedings of the Third International Conference, Pont-à-Mousson, France, April 19-25, 1970. v. 1. Amsterdam, North-Holland Pub., 1971, pp. 510-515).

**1994**
"Gold Mine in the Sky." Newsweek 77, April 26, 1971, p. 56.

**1995**
"Goddard Space Flight Center's Dr. R. A. Hanel Says That Although Prospect of Life on Mars is not Certain, Evidence Gathered by Mariner 9 Supports Prospect of Life Elsewhere in Solar System." New York Times, June 15, 1971, p. 1, col. 1.

**1996**
Luchitsky, I.
 "Space Aspect of Geology." (Tr. into English from Pravda, Moscow, July 15, 1971, 3 pp. NLL-M-20759-(5828.4F). Available from National Lending Library for Science & Technology, Boston Spa, Yorkshire, England.

**1997**
Penzias, A. A., & P. Encrenaz
 "Les Molecules de l'Espace." Paris, La Recherche 2, Oct. 1971, pp. 817-823.

**1998**
"Columbia University Scientists Drs. G. Wollin and D. B. Ericson Hold...That Life Might Arise on Waterless Planet, Article in Nature." New York Times, Nov. 17, 1971, p. 55, col. 6.

**1999**
"Nature's Way." Time 98, Nov. 29, 1971, pp. 64-65.

Notes:

# LIFE ON OTHER PLANETS IN GENERAL AND IN SPACE (EXCLUDING INTELLIGENT LIFE)

**2000**
"Is There Life on Mars or Beyond?" Time 98, Dec. 13, 1971, pp. 50-52.

**2001**
Astapovich, I. S.
"Organika Meteoritov." Kiev, Problemy Kosmicheskoi Fiziki 7, 1972, pp. 79-99.

**2002**
Sagan, Carl
"Life Beyond the Solar System." (In Ponnamperuma, Cyril, ed. Exobiology. Amsterdam, North-Holland Pub., 1972, pp. 465-477).

**2003**
Ehricke, Krafft A.
"Astrogenic Environments. The Effect of Stellar Spectral Classes on the Evolutionary Pace of Life." London, Spaceflight 14, Jan. 1972, pp. 2-14.

**2004**
Sagan, Carl
"Interstellar Organic Chemistry." London, Nature 238, July 14, 1972, pp. 77-80.

**2005**
Derpgolts, Vladimir, & Gennady Katterfeld
"Water on Distant Planets." Space World I-8-104, Aug. 1972, pp. 43-45.

**2006**
"Water on Mars Might not Support Life." Space World I-8-104, Aug. 1972, p. 42

**2007**
Foster, G. V.
"Non-Human Artifacts in the Solar System." London, Spaceflight 14, Dec. 1972, pp. 447-453.

**2008**
Crick, F. H. C., & L. E. Orgel
"Directed Panspermia." Icarus 19, July 1973, pp. 341-346.

NOTES:

# LIFE ON INDIVIDUAL PLANETS (EXCLUDING INTELLIGENT LIFE)

## BOOKS

**2009**
Lowell, P.
　New Observations of the Planet Mercury. Brookline, American Academy of Arts & Sciences, 1898, 46p.

**2010**
Mercier, A.
　Conférence Astronomique sur la Planète Mars. Projet d'Etudes sur les Moyens Pratiques d'Exécution de Signaux Lumineux de la Terre à Mars. Orléans, M. Marron, 1902.

**2011**
Lowell, P.
　Mars and its Canals. 2d ed. New York, Macmillan, 1906, 393p.

**2012**
Morse, E. S.
　Mars and its Mystery. Boston, Little, Brown, 1906, 16p.

**2013**
Lowell, P.
　The Temperature of Mars. Brookline, American Academy of Arts & Sciences, 1907, 66p.

**2014**
Lowell, P., & A. R. Wallace
　Is Mars Habitable? New York, Macmillan, 1907, 179p.

**2015**
Lowell, P.
　Mars as the Abode of Life. New York, Macmillan, 1908, 288p.

**2016**
Abbot, C. G.
　Habitability of Venus, Mars, and Other Worlds. Smithsonian Institution, Annual Report for 1920; Washington, 1921, pp. 165-171.

**2017**
Pickering, W. H.
　Mars. Boston, Gorham Press, 1921, 173p.

**2018**
Arrhenius, Svante
　Die Sternenwelt...Trans. by Alexis Finkelstein. Leipzig, Akademische Verlagsgesellschaft, 1931, 359p.

**2019**
Rudaux, L.
　La Lune et son Histoire. Paris, Nouvelles Eds. Latines, 1947.

NOTES:

## LIFE ON INDIVIDUAL PLANETS (Excluding Intelligent Life)

2020
De Vaucouleurs, G.
 Physics of the Planet Mars: an Introduction to Aerophysics. New York, Macmillan, 1954, pp. 99-127. ($10.00)

2021
Strughold, H.
 The Green and red Planet: a Physiological Study of the Possibility of Life on Mars; With Editorial Assistance and a Foreword by Green Peyton. London, Sidgwick & Jackson, 1954, 96p. (7 shillings, 6 pence)

2022
Nelson, Albert F.J.H.N.
 There is Life on Mars. London, Werner Laurie, 1955, 152p.

2023
Moore, P.
 Guide to Mars. London, Frederick Muller, 1956, 124p. (10 shillings, 6 pence)

2024
Wilkins, H. P.
 The Mystery of Mars. (In U.S. Smithsonian Institution. Annual Report of the Board of Regents, 1955; Publication 4272. Washington, Government Printing Office, 1956, pp. 229-244).

2025
Georgetown College Observatory
 The Known Physical Characteristics of the Moon and the Planets, by Carl C. Kiess, and K. Lassovsky. ARDC-TR-58-41; AD-115617; N63-83268. Washington, 1958, 43p.

2026
Dollfus, A.
 Resultats d'Observations Indiquant la vie sur la Planète Mars. (In Proceedings of the International Space Science Symposium, 1st. Nice, France, 1960. New York, Interscience, 1960. ($26.00)

2027
Hess, L.
 Mars as an Astronautical Objective. (In Ordway, F. I., editor. Advances in Space Science and Technology. 3v. New York, Academic Press, 1961, pp. 151-192).

2028
Kellogg, W. W., & C. Sagan
 The Atmospheres of Mars and Venus: a Report by the ad hoc Panel on Planetary Atmospheres. Washington, National Academy of Sciences, Space Science Board, 1961, 151p.

2029
Kuiper, G. P., & B. M. Middlehurst, editors
 Planets and Satellites. Chicago, University of Chicago Press, 1961, 601p. ($12.50)

2030
Newburn, R. L.
 The Exploration of Mercury, the Asteroids, the Major Planets and Their Satellite Systems and Pluto. (In Ordway, F. I., ed. Advances in Space Science and Technology 3. New York, Academic Press, 1961, pp. 196-266).

2031
Ramo, S., ed.
 Peacetime Uses of Outer Space. New York, McGraw-Hill, 1961, 279p. ($6.95)

2032
Sagan, Carl
 Organic Matter and the Moon. By Carl Sagan, Panel on Extraterrestrial Life for the Armed Forces--NRC Committee on Bioastronautics. Washington, National Academy of Sciences, National Research Council, 1961, 49p.

2033
Douglas Aircraft Company
 Physical Properties of the Planet Mars. SM-43634; N63-22371. Santa Monica, 1963, 98p.

Notes:

## LIFE ON INDIVIDUAL PLANETS (Excluding Intelligent Life)

**2034**
U.S. National Aeronautics & Space Administration
Experiments From a Small Probe Which Enters the Atmosphere of Mars, by R. A. Hanel, et al. NASA-TN-D-18991; N64-11234. Washington, 1963, 23p.

**2035**
U.S. National Aeronautics & Space Administration
Studies on Martian Biology, by Paul Deal. (In Naval Reserve Research Co. 12-5, Berkeley, California. Research Reserve Space Science Seminar, Treasure Island, San Francisco, Oct. 20-Nov. 2, 1963. N64-14993. Washington, 1963, 13p.).

**2036**
Yagoda, Herman
Interaction of Cosmic and Solar Flare Radiations With the Martian Atmosphere and Their Biological Implications. Paper Presented at COSPAR International Space Science Symposium. 4th, Warsaw, June 3-11, 1963. Bedford, Mass., U.S. Air Force Cambridge Research Laboratories, 1963, 4p.

**2037**
Dole, S. H.
Habitable Planets for man. Waltham, Blaisdell Pub. Co., 1964, 158p. ($5.75)

**2038**
Dole, S. H., & I. Asimov
Planets for man. New York, Random House, Inc., 1964, 242p. ($4.95)

**2039**
Hess, H. H. et al
Biology and the Exploration of Mars; Summary and Conclusions of a Study by the NAS/NRC Space Science Board, Stanford University, June 1964. Washington, National Academy of Sciences, 1965, 19p.

**2040**
Pittendrigh, Colin S., et al, eds.
Biology and the Exploration of Mars. Report of a study held at Stanford University and New York, 1964 - Oct. 1965. Pub. 1286. (NASA-CR-77938). Washington, National Academy of Sciences - National Research Council, Space Sciences Board, 1966, 528p. (Available from NTIS: $8.27).

**2041**
Three Undiscovered Planets. Los Angeles, DeVorss & Co., 1967, 15p.

**2042**
Grumman Aircraft Engineering Corp.
A Review of the Environments of Mercury, Venus and Mars, by K.M. Foreman, et al. Bethpage, N.Y., July 1968, 245p. (RM-420).

**2043**
Denaerde, Stefan
Buitenaardse Beschaving. De Planeet Iarga. Naar Aanwijzingen van de Schrijver Geillustreerd Door Rudolf Das. Deventer, N. Kluwer, 1969, 215p.

**2044**
Chernigovskii, V. N., ed.
Problemy Kosmicheskoi Biologii. v. 16. Moscow, Izd. Nauka, 1971, 352p.

**2045**
Cole, George H. A.
On Inferring Elastic Properties of Deep Planetary Interiors: Moon and Mars. Washington, 1971, 46p. (Available from NTIS).

**2046**
Schopf, J. W.
Micropaleontological Studies of Lunar Samples. A Search for Biogenic Structures in the Apollo 12 Lunar Samples. NASA-CR-114839. Washington, 1971, 37p. (Available on microfilm from NTIS).

Notes:

## LIFE ON INDIVIDUAL PLANETS (EXCLUDING INTELLIGENT LIFE)

**2047**
Tass Report: Mars: Complex Investigations. Soft Landing on the Planet Surface. Tr. into English from Pravda, #353 (19496), Dec. 19, 1971, 14p. (Available from National Translations Center, John Crerar Library, Chicago, Ill. 60616).

**2048**
California. University at Berkeley. Dept. of Soils and Plant Nutrition
Enzyme Activity in Terrestrial Soil in Relation to Exploration of the Martian Surface: Semiannual Progress Report, by M. S. Ardakani, et al. Berkeley, 1972, 26p. ($3.50) (NASA-CR-128399; SAPR-16; SSL-Ser-13- Issue-76). Available from NTIS.

**2049**
The Planet Mercury, 1971. NASA Space Vehicle Design Criteria (Environment). (NASA-SP-8085), Washington, NASA, 1972, 63p. (Available from NTIS).

**2050**
Moore, Patrick, & Charles A. Cross
Mars. New York, Crown, 1973, 48p.

## PERIODICAL ARTICLES

**2051**
Newcomb, S.
"Interpretation of the So-called Canals of Mars." Astrophysical Journal 26, 1907, pp. 1-17.

**2052**
Lowell, Percival
"Proofs of Life on Mars." Century (2:76), June 1908, pp. 292-303.

**2053**
Henkel, F. W.
"Venus as the Abode of Life." Popular Astronomy 17, Aug. 1909, pp. 412-417.

**2054**
"Editorial Denounces E. Perrier Claim That Huge Animals and Plants Exist on Mars." New York Times, Jan. 23, 1911, p. 6, col. 4.

**2055**
Perrier, E.
"What is Life on Mars Like?" North American Review 197, Jan. 1913, pp. 105-111.

**2056**
"Flammarion, Camille, on Canals of Mars." New York Times, Nov. 9, 1913, sec. V, p. 6, col. 1.

**2057**
"Dr. P. Lowell Presents Arguments to Prove Mars Canals Artificially Made." New York Times, Dec. 1, 1915, p. 24, col. 4.

NOTES:

## LIFE ON INDIVIDUAL PLANETS (Excluding Intelligent Life)

**2058**
"Latest Theory Regarding Life on the Planet Mars." Current Opinion 63, Oct. 1917, p. 254.

**2059**
"Prof. A. Arrhenius Declares Life Possible on Planet." New York Times, March 8, 1919, p. 7, col. 6.

**2060**
Beard, D. P.
"Life on Mars?" Scientific American 123, Sept. 18, 1920, p. 277.

**2061**
Abbot, C. G.
"The Habitability of Venus, Mars, and Other Worlds." (In Annual Report of the Board of Regents of the Smithsonian Institution for the Year Ending June 1920. Washington, U.S. Govt. Printing Office, 1922, pp. 165-171).

**2062**
"C. G. Abbott Says Venus not Mars, is Abode of Life." New York Times, May 26, 1922, p. 21, col. 7.

**2063**
Coblentz, W. W.
"Can Life Exist on Mars?" Scientific Monthly 20, April 1925, pp. 337-340.

**2064**
Aitken, R. G.
"Why Popular Interest in Mars?" Leaflets of the Astronomical Society of the Pacific (1:2), July 1925, pp. 3-6.

**2065**
Kaempffert, W.
"Habitability of Venus and Mars." New York Times, April 10, 1927, sec. X, p. 4.

**2066**
Slipher, E. C.
"The Planet Jupiter." Leaflets of the Astronomical Society of the Pacific (1:12), Sept. 1927, pp. 47-50.

**2067**
Gray, G. W.
"Life on Mars is Almost Certain. Interview With W. H. Pickering." American Magazine 105, Feb. 1928, pp. 54-55.

**2068**
"Oxygen on Mars as an Indication of Life." Science, Supplement 10, April 6, 1928, p. 67.

**2069**
Lowell, P.
"Life on Mars." (In Shapley, Harlow, & Helen E. Howorth, eds. A Source Book in Astronomy. New York, McGraw-Hill, 1929, pp. 388-393).

**2070**
Leonard, F. C.
"The New Planet Pluto." Leaflets of the Astronomical Society of the Pacific (1:30), Aug. 1930, pp. 121-124.

**2071**
Aitken, R. G.
"Venus: the Earth's Twin Planet." Leaflets of the Astronomical Society of the Pacific (1:48), Jan. 1933, pp. 194-198.

**2072**
Lewis, I. M.
"Life on Venus and Mars? Why These Planets may Sustain Life." Nature Magazine 24, Sept. 1934, pp. 133-134.

**2073**
"Dr. L. B. Andrews on Evidence That There is no Life on Mars." New York Times, Oct. 31, 1935, p. 21, col. 3.

Notes:

## LIFE ON INDIVIDUAL PLANETS (Excluding Intelligent Life)

**2074**
"Comment on Possibilities of Life on Mars."
New York Times, Jan. 12, 1936, sec. X, p. 8, col. 2.

**2075**
Pickering, W. H.
"Life on the Moon." Popular Astronomy 45, June 1937, pp. 317-319.

**2076**
Bennett, Dorothy A.
"Men From Mars." The Sky 3, Dec. 1938, pp. 8-9.

**2077**
Fournier, G.
"La Planète Mars et la vie." Paris, Astronomie 53, 1939, pp. 348-352.

**2078**
Antoncadi, E. M.
"The Planet Mars." Sky, Magazine of Cosmic News 31, July 1939, pp. 6-7, 22.

**2079**
Duckwall, W. E.
"Life on the Moon." Popular Astronomy 47, Nov. 1939, pp. 517-518.

**2080**
"Life on the Planet Venus now Seems Impossible." Science, Supplement 10, Sept. 27, 1940, p. 92.

**2081**
"Dr. S. Jones Speculates on Life on Venus."
New York Times, Oct. 27, 1940, p. 14, col. 1.

**2082**
"Comment on Dr. W. S. Adams's Theory of no Life on Mars." New York Times, Feb. 2, 1941, sec. IV, p. 2, col. 1.

**2083**
"Any Life on Venus is Declared Impossible; Clouds on Surface of Planet are Believed to be Composed of Solid Formaldehyde."
Science News Letter 40, July 12, 1941, p. 26.

**2084**
Duckwall, W. E.
"Life in Mars." Popular Astronomy 49, Nov. 1941, pp. 100-103.

**2085**
Barton, W. H.
"Mystery of Mars." Science Digest 11, Jan. 1942, pp. 75-77.

**2086**
Maude, A. P.
"Vegetation on Mars." London, Journal of the British Astronomical Association 52, Sept. 1942, pp. 264-265.

**2087**
Menzel, Donald H.
"New Light on the Mystery of Mars; Does Life Exist on our Neighbouring Planet?" Popular Science 143, Oct. 1943, pp. 125-127.

**2088**
"Flu From Venus?" Time 43, Feb. 21, 1944, p. 90.

**2089**
"Evergreens Will Greet Visitors to Mars."
Science Digest 18, Nov. 1945, p. 88.

**2090**
Lewis, I. M.
"Does Life Exist on Mars?" Nature Magazine 41, Feb. 1948, pp. 99-100.

**2091**
"U.S. and USSR Scientists Study Plant Life on Mars; Possibility of Animal Life Discussed." New York Times, Feb. 22, 1948, sec. IV, p. 9, col. 6.

Notes:

## LIFE ON INDIVIDUAL PLANETS (Excluding Intelligent Life)

**2092**
Mather, S. M.
"The Planet Mercury." London, Journal of the British Interplanetary Society 8, May 1949, pp. 119-120.

**2093**
"Dr. H. C. Urey Sees Life on Other Planets." New York Times, Dec. 9, 1950, p. 18, col. 6.

**2094**
Dauvillier, A.
"Sur la Nature de Pluton et de Triton." Paris, Comptes Rendus Hebdomadaires des Séances, Académie des Sciences 233, 1951, pp. 901-903.

**2095**
Strughold, Hubertus
"Life on Mars in View of Physiological Principles." (In U.S. Air Force School of Aviation Medicine. Epitome of Space Medicine. Randolph Air Force Base, Texas, 1951).

**2096**
"Plant Life on Mars." Science Digest 32, Aug. 1952, p. 91.

**2097**
"H. K. Kaiser Holds Mars has Primitive Vegetation and Venus may Have Organic Life in Millions of Years." New York Times, Sept. 2, 1952, p. 25, col. 8.

**2098**
Strughold, H.
"Das Leben auf dem Mars." Frankfurt-am-Main, Weltraumfahrt 4, 1953, pp. 24-26.

**2099**
"Moscow Radio Reports Prof. Tikhov Holds Life Exists on Mars and Vegetation is Blue." New York Times, June 12, 1953, p. 9, col. 5.

**2100**
Bowman, N. J.
"Mars, Planet of Mystery." Journal of Space Flight (6:1), 1954, pp. 1-7.

**2101**
Richardson, R. S.
"Life on Mars." (In his Exploring Mars. New York, McGraw-Hill, 1954, chapter 8).

**2102**
Strughold, H.
"What Kind of Life on Mars? Excerpt From The Green and red Planet." Science Digest 35, March 1954, pp. 58-62.

**2103**
"Venus, Near Neighbor of the Earth." Hobbies 59, March 1954, p. 158.

**2104**
Whipple, F. L.
"Is There Life on Mars?" Collier's 133, April 30, 1954, p. 21.

**2105**
Whipple, Fred L.
"Hay Vida Alguna en Marte?" Madrid, Revista Oficial Compl. (11: 124), Aug. 1954, pp. 31-32.

**2106**
Deville, Gérard
"Nos va, por fin, a Revelar su Secreto el Planeta Marte?" Bogotá, Lecturas 362, Dec. 1954, pp. 89-91.

**2107**
Moore, P.
"Mercury: the Barren Planet." Journal of the Astronautical Sciences (2:4), 1955, pp. 145-148.

**2108**
David, H. M.
"AF Lectures Cover Host of Lunar Problems." Missiles and Rockets (10:4), Jan. 22, 1962, p. 28.

NOTES:

## LIFE ON INDIVIDUAL PLANETS (EXCLUDING INTELLIGENT LIFE)

2109
"Dr. D. B. McLaughlin Says Mars is Barren, no Life Exists." New York Times, April 6, 1955, p. 31, col. 8.

2110
Keenan, C.
"What's new on Mars?" America 93, July 16, 1955, p. 383.

2111
Keyhoe, Donald E.
"Enigma on the Moon." Fate 9, Aug. 1956, pp. 36-43.

2112
Moore, P.
"Venus, the Nearest Planet." London, New Scientist (2:52), 1957, pp. 12-14.

2113
Salisbury, F. B.
"A Discussion of Some Current Theories Regarding the Markings on Mars." Publications of the Astronomical Society of the Pacific 69, 1957, pp. 396-397.

2114
"Life on Mars Hardy." Science Newsletter 71 Jan. 12, 1957, p. 31.

2115
"Vegetation on Mars?" Sky and Telescope 16, April 1957, p. 275.

2116
"Mars has Plant Life, but no Canals." Science Digest 41, May 1957, p. 66.

2117
"Randolph School's Test Project Proves Life can Exist on Mars." Air Force Times 17, June 22, 1957, p. 6.

2118
"Life on Mars?" Time 70, July 1, 1957, p. 58.

2119
Sinton, W. M.
"Spectroscopic Evidence for Vegetation on Mars." Astrophysical Journal (126: 2), Sept. 1957, p. 231.

2120
"Studies Show Martian Life." Science Newsletter 72, Nov. 2, 1957, p. 275.

2121
Lederberg, Joshua, & D. B. Cowie
"Moondust." Science (127: 3313), June 27, 1958, p. 1473.

2122
"Conserving Moondust." Scientific American 199, Aug. 1958, p. 48.

2123
Edwards, Frank
"Frank Edwards' Report: Keep Your eye on Venus." Fate 12, 1959, pp. 57-63.

2124
"Life on Mars?" America 101, May 23, 1959, p. 355.

2125
Drake, Walter R.
"Is our sun Inhabited?" London, Flying Saucer Review 5, Nov.-Dec. 1959, pp. 15-17.

2126
Sinton, W. M.
"Further Evidence of Vegetation on Mars." Science (130: 3384), Nov. 6, 1959, p. 1234.

NOTES:

## LIFE ON INDIVIDUAL PLANETS (Excluding Intelligent Life)

2127
Macvey, John
"Ultimate Objective: Pluto." London, Spaceflight 2, Jan. 1960, pp. 159-160.

2128
Parry, A.
"Three Billion Year old Sputniks? Martian's Moon." Science Digest 47, Jan. 1960, pp. 6-11.

2129
Dollfus, A.
"La vie sur la Planète Mars." Paris, Comptes Rendus Hebdomadaires des Séances, Académie des Sciences (250:3), Jan. 18, 1960, pp. 463-465.

2130
"Life on Mars?" Missiles and Rockets (6:3), Jan. 18, 1960, pp. 32-33.

2131
Marayo Magdaleno, Feliciano
"Podremos Vivir en Marte?" Bogotá, Revista Aeronautica 236, July 1960, pp. 557-564.

2132
Drake, Walter R.
"Mercury, Jupiter, and Others: can Life Exist?" London, Flying Saucer Review 6, Sept.-Oct. 1960, pp. 18-22.

2133
Sagan, Carl
"Biological Contamination of the Moon." (In Proceedings of the National Academy of Sciences: v. 46, #4, June 15, 1960. Washington, 1961, p. 396).

2134
"Dr. Shapley Sure no Life Above Vegetal Exists on Mars." New York Times, March 2, 1961, p. 29, col. 8.

2135
"Space Life on Earth; Bacteria-like Cells From Meteorites." Science Newsletter 79, April 15, 1961, p. 227.

2136
Binder, Otto O.
"Mars Colony: Support of Human Life Will be Relatively Simple on Mars." Space World 1, June 1961, pp. 30-31.

2137
"Life on Venus." The Airman (5:6), June 1961, p 44.

2138
"Prof. W. W. Howells Speculates Beings on Other Planets Resemble Mythical Centaurs." New York Times, July 4, 1961, p. 21, col. 8.

2139
"Dr. Sagan Sees Jupiter Better Possibility Than Venus for Life." New York Times, Aug. 20, 1961, sec. IV, p. 10, col. 8.

2140
Colthup, N. B.
"Identification of Aldehyde in Mars Vegetation Regions." Science 134, Aug. 25, 1961, p. 529.

2141
"Life may Exist on Jupiter." Science Newsletter 80, Aug. 26, 1961, p. 134.

2142
Barricelli, N. A.
"Prospects and Physical Conditions for Life on Venus and Mars." Milan, Scientia (96:11), Nov. 1961, pp. 337-343.

NOTES:

# LIFE ON INDIVIDUAL PLANETS (EXCLUDING INTELLIGENT LIFE)

2143
Rublowsky, J.
"Is There Life on the Moon?" Space World, Nov. 1961, pp. 12-15, 49-50.

2144
Rice, R. V.
"Martian Life." Science (138:3545), 1962, pp. 1197-1198.

2145
Strughold, H.
"Ecological Aspects of Planetary Atmospheres With Special Reference to Mars." Journal of Aviation Medicine (23:2), 1962, pp. 130-140.

2146
Sagan, Carl
"Earth Life not From Space." Science Newsletter 81, Jan. 6, 1962, p. 4.

2147
"Bacteria Able to Survive Mars-like Conditions." Science Newsletter 81, Feb. 17, 1962, p. 105.

2148
"Dr. F. B. Salisbury Holds Flourishing Plant Life Probable, Some Animal Life Possible on Mars." New York Times, April 6, 1962, p. 46, col. 7.

2149
Kilson, Walter K.
"Exploration on the Moon." Army Information Digest 17, June 1962, pp. 32-37.

2150
"Moon Life Seen Possible." Science Newsletter 81, June 9, 1962, p. 359.

2151
Ewing, A.
"Life on Mars Seen Possible." Science Newsletter 81, June 16, 1962, pp. 378-379.

2152
Schindler, Charles A., et al
"Feasibility of Laboratory Studies Concerning Life on Venus." Aerospace Medicine 33, July 1962, pp. 859-861.

2153
"Proteins Could Give Clue to Unearthly Life on Mars." Science Newsletter 82, July 28, 1962, p. 56.

2154
Lederberg, J., & Carl Sagan
"Microenvironments for Life on Mars." Proceedings of the National Academy of Sciences (48:9), Sept. 1962, pp. 1473-1475.

2155
Levin, G. V., & A. W. Carriker
"Life on Mars." Nucleonics (20:10), Oct. 1962, pp. 71-72.

2156
"Life Forms may Exist in hot Spots on Mars." Science Newsletter 82, Oct. 13, 1962, p. 240.

2157
"No Chance of Life on Venus; Scientific Data From Mariner II." Business Week, Dec. 1962, p. 22.

2158
Cross, C. A.
"Conditions on Mars." (In Gatland, Kenneth W., ed. Spaceflight Today. London, Iliffe Books, Ltd., 1963, pp. 190-203)

2159
de Vaucouleurs, G.
"Optical Studies of the Surface and Atmosphere of Mars." Advances in the Astronautical Sciences 15, 1963, pp. 519-532.

NOTES:

# LIFE ON INDIVIDUAL PLANETS (Excluding Intelligent Life)

**2160**
Kleczek, Josip
"Life on the Planets of Other Suns." (In To the Near and Distant Universe. Tr. into English of the manuscript Do Blizkeho i Vzdaleneho Vesmiru. Prague, Rozhlasova University, 1960. Dayton, Wright-Patterson Air Force Base, 1963, pp. 411-427). (FTD-TT-62-7) 1 & 2 & 3 & 4; AD-413009).

**2161**
Miller, S. L.
"The Possibility of Life on Mars." (In North American Aviation, Inc. Proceedings of the Lunar and Planetary Exploration Colloquium. Santa Monica, Calif., May 23-25, 1962, v. 3, #2, ed. by E. M. Fallone. Publication #513-W-12. Downey, Calif., 1963, pp. 1-7).

**2162**
Sagan, Carl
"Biological Exploration of Mars." (In Morgenthaler, George W., ed. Exploration of Mars. North Hollywood, Calif., Western Periodicals Co., 1963, pp. 571-581). (Advances In the Astronautical Sciences 15).

**2163**
Sinton, William M.
"Evidence of the Existence of Life on Mars." (In Morgenthaler, George W., ed. Exploration of Mars. North Hollywood, Calif., Western Periodicals Co., 1963, pp. 543-551. (Advances In the Astronautical Sciences 15).

**2164**
Siegel, S. M., et al
"Martian Biology: the Experimentalist's Approach." London, Nature 197, Jan. 1963, pp. 329-331.

**2165**
Siegel, S. M., et al
"Martian Biology: The Experimentalist's Approach." London, Nature (197: 4865), Jan. 26, 1963, p. 329.

**2166**
Croome, A.
"Venus, an Inferno." London, Discovery (24:4), April 1963, p. 5.

**2167**
Gossner, Simone D.
"Life may Exist on Mars, but Mistranslation Made the Canals." Natural History 72, April 1963, pp. 56-57.

**2168**
"Martian Environment." Science Newsletter 83, June 22, 1963, p. 386.

**2169**
Bongers, Leonard H.
"Is There Life on Mars?" Space/Aeronautics 40, Aug. 1963, pp. 86-88.

**2170**
Von Braun, W.
"Mars: are its Canals Full of Water; is There Really Life There?" Popular Science 183, Aug. 1963, p. 18.

**2171**
Fesenkov, V. G.
"Mars and Organic Life." Moscow, Priroda 2, 1963, pp. 11-17. English Trans. by Joint Publications Research Service, Washington, Sept. 11, 1963: JPRS-21026; OTS-63-3173B; N63-21489.

**2172**
"Life on Mars Talk Dealt a Setback." Science Newsletter 84, Sept. 21, 1963, p. 185.

**2173**
Rea, D. G.
"Evidence for Life on Mars." London, Nature (200: 4902), Oct. 12, 1963, pp. 114-116.

Notes:

## LIFE ON INDIVIDUAL PLANETS (Excluding Intelligent Life)

**2174**
Watkins, H. D.
"Ames Study Supports Mars Life Theory." Aviation Week and Space Technology 79, Nov. 18, 1963, p. 61.

**2175**
Horowitz, Norman H.
"The Design of Martian Biological Experiments." (In Life Sciences and Space Research II. International Space Science Symposium. 4th, Warsaw, June 3-12, 1963. Ed. by M. Florkin and A. Dollfus. New York, Interscience. 1964, pp. 133-138).

**2176**
Kellogg, W. W.
"Mars." International Science and Technology, Feb. 1964, pp. 40-48, 114.

**2177**
Rea, D. G.
"The Darkening Wave on Mars." London, Nature 201, March 7, 1964, pp. 1014-1015.

**2178**
"Prof. J. Strong Sees Discovery of Water Vapor in Atmosphere Increasing Chances of Existence of Life (on Venus)". New York Times, April 12, 1964, p. 39, col. 1.

**2179**
Ewing, A.
"Water Vapor on Venus." Science Newsletter 85, April 25, 1964, p. 261.

**2180**
"Signs of Life on Mars: Life on Mars Hard to see." Science Newsletter 85, May 30, 1964, pp. 339-340.

**2181**
"Dr. B. Commoner Says Possibilities of Life on Other Planets are too Small to Justify Scientific Exploration of Them (Speech at Hahneman Medical College)". New York Times, June 9, 1964, p. 21, col. 1.

**2182**
"Mars Biological lab Designs are Sought." Aviation Week and Space Technology 80, June 15, 1964, p. 37.

**2183**
"W. L. Laurence on Dr. C. Sagan Discussion of Possibility That Life Exists on Other Planets." New York Times, June 23, 1964, sec. IV, p. 7, col. 7.

**2184**
Wetmore, W. C.
"Microbe Survivability on Mars Simulated." Aviation Week and Space Technology 80, June 29, 1964, pp. 75-76.

**2185**
"How to Find Life on Mars." Science Digest 56, Aug. 1964, p. 90.

**2186**
Abelson, P. H.
"Martian Environment." Science 147, Feb. 12, 1965, p. 683.

**2187**
"Dr. Abelson Decries Search for Life on Mars...Doubts Life Exists There..." New York Times, Feb. 13, 1965, p. 23, col. 7.

**2188**
"Finding Life on Mars may Depend on Season." Science Newsletter 87, March 13, 1965, p. 168.

**2189**
Hibben, R. D.
"Some Earth Life Seen Adaptable to Mars." Aviation Week and Space Technology 82, April 12, 1965, p. 71.

**2190**
"Does Life Exist on Mars?" Science Newsletter 87, May 8, 1965, p. 292.

NOTES:

## LIFE ON INDIVIDUAL PLANETS (Excluding Intelligent Life)

2191
Armagnac, A. P.
"Big Question About Mars: is There Life on it?" Popular Science 187, July 1965, p. 86.

2192
Frisch, B. H.
"What We'll see on Mars." Science Digest 58, July 1965, pp. 44-53.

2193
Sullivan, W.
"Mars, Tantalizing Question Mark in the sky." New York Times Magazine, July 11, 1965, pp. 12-13.

2194
"Martian Life: a Darkening Wave." Newsweek 66, July 26, 1965, pp. 56-57.

2195
Eberhart, J.
"Is There Life on Mars?" Science Newsletter 88, July 31, 1965, pp. 74-75.

2196
Watkins, H. D.
"Automated lab with Search for Mars Life." Aviation Week and Space Technology 83, Aug. 2, 1965, p. 61.

2197
Salisbury, Frank B.
"The Possibilities of Life on Mars." (In Virginia Polytechnic Institute. Conference on the Exploration of Mars and Venus, Virginia Polytechnic Institute, Blacksburg, Va., Aug. 23-27, 1965. Proceedings. Blacksburg, 1965, pp. VI-1 to VI-16).

2198
Maney, Charles A.
"Is Venus Inhabited?" London, Flying Saucer Review 11, Sept.-Oct. 1965, pp. 6-8.

2199
Ponnamperuma, Cyril
"The Search for Extraterrestrial Life." Science Teacher 32, Oct. 1965, 6p.

2200
"Where There's Hope There may be Life: Mars." Time 86, Oct. 8, 1965, p. 67.

2201
Weston, Charles R.
"A Strategy for Mars." American Scientist 53, Dec. 1965, pp. 495-507.

2202
"Is There Life on Mars or Earth? Comparison of Photographs." Time 87, Jan. 7, 1966, p. 44.

2203
"Ames Researchers Find Mars Produces Life Forms." Missiles and Rockets 18, Feb. 14, 1966, p. 23.

2204
Ladin, V.
"Life on the Moon." Space World C-4-30, April 1966, pp. 11-12.

2205
Mumford, G. S.
"Is There Life on Earth?" Sky and Telescope 31, April 1966, pp. 213-214.

2206
"Drs. Plummer and Strong Believe Extensive Areas of Venus Have Temperature Comfortable to man." New York Times, April 18, 1966, p. 20, col. 1.

Notes:

# LIFE ON INDIVIDUAL PLANETS (Excluding Intelligent Life)

**2207**
Radunskaya, Irina
"A Mirror for Venus." Space World C-6-32, June 1966, pp. 40-42.

**2208**
"From Surveyor I Findings, Dr. J. J. Gilvarry Believes Moon Once had Seas and Life." New York Times, June 19, 1966, sec. IV, p. 14, col. 1.

**2209**
Evans, Gordon H.
"Image Orthicon Photographs of Martian Canals." London, Flying Saucer Review 12, July-Aug. 1966, pp. 7-9.

**2210**
"Dr. E. J. Opik Sees Craters (on Mars) Dating from Earliest Period...Doubts Advanced Form of Life has Ever Existed ...but Sees Likelihood of Primitive Form of Vegetation." New York Times, July 17, 1966, p. 55, col. 1.

**2211**
Strughold, Hubertus
"A new Look at Mars." London, Spaceflight 8, Sept. 1966, pp. 302-306.

**2212**
"Marsh gas on Mars." Time 88, Nov. 4, 1966, p. 56.

**2213**
Haas, Ward J.
"The Biological Significance of the Space Effort." New York Academy of Science, Annals 140, Dec. 16, 1966, pp. 659-666.

**2214**
"A. Oparin Sees Primitive Life Forms Possible on Moon, Pravda Article." New York Times, Dec. 30, 1966, p. 7, cols. 4, 5.

**2215**
Strughold, Hubertus
"Synopsis of Martian Life Theories." (In Ordway, Frederick I., ed. Advances in Space Science and Technology 9. New York, Academic Press, 1967, pp. 105-122).

**2216**
"Lunar Orbiter 1 Photo of Earth Taken in August 1966 Suggests...That Venus Might Support Life if There are Holes in Cloud Cover as There are on Earth." New York Times, Jan. 14, 1967, p. 12, col. 3.

**2217**
"Life on Mars?" Space World D-2-38, Feb. 1967, pp. 14-15.

**2218**
"Dr. H. K. Debus Predicts Astronauts Will one day Find Living Things in Space." New York Times, Feb. 27, 1967, p. 10, col. 3.

**2219**
"Pre-Life on Jupiter?" Science News 91, April 22, 1967, p. 376.

**2220**
"Martian Wolf Trap in Action; Soil Samples to be Examined for Evidence of Living Organisms." Science Digest 61, June 1967, pp. 17-21.

**2221**
"Venus is Dead and too hot." Time 89, June 9, 1967, pp. 87-88.

Notes:

## LIFE ON INDIVIDUAL PLANETS (Excluding Intelligent Life)

**2222**
"Life on Jupiter?" London, Spaceflight 9, Aug. 1967, p. 267.

**2223**
Jastrow, Robert
"The Planet Jupiter." London, Science Journal 3, Sept. 1967, pp. 50-54.

**2224**
"Dr. J. W. Schopf's Search for Primitive Life on Earth...Connected With Work at NASA's Ames Research Center on Developing Methods of Detecting Possible Life on Other Planets." New York Times, Sept. 5, 1967, p. 45, col. 8.

**2225**
"Life in the Clouds: Conditions on Venus." Science News 92, Sept. 30, 1967, p. 320.

**2226**
Sagan, Carl, & Joseph Veverka
"The Martian Ionosphere: a Component due to Solar Protons." Science 158, Oct. 6, 1967, pp. 110-112.

**2227**
"Gasbags of Venus." Time 90, Oct. 20, 1967, p. 60.

**2228**
Sagan, Carl
"Mars: a new World to Explore." National Geographic Magazine 132, Dec. 1967, pp. 820-841.

**2229**
Sagan, Carl
"Life on the Surface of Venus?" London, Nature 216, Dec. 23, 1967, pp. 1198-1199.

**2230**
"Mars Surface Multicratered: Mariner IV Photographs." Space World E-1-49, Jan. 1968, pp. 28-29.

**2231**
"Venus, a Hellhole." Science Digest 63, Jan. 1968, pp. 16-18.

**2232**
Greenspan, Jack A., & Tobias Owen
"Jupiter's Atmosphere: its Structure and Composition." Science 159, Jan. 26, 1968, pp. 448-450.

**2233**
"Green Salad on the red Planet: What Next?" Space World E-2-50, Feb. 1968, pp. 34-35.

**2234**
Hide, Raymond
"Jupiter's Great red Spot." Scientific American 218, Feb. 1968, pp. 74-82.

**2235**
"Exploring the Planets; Mariner V Probes Venus Secrets." Space World E-3-51, March 1968, pp. 18-23.

**2236**
Fesenkov, V.
"The Mystery Planet." Space World E-31-51, March 1968, pp. 38-39.

**2237**
Wend, Richard E.
"Jupiter, the Active Planet." The Review of Popular Astronomy 62, March 1968, pp. 4-7.

**2238**
Smoluchowski, R.
"Mars: Retention of Ice." Science 159, March 22, 1968, pp. 1348, 1350.

Notes:

# LIFE ON INDIVIDUAL PLANETS (Excluding Intelligent Life)

2239
Adams, John B.
"Lunar and Martian Surfaces: Petrologic Significance of Absorption Bands in the Near-Infrared." Science 159, March 29, 1968, pp. 1348-1350.

2240
"Venus Lives?" Newsweek 71, April 8, 1968, pp. 113-114.

2241
Chapman, Clark R.
"The Discovery of Jupiter's red Spot." Sky & Telescope 35, May 1968, pp. 276-278.

2242
Hodgson, Richard G.
"Mercury, the Elusive Planet." The Review of Popular Astronomy 62, May 1968, pp. 4-6.

2243
"Mars Surface Laboratory." Space World E-5-53, May 1968, pp. 36-37.

2244
Ash, M. E., et al
"The Case for the Radar Radius of Venus." Science 160, May 31, 1968, pp. 985-987.

2245
Marcus, A. H.
"Martian Craters: Number Density." Science 160, June 21, 1968, pp. 1333-1335.

2246
Levin, V. L.
"The Possibility of Survival of Terrestrial Organisms Under 'Martian' Conditions, a Review of the Foreign Literature." (In Space Biology and Medicine, Vol. 2, #2). Washington, Joint Publications Research Service, June 27, 1968, pp. 131-137. (JPRS-45798), Available from NTIS.

2247
Jastrow, Robert
"Planet Venus: Comparison of Mariner V and Venera 4 Data." Science 160, June 28, 1968, pp. 1403-1410.

2248
Sagan, Carl, et al
"Contamination of Mars." Science 159, March 15, 1968, pp. 1191-1196; Reply by D. L. Amsbury, 161, July 19, 1968, p. 298.

2249
Salisbury, John W., & Graham R. Hunt
"Martian Surface Materials: Effect of Particle size on Spectral Behavior; Presence of Limonite." Science 161, July 26, 1968, pp. 365-366.

2250
Libby, Willard F.
"Ice Caps on Venus?" Science 159, March 8, 1968, pp. 1097-1098; discussion, 161, Aug. 30, 1968, pp. 915-916.

2251
Wood, A. T., Jr., et al
"Venus: Estimates of the Surface Temperature and Pressure From Radio and Radar Measurements." Science 162, Oct. 4, 1968, pp. 114-116.

2252
Owen, Tobias
"Jupiter and the Outer Planets: AAAS Symposium." Science 162, Nov. 1, 1968, pp. 588-589.

2253
Jurgens, R. F.
"Radar Scattering From Venus at Large Angles of Incidence and the Question of Polar Ice Caps." Science 162, Dec. 20, 1968, pp. 1388-1390.

Notes:

## LIFE ON INDIVIDUAL PLANETS (Excluding Intelligent Life)

**2254**
Capen, Charles F.
   "Mars: a Dynamic World." *The Review of Popular Astronomy* 63, Feb. 1969, pp. 4-7.

**2255**
Eshleman, Von R.
   "The Atmospheres of Mars and Venus." *Scientific American* 220, March 1969, pp. 78-88.

**2256**
"Life on Venus." *Science Digest* 65, March 1969, p. 80.

**2257**
Mueller, Robert F.
   "Planetary Probe: Origin of Atmosphere of Venus." *Science* 163, March 21, 1969, pp. 1322-1324.

**2258**
"McDonald Observatory Astronomers Report They Have Obtained First 'Absolutely Conclusive Proof' That Water Exists in Martian Atmosphere." *New York Times*, March 25, 1969, p. 1, col. 4.

**2259**
Liubarskii, K. A.
   "The Hypothetical Biosphere of Mars." *Environmental Space Sciences* 3, May-June 1969, pp. 167-172.

**2260**
Jastrow, Robert
   "Venus and Mars: Statements." *The New Yorker* 45, May 10, 1969, pp. 29-31.

**2261**
"Animal Life on Jupiter?" *Science Digest* 66, July 1969, pp. 30-31.

**2262**
Liubarskii, K. A.
   "Dykhanie i Mineral'noe Pitanie Gipoteticheskikh Marsianskikh Organizmov i Drugie Voprosy Biologii Marsa." Moscow, *Kosmicheskaia Biologiia i Meditsina* 3, July-Aug. 1969, pp. 12-17.

**2263**
Himmel, N. S.
   "Contradictions in Temperature Data Spark Dispute About Possibility of Life on Mars." *Aviation Week* 91, Aug. 18, 1969, pp. 90-91.

**2264**
"New Debate Over Life on Mars." *U.S. News & World Report* 67, Aug. 18, 1969, p. 9.

**2265**
Gale, W. A., & A. C. E. Sinclair
   "The Polar Temperature of Venus." *Science* 165, Sept. 26, 1969, pp. 1356-1357.

**2266**
Hunten, Donald M., & Richard M. Goody
   "Venus: the Next Phase of Planetary Exploration." *Science* 165, Sept. 26, 1969, pp. 1317-1323.

**2267**
Wells, R. A.
   "Martian Topography: Large Scale Variations." *Science* 166, Nov. 14, 1969, pp. 862-865.

**2268**
Nikander, Jukka
   "Some Problems Posed by the Planet Venus." London, *Spaceflight* 12, April 1970, pp. 180-183.

**2269**
Seckbach, Joseph, & Willard F. Libby
   "Vegetative Life on Venus." Dordrecht, *Space Life Sciences* 2, Sept. 1970, pp. 121-143.

NOTES:

## LIFE ON INDIVIDUAL PLANETS (EXCLUDING INTELLIGENT LIFE)

2270
"NASA Studies Planetary Habitation Methods."
Aviation Week 93, Nov. 30, 1970, pp. 62-63.

2271
Soffen, G. A. & James S. Martin, Jr.
"How We'll Search for Life on Mars."
Popular Science 198, Feb. 1971, pp. 51-53.

2272
"Jet Propulsion Lab Scientists...Report Lab Experiments Conducted With Soil, Atmospheric Gases and Ultraviolet Radiation Thought to be Present on Mars Reveal Primitive Form of Life Could Exist on Planet..." New York Times, March 23, 1971, p. 17, col. 1.

2273
"Organic Production on Mars." Science News 99, March 27, 1971, p. 210.

2274
Jethon, Zbigniew
"Egzobiologia w Astronautyce." Warsaw, Technika Lotnicza i Astronautyczna 26, June 1971, pp. 4-7.

2275
Moore, Patrick
"The Approach of Mars." Astronomy and Space 1, June 1971, pp. 32-36.

2276
"Primitive Life on Mars?" Space World H-7-91, July 1971, p. 32.

2277
Pollard, William G.
"The Uniqueness of the Earth." Air University Review 22, July-Aug. 1971, pp. 34-44.

2278
Hieronymus, W. S.
"Mariner Seeks Clues to Martian Life Forms."
Aviation Week 95, Nov. 8, 1971, pp. 49-50.

2279
Heinrich, M. R.
"Solvent Effect on Enzymes; Implications for Extraterrestrial Life." (In Molecular Evolution, Prebiological and Biological. New York, Plenum Press, 1972, pp. 331-339).

2280
Kahn, F. D.
"Life in the Universe." (In The Emerging Universe. Essays on Contemporary Astronomy. Charolettesville, University of Virginia Press, 1972, pp. 71-89).

2281
Koval, I. K.
"The Martian Crest." (Tr. into English from Priroda, Moscow, #4, 1972, pp. 2-9. (NASA-TT-F-13919). (Available from NTIS: $3.00).

2282
Molton, P.
"Exobiology, Jupiter and Life." London, Spaceflight 14, June 1972, pp. 220-223.

2283
"Water on Mars Might not Support Life."
Space World I-8-104, Aug. 1972, p. 42.

2284
"Dr. C. Sagan Says Future Spacecraft That Land on Mars are Likely to Find Evidence of Some Forms of Life, Past or Present..."
New York Times, Oct. 8, 1972, p. 68, col. 3.

NOTES:

## LIFE ON INDIVIDUAL PLANETS (Excluding Intelligent Life)

**2285**
Caidin, Martin
 "The Case for Life on Mars; Excerpts from
 Destination Mars." Science Digest 73, Jan.
 1973, pp. 26-31.

**2286**
Molton, P.
 "Limitations of Terrestrial Life." London,
 Spaceflight 15, Jan. 1973, pp. 27-30.

**2287**
Ponnamperuma, Cyril, & P. Molton
 "The Prospect of Life on Jupiter."
 Dordrecht, Holland, Space Life Sciences 4,
 Jan. 1973, pp. 32-44.

**2288**
Lozina-Lozinsky, L. K., et al
 "Effect on Infusoria of Physical Conditions
 Which Simulate the Environment on the Surface
 of the Planet Mars." (In Problems of Space
 Biology 16, Feb. 1973, pp. 351-365. Tr.
 into English of book in Russian). (NASA-
 TT-F-719). Available on microfilm from NTIS:
 $6.00.

**2289**
Marov, Michael
 "Venus: How Much do we Know?" London,
 Spaceflight 15, Feb. 1973, pp. 48-50.

**2290**
Thomsen, D. E.
 "Extraterrestrial Life, if There is any,
 and us." Science News 104, July 14, 1973,
 pp. 29-31.

**2291**
 "Earth: zoo or Petri Dish?" Scientific
 American 229, Oct. 1973, p. 51.

**2292**
 "Bacteria Thrive in Jupiter-like Atmosphere."
 Science News 104, Nov. 17, 1973, p. 309.

**2293**
Bozhich, Serge P.
 "Jupiter, Planète Habitée? ou l'Enigme
 de son Emission sur Ondes Décamétriques."
 Paris, Phénomènes Spatiaux (38: 4), Dec.
 1973, pp. 3-6.

**2294**
 "NASA Biologists Discover Rare Earth Organism."
 Space World K-1, Jan. 1974, p. 33.

**2295**
 "Scientific Opinion Divided About Life on
 Jupiter." Space World K-1, Jan. 1974, pp.
 21-23.

NOTES:

## OCCULT ASPECTS, INCLUDING ESP, MENTAL TELEPATHY, PSYCHOLOGICAL AND PHYSICAL EFFECTS, ETC.

**BOOKS**

**2296**
Gaston, Henry A.
  Mars Revealed; or, Seven Days in the Spirit World. San Francisco, 1880, 208p.

**2297**
Buck, Richard M.
  Cosmic Consciousness. New York, E. P. Dutton and Co., 1901, 384p.

**2298**
Oliver, Frederick S.
  A Dweller on two Planets, or The Dividing of the way, by Phylos the Tibetan. Los Angeles, Baumgardt Publishing Co., 1905, 423p.

**2299**
Flammarion, Camille
  Mysterious Psychic Forces. Boston, Small Maynard and Co., 1907, 466p.

**2300**
Fort, Charles
  The Book of the Damned. New York, Boni & Liveright, 1919, 298p.

**2301**
Fort, Charles
  New Lands. Introduction by Booth Tarkington. New York, Boni & Liveright, 1923, 249p.

**2302**
Fort, Charles
  Lo! Introduction by Tiffany Thayer. Illustrated by Alexander King. New York, C. Kendall, 1931, 411p.

**2303**
Fort, Charles
  The Books of Charles Fort, With an Introduction by Tiffany Thayer. New York, Published for the Fortean Society by H. Holt & Co., 1941, 1125p.

**2304**
Miller, Will, & Evelyn Miller
  We of the new Dimension. Los Angeles, 195-, 115p.

**2305**
Pelley, William D.
  Star Guests...Design for Mortality. Noblesville, Ind., Soulcraft Press, 1950, 318p.

**2306**
  The Mystery of Other Worlds Revealed. Greenwich, Conn., Fawcett Pubs., 1952, 144p.

NOTES:

## OCCULT ASPECTS, INCLUDING ESP, MENTAL TELEPATHY, PSYCHOLOGICAL AND PHYSICAL EFFECTS, ETC.

**2307**
Ley, Will, et al
  The Mystery of Other Worlds Revealed. New York, Sterling, 1953, 144p. ($2.95)

**2308**
Wilkins, H. P.
  Mysteries of Space and Time. London, Frederick Muller, 1955, 208p.

**2309**
Barton, Michel X.
  Flying Saucer Revelations. Los Angeles, Futura Press, 1957, 38p.

**2310**
Herschel, W.
  On the Construction of the Heavens. (In Munitz, M. K., editor. Theories of the Universe From Babylonian Myth to Modern Science. New York, Free Press of Glencoe, c/o Free Press, 1957, pp. 264-268). ($7.95)

**2311**
Howard, Dana
  Over the Threshold. Los Angeles, Llewellyn Pubs., 1957, 140p.

**2312**
Sumner, F. W.
  The Coming Golden age. Los Angeles, New Age Pub. Co., 1957, 206p.

**2313**
King, George
  Life on the Planets. Hollywood, Calif., The Aetherius Society, 1958, 29p.

**2314**
Layne, Stan
  I Doubted Flying Saucers. Boston, Meador Press, 1958, 177p.

**2315**
Thomas, Dorothy
  Life on Mars According to the Great Mystics. Los Angeles, New Age Pub. Co., 1958, 9p.

**2316**
Williamson, George H.
  Secret Places of the Lion. Amherst, Wis., Amherst Press, 1958, 230p.

**2317**
Barton, Michael X.
  Secrets of Higher Contact. Los Angeles, Futura Press, 1959, 30p.

**2318**
Norman, Mark A.
  Many Shall be Called. El Monte, Calif., Understanding Pub. Co., 1959, 103p.

**2319**
Barton, Michael X.
  The Spacemasters Speak. Los Angeles, Futura Press, 1960, 34p.

**2320**
Barton, Michael X.
  Your Part in the Great Plan. Los Angeles, Futura Press, 1960, 30p.

**2321**
Howard, Dana
  The Keys to the Citadel of Space. Los Angeles, Llewellyn Pubs., Ltd., 1960, 203p.

**2322**
Philip, Brother .
  Secret of the Andes. Clarksburg, W. Va., Saucerian Books, 1961, 151p.

NOTES:

## OCCULT ASPECTS, INCLUDING ESP, MENTAL TELEPATHY, PSYCHOLOGICAL AND PHYSICAL EFFECTS, ETC.

2323
Herrmann, J.
  Das Falsche Weltbild; Astronomie und Aberglaube. Eine Kritische Untersuchung Uber Astrologie, Welteislehre, Hohlwelttheorie, Bewohnbarkeit der Sonne, Fliegende Untertassen und Andere Astronomische Irrlehren. Stuttgart, Franckhsche Verlags., 1962, 162p. (6.80 marks)

2324
Going Up. Practical Methods of Astral
  Projection. Azusa, Calif., Galaxy Press, 1963.

2325
King, George
  The Nine Freedoms. Los Angeles, The Aetherius Society, 1963, 200p.

2326
Mustapa, Margit
  Book of Brothers. New York, Vantage Press, 1963, 196p.

2327
King, George
  A Cosmic Message of Divine Opportunity. Hollywood, Calif., The Aetherius Society, 1964, 9p.

2328
King, George
  Join Your Ship. Hollywood, Calif., The Aetherius Society, 1964, 16p.

2329
Fry, Daniel W.
  The Curve of Development. Lakemont, Ga., CSA Printers and Publishers, 1965, 75p.

2330
Gould, Rupert R.
  Enigmas. New Hyde Park, N.Y., University Books, 1965, 248p.

2331
Halsey, Wallace C.
  Cosmic End-time Secrets. Los Angeles, Futura Press, 1965, 102p.

2332
King, George
  The day the Gods Came. Los Angeles, The Aetherius Society, 1965, 71p.

2333
The Book of Spaceships and Their Relationship
  With Earth, by the god of a Planet Near the Earth and Others. Clarksburg, W. Va., Saucerian Pubs., 1966, 70p.

2334
Carter, Joan F.
  Fourteen Footsteps From Outer Space. Dallas, Royal Pub. Co., 1966, 168p.

2335
Dewey, Mark
  A man From Space Speaks. Houston, 1966, 38p.

2336
Lee, Gloria
  The Going and the Glory. Auckland, New Zealand, Heralds of the New Age, 1966, 73p.

2337
Pritchett, E. B.
  Transcripts of '44.' Arlington, Va., Marcap Council, 1966, 66p.

2338
From Jupiter, Planet of joy. Los Angeles,
  DeVorss & Co., 1967, 15p.

2339
From Planet Pluto With Brotherly Love. Los
  Angeles, DeVorss & Co., 1967, 32p.

NOTES:

## OCCULT ASPECTS, INCLUDING ESP, MENTAL TELEPATHY, PSYCHOLOGICAL AND PHYSICAL EFFECTS, ETC.

**2340**
Invitation From the Planet Venus. Los Angeles, DeVorss & Co., 1967, 20p.

**2341**
Planet Mercury Sends Greetings. Los Angeles, DeVorss & Co., 1967, 32p.

**2342**
Uranus. Lover of man, Speaks. Los Angeles, DeVorss & Co., 1967, 29p.

**2343**
Sprinkle, R. L.
   Some Uses of Hypnosis in UFO Research. Laramie, Wyoming, 1968, 18p.

**2344**
Belle, Jack van
   Zienswijze. Op Occultisme, Meditatie, Spiritisme, Reincarnatie, UFO's. Laren, Holland, A. J. Luitingh, 1972, 208p. (9.90 florins)

**2345**
Charroux, Robert
   The Mysterious Unknown. London, Spearman, 1972, 236p.

**2346**
Godwin, John
   Occult America. Garden City, N.Y., Doubleday, 1972, 314p.

**2347**
Stranges, Frank E.
   My Friend from Beyond the Earth. Kitchener, Canada, Galaxy Press, 1972, 24p.

**2348**
Tarade, Guy
   Les Archives du Savoir Perdu. Paris, R. Laffont, 1972, 345p.

**2349**
Le Poer Trench, Brinsley
   The Eternal Subject. London, Souvenir Press, 1973, 200p.

**2350**
In Search of Ancient Mysteries. NBC Television Network Program, Jan. 31, 1974, 8-9 P.M., Eastern Daylight Time.

## PERIODICAL ARTICLES

**2351**
Stumbough, Virginia
   "The Anatomy of Mirages." Fate 6, April 1953, pp. 85-93.

**2352**
Hervey, Michael
   "The Strange Growing in Martinique." Fate 9, Sept. 1956, pp. 27-41.

**2353**
Hampton, Wade T.
   "Mystery of Hypnotic 'Ecstacy'." Fate 10, Jan. 1957, pp. 94-99.

**2354**
Laine, Juliette
   "Gullivers' two Moons on Mars." Fate 10, Aug. 1957, pp. 43-44.

**2355**
Michel, Aimé
   "Flying Saucers in Europe, Saucers--or Delusions?" Fate 11, Jan. 1958, pp. 73-79.

NOTES:

## OCCULT ASPECTS, INCLUDING ESP, MENTAL TELEPATHY, PSYCHOLOGICAL AND PHYSICAL EFFECTS, ETC.

**2356**
"The People who see 'Flying Saucers'." The UFO Investigator 1, Jan. 1958, pp. 23-24.

**2357**
"Are 'Contact Group' Sightings Metaphysical?" Flying Saucers, July-Aug. 1958, pp. 12-15, 19.

**2358**
Larson, Kenneth L.
"Charles Fort and the UFO." Flying Saucers, July 1959, pp. 20-23, 42.

**2359**
James, Trevor
"Saucers and Psychism." London, Flying Saucer Review 5, Nov.-Dec. 1959, pp. 24-27.

**2360**
James, Trevor
"Scientists, Contactees and Equilibrium." London, Flying Saucer Review 6, Jan.-Feb. 1960, pp. 19-21.

**2361**
Foght, Paul
"Guilty: the Mystery ray That Kills." Fate 14, March 1961, pp. 31-33.

**2362**
Schoenherr, Luis
"UFO's and Fourth Dimension." London, Flying Saucer Review 9, March-April 1963, pp. 10-12.

**2363**
Cox, Adrian
"Thoughts on Extended Dimensions." London, Flying Saucer Review 9, July-Aug. 1963, pp. 8-9.

**2364**
Liss, Jeffrey
"The Problem of Life on Mars." Fate 16, Aug. 1963, pp. 39-47.

**2365**
Wellman, Wade
"The Psychology of Scepticism." London, Flying Saucer Review 9, Sept.-Oct. 1963, pp. 32-34.

**2366**
Schoenherr, Luis
"UFO's and the Fourth Dimension." London, Flying Saucer Review 10, Jan.-Feb. 1964, pp. 16-20.

**2367**
Cleary-Baker, John
"The 'Psychological' Saucer." London, BUFORA Journal and Bulletin 1, Winter 1965, pp. 17-18.

**2368**
Haythornthwaite, P. K.
"Radioactivity and the UFO." London, BUFORA Journal and Bulletin 1, Summer 1965, pp. 13-14.

**2369**
Schoenherr, Luis
"UFO's and the Fourth Dimension." London, Flying Saucer Review 11, Nov.-Dec. 1965, pp. 12-13, 18.

**2370**
"Sur les Effets Biologiques des Champs Magnétiques Intenses." Paris, Phénomènes Spatiaux, Dec. 1965, pp. 8-10.

**2371**
Asimov, Isaac
"UFO's--What I Think." Science Digest 59, June 1966, pp. 44-47.

**2372**
Kor, Peter
"Perspective, Flying Saucers--Physical or Psychic?" Saucer News 13, Fall 1966, pp. 10-12.

NOTES:

## OCCULT ASPECTS, INCLUDING ESP, MENTAL TELEPATHY, PSYCHOLOGICAL AND PHYSICAL EFFECTS, ETC.

2373
"Perturbations Psychiques." Paris, Phénomènes Spatiaux, Dec. 1966, pp. 21-23.

2374
Creighton, Gordon
"Mysterious Psychological Effects of Flying Saucers." London, Flying Saucer Review 13, July-Aug. 1967, pp. 5-6.

2375
Markowitz, William
"The Physics and Metaphysics of Unidentified Flying Objects." Science 157, Sept. 15, 1967, pp. 1274-1279.

2376
Finch, Bernard E.
"Physiological Effect on Witness at Hook." London, Flying Saucer Review 13, Nov.-Dec. 1967, pp. 7, 27.

2377
Meerloo, Joost A.
"Flying Saucer Syndrome and the Need for Miracles." Journal of the American Medical Association 203, March 18, 1968, p. 170.

2378
Dick, William
"U.S. Air Force Project Uses Doctor to Prove Cop's Report of Flying Saucer by Hypnotism." National Enquirer 42, May 19, 1968, pp. 1, 3, 4.

2379
Dawson, William F.
"So What's new With Charles Fort?" Fate 230, May 1969, pp. 76-80.

2380
Kelly, Maryellen
"UFO Made me Sick, Says Housewife." Fate 230, May 1969, pp. 34-39.

2381
Bowen, Charles
"UFO's and Psychic Phenomena." London, Flying Saucer Review (15: 4), July-Aug. 1969, pp. 22-25.

2382
Chibbett, H. S. W.
"UFO's and Parapsychology." London, Flying Saucer Review, special issue #3, Sept. 1969, pp. 33-38.

2383
Sprinkle, R. L.
"Some Uses of Hypnosis in UFO Research." London, Flying Saucer Review, Sept. 1969, pp. 17-19.

2384
Creighton, Gordon
"Healing from UFO's." London, Flying Saucer Review (15: 5), Sept.-Oct. 1969, pp. 20-23.

2385
Mackay, Ivar
"UFO's and the Occult." London, Flying Saucer Review (16:4), July-Aug. 1970, pp. 27-29; (16:5), Sept.-Oct. 1970, pp. 24-25.

2386
Dick, William
"Research Says Hundreds Have Been Injured, Temporarily Blinded and Burned by UFO's." National Enquirer, Nov. 1, 1970, back page.

2387
Edwards, P. M. H.
"UFO's and ESP." London, Flying Saucer Review (16: 6), Nov.- Dec. 1970, pp. 18-20, 26.

NOTES:

## OCCULT ASPECTS, INCLUDING ESP, MENTAL TELEPATHY, PSYCHOLOGICAL AND PHYSICAL EFFECTS, ETC.

**2388**
Schwartz, Dr. Berthold E.
  "Possible UFO-Induced Temporary Paralysis."
London, *Flying Saucer Review* (17:2), March-April 1971, pp. 4-9.

**2389**
"Hoax and Hallucination: the Evidence."
  *Flying Saucers* 73, June 1971, pp. 26-27.

**2390**
"Jung and the UFO's." *Flying Saucers* 73, June 1971, pp. 10-11.

**2391**
Saunders, Alex
  "Are Saucer Sighters Hypnotized." *Search* 101, Jan. 1972, pp. 19-24.

**2392**
Saunders, Alex
  "Exploring Space With the Mind." *Flying Saucers* 76, March 1972, pp. 18-19.

**2393**
Bord, Janet
  "Are Psychic People More Likely to See UFO's?" London, *Flying Saucer Review* (18:3), May-June 1972, pp. 20-22.

**2394**
Adell, Albert, and Père Redon
  "UFO Enters and Inspects a Room." London, *Flying Saucer Review* (19:2), March-April 1973, pp. 10-13+.

**2395**
Mackay, E.A.I.
  "UFO Entities: Occult and Physical." London, *Flying Saucer Review* (19:2), March-April 1973, pp. 26-29.

**2396**
Lewin, Harold
  "Beings from Outer Space are Communicating with People on Earth Through ESP."
*National Enquirer*, Oct. 2, 1974, p.3.

NOTES:

## ORGANIZATIONS

### BOOKS

**2397**
U.S. Air Force. Air Technical Intelligence Center
 <u>Project Sign</u>. Washington, G.P.O., 1949? 46p.

**2398**
U.S. Air Force. Air Technical Intelligence Center
 <u>Project Saucer</u>. Washington, G.P.O., 1950? 22p.

**2399**
U.S. Air Force
 <u>Report on the Meetings of the Scientific Advisory Board Panel on Unidentified Flying Objects</u>. Washington, G.P.O., 1953, 30p.

**2400**
Archers' Court Research Group. <u>Biometric Analysis of the 'Flying Saucer' Photographs</u>. Archers' Court, Hastings, Sussex, 1954, 27p.

**2401**
Barabashev, G.
 <u>They Knew too Much About Flying Saucers</u>. New York, University Books, 1956.

**2402**
Congress of Scientific UFOlogists, New York, 1967
 <u>New York's First Flying Saucer Convention</u>. Reported by: Bessie J. Gibbs and Opal Smith. Winchester, Va., B. J. Gibbs, 1967, 106p.

### PERIODICAL ARTICLES

**2403**
Rougeron, C.
 "Soucoupistes et Antisoucoupistes." Paris, <u>France Illustration</u> 6, April 29, 1950, p. 422.

**2404**
"Aero Club Asks U.S. Air Force to Reopen Probe." <u>New York Times</u>, Jan. 22, 1951, p. 19, col. 6.

**2405**
"M. Berger on Societies and Publications in U.S., Devoted to Saucer News." <u>New York Times</u>, Feb. 7, 1955, p. 23, col. 5.

---

NOTES:

## ORGANIZATIONS

**2406**
Kobler, John
"He Runs Flying-saucer Headquarters."
Saturday Evening Post 228, March 10, 1956, pp. 26-27, 69, 72.

**2407**
"National Investigations Committee on Aerial Phenomena Organized to Provide 'Honest Information'." New York Times, Nov. 4, 1956, p. 48, col. 1.

**2408**
"Saucer Session for Spaceship Sighters. Interplanetary Spacecraft Convention." Life 42, May 27, 1957, pp. 117-118.

**2409**
"Committee on Space Research (COSPAR): Organization Meeting, London...Executive Committee Elected; Prof. H. C. van der Hulst President." New York Times, Nov. 16, 1958, p. 14, col. 1.

**2410**
Priestly, Lee
"Inside APRO...a Saucer fan Club." Fate 12, Jan. 1959, pp. 59-65.

**2411**
"Out-of-the-Blue Believers: Civilian Saucer Intelligence of New York." New Yorker 35, April 18, 1959, pp. 36-37.

**2412**
Szachnowski, Antoni
"The Necessity of a Global International Federation of UFO Groups." London, BUFORA Journal and Bulletin 1, Spring 1965, pp. 5-9.

**2413**
Van Sommers, T.
"These are our Ufologists." Sydney, Pix 78, July 31, 1965, p. 14.

**2414**
Buckner, H. T.
"The Flying Saucerians: a Lingering Cult."
London, New Society 9, Sept. 1965, pp. 14-16.

**2415**
"Flying Saucer Buffs Discuss Ways to Improve Communications with Objects, Convention, Los Angeles." New York Times, Feb. 7, 1966, p. 6.

**2416**
"Out of This World: Convention of the Amalgamated Flying Saucer Clubs of America." Newsweek 68, Nov. 7, 1966, p. 38.

**2417**
"...ed ora Parliamo di Allucinazioni Collettive. Centro Unico Nazionale per lo Studio del Fenomeni Ritenuti di Natura Extraterrestre." Milan, Notiziario 4, 1967, pp. 5-25.

**2418**
"I Nostri Aderenti ci Hanno Chiesto il Parere Sull' AIAS. Centro Unico Nazionale per lo Studio del Fenomeni Ritenuti di Natura Extraterrestre." Milan, Notiziario 4, 1967, pp. 1-4.

**2419**
Adamski, George
"Centro Unico Nazionale per lo Studio del Fenomeni Ritenuti di Natura Extraterrestre." Milan, Notiziario 5, 1967, pp. 2-7.

**2420**
"Gli UFO Debattuti all' ONU. Centro Unico Nazionale per lo Studio del Fenomeni Ritenuti di Natura Extraterrestre." Milan, Notiziario 6, 1967.

**2421**
"Vonkeviczky e gli UFO all'UNU. Centro Unico Nazionale per lo Studio del Fenomeni Ritenuti di Natura Extraterrestre." Milan, Notiziario 5, 1967, pp. 8-14.

NOTES:

## ORGANIZATIONS

**2422**
Vallée, Jacques
  "A Survey of French UFO Research Groups."
London, Flying Saucer Review 13, Sept.-Oct.
1967, pp. 22-24.

**2423**
Veit, Karl
  "Mainzer Weltkongres der UFO-Forscher."
Wiesbaden, UFO-Nachrichten, Dec. 1967, pp. 1, 3.

**2424**
"Prof. F. Ziegel of Moscow Aviation Institute
  Urges 'Joint Effort' of World Scientists
  to Determine Nature of UFO's..." New
  York Times, Dec. 10, 1967, p. 70, col. 1.

**2425**
"Proposal for Coordinated International
  Effort to Deal with Sightings, Report
  International Astronautical Federation
  Congress." New York Times, Oct. 16,
  1968, p. 12, col. 2.

**2426**
"Great Lakes Identified Flying Objects
  Association Meets in Ill." Chicago
  Tribune, April 12, 1973, sec. N4A,
  p. 18, col. 1.

**2427**
"Convegno Nazionale di Studi fra i Gruppi
  di Ricera Ufologica." Milan, Notiziario
  UFO 62, April-June 1974, pp. 2-5.

NOTES:

# RELIGIOUS AND PHILOSOPHICAL ASPECTS

## BOOKS

**2428**
Of the Plurality of Worlds: an Essay; Also Dialogue on the Same Subject. 2d ed. London, John W. Parker & Son, 1854, 395p.

**2429**
Brewster, D.
More Worlds Than one: the Creed of the Philosopher, and the Hope of the Christian. London, John C. Hotten, 1870, 262p.

**2430**
Charles, R. H. editor
The Book of the Secrets of Enoch. Trans. From Slavonic by W. R. Morfill. London, Oxford, 1896, (Chapters 1 & 3).

**2431**
Arrhenius, Svante
The Life of the Universe as Conceived by man From the Earliest Ages to the Present Time. Trans. by H. Borns. 2v. New York, Harper & Brothers, c/o Harper and Row, 1909.

**2432**
Sewall, Frank
Life on Other Planets as Described by Swedenborg. Philadelphia, Swedenborg Scientific Association, 1911, 20p.

**2433**
Townsend, L. T.
Stars not Inhabited; Scientific and Biblical Points of View. Nashville, Methodist Book Concern, c/o Methodist Pub. Co., 1914, 254p. ($1.00)

**2434**
Arrhenius, Svante
The Destinies of the Stars. New York, Putnams, 1918, 256p.

**2435**
Schuster, O. J.
Other Worlds. Boston, Christopher Publishing House, 1927, 104p. ($1.50)

**2436**
Swedenborg, E.
The Earths in our Solar System, Which are Called Planets, and the Earths in the Starry Heavens, Their Inhabitants and Spirits and Angels Thence, From Things Heard and Seen. Originally Published in Latin at London, A.D. 1758. Rotch Edition, With an Introduction by Garrett P. Serviss. Boston, B. A. Whittemore, 1928, 124p.

NOTES:

# RELIGIOUS AND PHILOSOPHICAL ASPECTS

2437
Inge, W. R.
   God and the Astronomers. London, Longmans, Green, 1933, 308p.

2438
Swedenborg, E.
   The Apocalypse Explained, According to the Spiritual Sense: in Which are Revealed the Arcana Which are There Predicted, and Have Been Hitherto Deeply Concealed. Trans. From a Latin Posthumous Work, of Emanuel Swedenborg. v.6. London, J. S. Hodson, 1834-40.

2439
Swedenborg, E.
   Heaven and its Wonders and Hell, From Things Heard and Seen. First Published in Latin, London, 1758. Standard Edition. New York, Swedenborg Foundation, 1938, 455p.

2440
Swedenborg, E.
   The World of Spirits and Man's State After Death. First Published in Latin in 1758, and trans. by John C. Ager. New York, Swedenborg Foundation, 1940, 112p.

2441
Grant, W. V.
   Men From the Moon in America. Dallas, Texas, 195-, 32p.

2442
Cove, Gordon
   Who Pilots the Flying Saucers? London, 1955, 80p.

2443
Fontenelle, B. Le Bovier de
   Entretiens sur la Pluralité des Mondes. Digression sur les Anciens et les Modernes. Edited by Robert Shackleton. Oxford, Clarendon Press, 1955, 218p. (30 shillings)

2444
Goff, Kenneth
   The Flying Saucers. Englewood, Colorado, 1955, 32p.

2445
Festinger, Leon
   When Prophecy Fails. Minneapolis, University of Minnesota Press, 1956, 256p.

2446
Jessup, Morris K.
   UFO and the Bible. New York, Citadel Press, 1956, 126p. ($2.50)

2447
Mascall, E. L.
   Christian Theology and Natural Science; Some Questions on Their Relations. London, Longmans, Green, 1956, 328p. ($6.00)

2448
Unger, George
   Flying Saucers: Physical and Spiritual Aspects. East Grinstead, Sussex, Eng., New Knowledge Books, 1958, 43p.

2449
Heinecken, M. J.
   God in the Space age. New York, Winston Co., c/o Holt, Rinehart & Winston, Inc., 1959, 216p. ($3.50)

2450
Asimov, Isaac
   The Wellsprings of Life. New York, Abelard-Schuman, 1960, 233p.

2451
Adamski, George
   Cosmic Philosophy. San Diego, 1961, 87p.

2452
Klotz, J. W.
   The Challenge of the Space age. St. Louis, Concordia Publishing House, 1961, 112p. ($1.00)

2453
Cheville, R. A.
   Spirituality in the Space Age. Independence, Herald House, 1962, 264p. ($2.75)

NOTES:

# RELIGIOUS AND PHILOSOPHICAL ASPECTS

**2454**
Gaspa, Pietro
  Monito All'umanita. Sassari, Italy, Arti Grafiche Editoriali S.p.A., 1962, 127p.

**2455**
Ostlin, M. T.
  Thinking out Loud About the Space age: is the Christian Faith Adequate for a Space age? Philadelphia, Dorrance, 1962, 144p. ($3.00)

**2456**
Brasington, Virginia F.
  Flying Saucers in the Bible. Clarksburg, W. Va., Saucerian Press, 1963, 78p.

**2457**
Hansen, L. T.
  He Walked the Americas. Amherst, Wis., Amherst Press, 1963, 256p.

**2458**
Dean, John W.
  Flying Saucers and the Scriptures. New York, Vantage Press, 1964, 173p.

**2459**
Drake, Walter R.
  Gods or Spacemen? Amherst, Wis., Amherst Press, 1964, 176p.

**2460**
Giret, A.
  L'Astronomie et le Sentiment Religieux. Paris, Librairie Paillard, 1964, 200p. (7F)

**2461**
Mundo, L.
  Flying Saucers and the Father's Plan. Clarksburg, Saucerian Books, 1964, 801p. ($3.00)

**2462**
Korell, Federico
  Kant und die "Fliegenden Teller." Eine Zeitgemässe Betrachtung. Calw, Wurttemberg, Schätzkammerverlag H. Fändrich, 1965, 31p. (2.70 marks)

**2463**
Oberth, Hermann
  Katechismus der Uraniden. Haben Unsere Religionen e. Zukunft. Gedanken aus Philosoph. Wiesbaden-Schierstein, Ventla-Verlag, 1966, 160p.

**2464**
  Philosophical Speculations Concerning Life on Earth and in Outer Space. Tr. into English from Priroda (Moscow) #11, 1965, pp. 88-101. (JPRS-34259; TT-66-30700). Washington, Joint Publications Research Service, Feb. 23, 1966, 27p. (Available from NTIS: $1.00).

**2465**
Sendy, Jean
  Les Dieux Nous Sont nés. Paris, Bernard Grasset, 1966, 343p.

**2466**
Dinotos, Sábado
  A Antiguidade dos Discos Voadores. São Paulo, 1967, 173p. (8 cruzeiros)

**2467**
Downing, Barry H.
  The Bible and Flying Saucers. Philadelphia, J. B. Lippincott Co., 1967, 221p.

**2468**
Greenstein, Jesse L.
  Speculation on man and the Universe. Pasadena, 1967, 11p.

NOTES:

# RELIGIOUS AND PHILOSOPHICAL ASPECTS

**2469**
Downing, Barry H.
 *The Bible and Flying Saucers*. Philadelphia, Lippincott, 1968, 221p.

**2470**
Drake, Walter
 Spacemen in the Ancient East. London, Spearman, 1968, 240p. (30 shillings)

**2471**
Dione, R.
 *God Drives a Flying Saucer*. New York, Exposition, 1969, 94p. ($5.00)

**2472**
Leonard, R.
 *Flying Saucers, Ancient Writings and the Bible*. New York, Exposition Press, 1969, 282p. ($7.50)

**2473**
Sendy, Jean
 *Ces Dieux qui Firent le Ciel et la Terre, le Roman de la Bible*. Paris, R. Laffont, 1969, 285p.

**2474**
Bourquin, Gilbert A.
 *Die Däniken-Story: Dokumente, Meinungen*. Munich, F. A. Herbig, 1970, 183p.

**2475**
Moyer, Ernest P.
 God, Man, and the UFOs. New York, Carlton Press, 1970, 422p. ($6.50)

**2476**
Jacchieri, Carlos
 Os Deuses não Eram Astronautas. São Paulo, Ed. Ciência e Progresso, 1971, 223p.

**2477**
Pereira, Fernando Cleto Nunes
 *A Bíblia e os Discos Voadores*. São Paulo Bisordi, 1971, 280p.

**2478**
Dione, R. L.
 *God Drives a Flying Saucer*. London, Corgi, 1973, 131p.

**2479**
Ellwood, Robert S.
 *Religious and Spiritual Groups in America*. Englewood Cliffs, N.J., Prentice-Hall, 1973, 334p.

**2480**
Sendy, Jean
 *The Coming of the Gods*. New York, Berkeley Pub., 1973, 237p.

**2481**
Blumrich, Josef F.
 *The Spaceships of Ezekiel*. New York, Bantam Books, 1974, 179p.

**2482**
Drake, William R.
 *Gods and Spacemen in the Ancient West*. New York, New American Library, 1974, 230p.

**2483**
Landsburg, Allan, & Sally Landsburg
 *In Search of Ancient Mysteries*. New York, Bantam Books, 1974, 197p.

**2484**
Steinhäuser, Gerhard R.
 *Jesus Christ, Heir to the Astronauts*. New York, Abelard-Schuman, 1974, 152p.

**2485**
Von Däniken, Erich
 *Gold of the Gods*. New York, Bantam, 1974, 235p.

NOTES:

RELIGIOUS AND PHILOSOPHICAL ASPECTS

## PERIODICAL ARTICLES

2486
Appel, T.
 "Plurality of Worlds." American Presbyterian Review 3, 1855, p. 572.

2487
 "Habitability of Worlds." Christian Review 20, 1855, p. 202.

2488
Leavitt, J.
 "Plurality of Worlds." Methodist Quarterly Review 15, 1855, p. 356.

2489
 "Plurality of Worlds." London, St. Paul's Magazine 3, 1868, p. 47.

2490
 "Plurality of Worlds Inhabited." London, St. Paul's Magazine 3, 1868, p. 676.

2491
Searle, G.
 "Plurality of Worlds." Catholic World 37, 1883, p. 49.

2492
Concilio, J. B. de
 "Plurality of Worlds." American Catholic Quarterly 9, 1884, p. 193.

2493
Hughes, T.
 "Plurality of Worlds." American Catholic Quarterly 9, 1884, p. 452.

2494
Burr, E. F.
 "Other Inhabited Worlds." Presbyterian Review 6, 1885, p. 257.

2495
Searle, G. M.
 "Is There a Companion World to our own?" Catholic World 55, Sept. 1892, pp. 860-878.

2496
Coupe, Charles
 "Are the Planets Inhabited?" American Catholic Quarterly 124, Oct. 1906, pp. 699-720.

2497
McColley, G.
 "Theory of a Plurality of Worlds as a Factor in Milton's Attitude Toward the Copernican Hypothesis." Modern Language Notes 47, May 1932, pp. 319-325.

2498
Cornford, F. M.
 "Innumerable Worlds in Pre-Socratic Philosophy." London, Classical Quarterly 28, Jan. 1934, pp. 1-16.

2499
McColley, G., & H. W. Miller
 "St. Bonaventure, Francis Mayron, William Vorilong and the Doctrine of a Plurality of Worlds." Speculum 12, July 1937, pp. 386-389.

2500
 "The Theology of Saucers." Time 60, Aug. 1952, p. 62.

2501
 "The Saucer Question." The Catholic Digest 16, Oct. 1952, p. 121.

NOTES:

# RELIGIOUS AND PHILOSOPHICAL ASPECTS

2502
Williamson, A. A.
"Speculation on the Cosmic Function of Life." Journal of the Washington Academy of Sciences (43:10), 1953, pp. 305-311.

2503
"Biblical Flying Saucers." Science News Letter 63, March 7, 1953, p. 148.

2504
Jack, Homer A.
"Religion and the Saucers." Fate 8, March 1955, pp. 20-23.

2505
Cassens, Kenneth H.
"UFO's and the Modern Bible." Fate 8, Aug. 1955, pp. 51-53.

2506
"Space Theology." Time 66, Sept. 19, 1955, p. 81.

2507
Moseley, James W.
"Peruvian Desert: map for Saucers?" Fate 8, Oct. 1955, pp. 28-33.

2508
Viney, Basil
"Invasion From Space." London, The Contemporary Review 188, Oct. 1955, pp. 257-260.

2509
Pittenger, Norman W.
"Christianity and the man on Mars." Christian Century 73, June 1956, pp. 747-748.

2510
"No Room for Christian Faith." Sign 36, Nov. 1956, p. 14.

2511
Webster, G.
"Life on Mars?" Catholic Digest 21, April 1957, pp. 86-89.

2512
Richards, Sam
"Some Philosophical Implications of UFO's." London, Spacelink 4, Summer 1957, pp. 4-11.

2513
"Other-Worldly Faith." Newsweek 51, March 24, 1958, p. 64.

2514
Lewis, C. S
"Faith and Outer Space." Time 71, March 31, 1958, p. 37.

2515
"Are 'Contact Group' Sightings Metaphysical?" Flying Saucers, July-Aug. 1958, pp. 12-15, 19.

2516
"Dr. Jung and the Saucers." Time 72, Aug. 11, 1958, p. 38.

2517
Carey, G. C. S.
"Lewis and Space: Biology, Ecology and Theology." Commonweal 69, Oct. 24, 1958, pp. 100-101.

2518
Casserly, J. J.
"The Church and Space Conquest." Ave Maria 89, March 21, 1959, pp. 5-8.

2519
Jung, Carl G.
"Flying Saucers." (Review by P. Wylie). Saturday Review 42, Aug. 8, 1959, p. 17.

NOTES:

# RELIGIOUS AND PHILOSOPHICAL ASPECTS

**2520**
James, Trevor
"Scientists, Contactees and Equilibrium." London, *Flying Saucer Review* 6, Jan.-Feb. 1960, pp. 19-21.

**2521**
"Missionaries to Space." *Newsweek* 55, Feb. 15, 1960, p. 90.

**2522**
Bruns, J. E.
"Cosmolatry." *Catholic World* 191, Aug. 1960, pp. 283-287.

**2523**
Kleinz, John P.
"Theology of Outer Space." *Columbia* 40, Oct. 1960, pp. 27-28, 36-37.

**2524**
Davis, C.
"The Place of Christ." London, *Clergy Review* 45, Dec. 1960, pp. 706-718.

**2525**
Easton, W. B.
"Space Travel and Space Theology." *Theology Today* 17, Jan. 1961, pp. 428-429.

**2526**
Edwards, Allan
"An Angel Unawares?" London, *Flying Saucer Review* 7, Jan.-Feb. 1961, pp. 7-10.

**2527**
Shapley, Harlow
"Riddle of god, man and Outer Space." *Coronet* 49, Feb. 1961, pp. 40-44.

**2528**
Conniff, James G.
"Who's out There: Meteorite Evidence." *Columbia* 41, Aug. 1961, pp. 3-5, 35, 37.

**2529**
McHugh, L.
"Life in Outer Space." (Interview). *Sign* 41, Dec. 1961, pp. 26-29.

**2530**
Zubek, T. J.
"Theological Questions on Space Creatures." *American Ecclesiastical Review* 145, Dec. 1961, pp. 393-399.

**2531**
"Like Nothing on Earth." *Lamp* 44, Summer 1962, pp. 16-21.

**2532**
Brandt, Ivan
"The Stumbling Block of Orthodoxy." London, *Flying Saucer Review* 8, July-Aug. 1962, pp. 26-29.

**2533**
Watson, S. H.
"The Secrets of Time and Space." (In Palmer, Ray. ed. *The Hidden World*, No. A-9. Amherst, Wis., Palmer Pubs., 1963, pp. 1536-1723).

**2534**
Lynch, J.
"Christians on Other Planets?" *Friar* 19, Jan. 1963, pp. 06-09.

**2535**
Lewis, C. S.
"Onward, Christian Spacemen." *Catholic Digest* 27, Aug. 1963, pp. 90-95.

NOTES:

## RELIGIOUS AND PHILOSOPHICAL ASPECTS

**2536**
Blaher, Damian J.
"Is Anybody There?" *Friar*, Sept. 1963, pp. 15-17.

**2537**
Dethier, V. G.
"Life on Other Planets." *Catholic World* 198, Jan. 1964, pp. 245-250.

**2538**
Conway, J.
"If There is Life on Other Planets Wouldn't There Have to be Polygenesis?" *Catholic Messenger* 82, Aug. 1964, p. 10.

**2539**
Carr, A.
"Take me to Your Leader." New York, *Homiletic and Pastoral Review* 65, Dec. 1964, pp. 255-256.

**2540**
Finch, Bernard
"The ark of the Israelites was an Electrical Machine." London, *Flying Saucer Review* 11, May-June 1965, pp. 18-19.

**2541**
Drake, Walter R.
"Space Gods of Ancient Britain." London, *Flying Saucer Review* 11, July-Aug. 1965, pp. 15-17.

**2542**
Gilman, Peter
"Do the Cherubim Come From Mars?" London, *Flying Saucer Review* 13, Sept.-Oct. 1967, pp. 19-21, 29.

**2543**
Stevens, C.
"Space Neighbors: Excerpt From *Essays in Astro-Theology*." *Family Digest* 24, Aug. 1969, pp. 20-25.

**2544**
Bylinsky, C.
"The Secrets of the Planets." *Catholic Digest* 34, March 1970, pp. 78-82.

**2545**
Guttilla, Peter
"The Godly Ones." *Flying Saucers* 72, March 1971, pp. 25-27.

**2546**
Ellwood, Robert S.
"UFO's and the Bible: a Review of the Literature." *APRO Bulletin*, Sept.-Oct. 1971.

**2547**
Stevens, C.
"God, man and Outer Space." *Liguorian* 59, Nov. 1971, pp. 13-15.

**2548**
"Theory of God as Extra-Terrestrial Visitor Presented." *Washington Post*, Feb. 20, 1972, sec. 1, p. 21, col. 1.

**2549**
Saunders, Alex
"Was Jesus Christ a Spaceman?" *Caveat Emptor* 5, Fall 1972, pp. 9, 23.

**2550**
Bord, Colin
"Angels and UFO's." London, *Flying Saucer Review* (18:5), Sept.-Oct. 1972, pp. 17-19.

**2551**
Michel, Aimé
"Of Gods, Genii, Heroes and Entities." London, *Flying Saucer Review* (19:2), March-April 1973, pp. 6-9.

**2552**
"*Crash go the Chariots*, by C. Wilson. Review." *Christianity Today* 17, Aug. 31, 1973, pp. 30-31.

NOTES:

## RELIGIOUS AND PHILOSOPHICAL ASPECTS

**2553**
Zullo, Allan
"Top NASA Scientist Says Bible Describes Spaceship Landing in Israel 2600 Years ago." National Enquirer, Feb. 3, 1974, p.45.

NOTES:

# THE SEARCH FOR LIFE

## BOOKS

**2554**
Clarke, A. C.
   The Exploration of Space.  New York, Harper, 1959, 199p.  ($4.95).

**2555**
NASA-University Conference on the Science and Technology of Space Exploration, Chicago, Nov. 1-3, 1962
   Proceedings.  v.2.  Washington, Office of Scientific and Technical Information, National Aeronautics and Space Administration, 1962; for Sale by Government Printing Office, Washington, (NAS 1.21; Item 830-I: $5.50)

**2556**
Meinschein, Warren G.
   Detecting Extraterrestrial Life.  (NASA Contract NAsw-508).  Linden, N.J., Esso Research & Eng. Co., 1963, 4p.

**2557**
Melpar, Inc.
   Detection of Extraterrestrial Life.  Method II: Optical Rotatory Dispersion.  First Quarterly Report, 20 Dec. 1962 to 19 March 1963-(NASA-CR-52073).  Falls Church, Va., 1963, 43p.  (Available from NTIS: $4.60).

**2558**
U.S.  National Aeronautics & Space Administration
   The Search for Extraterrestrial Life.  NASA EP-10; N63-20935.  Washington, 1963, 23p.

**2559**
Levin, Gilbert V., et al
   Radioisotopic Biochemical Probe for Extraterrestrial Life.  Second Annual Progress Report to NASA.  (NASA-CR-55318).  Washington, Resources Research, Inc., 26 March 1963, 152p.  (Available from NTIS: $11.50).

**2560**
Levin, Gilbert V., et al
   Gulliver, an Experiment for Extraterrestrial Life Detection and Analysis.  Presented at the COSPAR 4th International Space Science Symposium, Warsaw, 10 June, 1963.  (NASA-CR-55511).  Alexandria, Va., NTIS, 1963, 14p.  ($1.60)

**2561**
Stanford University.  School of Medicine
   The Search for Extra-terrestrial Life, by Paul Schneour.  (In Naval Reserve Research Co. 12-5, Berkeley, California.  Research Reserve Space Science Seminar, Treasure Island, San Francisco, Oct. 20-Nov. 2, 1963.  N64-14993. 1963, 12p.

**2562**
Melpar, Inc.
   Research on Detection of Extraterrestrial Life by Ultraviolet Spectrophotometry. (NASA-CR-55655).  Falls Church, Va., Jan. 1964, 60p.  (Available from NTIS: $5.60).

---

NOTES:

## THE SEARCH FOR LIFE

**2563**
Quimby, Freeman H.
  Concepts for Detection of Extraterrestrial Life. Washington, Scientific and Technical Information Division, National Aeronautics and Space Administration; for sale by the Superintendent of Documents, U.S. Govt. Print. Office, 1964, 53p. (NASA SP-56)

**2564**
Quimby, F. H.
  Some Criteria of Living Systems Useful in the Search for Extraterrestrial Life. (In Society for Industrial Microbiology. Developments in Industrial Microbiology 5, Washington, American Institute of Biological Sciences, 1964, pp. 224-234).

**2565**
Blei, Ira, & J.W. Liskowitz
  Review of Concepts and Investigations for the use of Optical Rotation as a Means of Detecting Extraterrestrial Life. Presented at 5th International Space Symposium (COSPAR), Florence, Italy, 8-20 May 1964. Falls Church, Va., Melpar, Inc., 1964, 15p. (Available from NTIS)

**2566**
Florkin, Marcel, ed.
  Life Sciences and Space Research, Vol. 3. International Space Science Symposium. 5th, Florence, Italy, May 12-16, 1964. New York, Wiley, 1965, 258p.

**2567**
Kok, B.
  A Study of the Feasibility of Detecting Extraterrestrial Life Based on the Exchange Between Water and Oxyanions and a Study of Energy Exchange in Autotrophic Life. Final Report. (NASA-CR-67924). Alexandria, Va., NTIS, 1965, 73p. ($3.00)

**2568**
Mutschall, Vladimir
  Soviet Long-Range Space Exploration Program. Surveys of Communist World Scientific and Technical Literature. (ATD-65-94). Washington, Library of Congress, Aerospace Technology Div., 1965, 37p.

**2569**
U.S. National Aeronautics & Space Administration
  Significant Achievements During 6 Years of Space Bioscience Research and Applications, 1958-1964. (NASA-TM-X-57051). Washington, 1965, 157p. ($5.00) Available from NTIS.

**2570**
Young, R. S., et al
  An Analysis of the Extraterrestrial Life Detection Problem. Washington, Scientific and Technical Information Division, National Aeronautics and Space Administration; for sale by the Clearinghouse for Federal Scientific and Technical Information, Springfield, Va., 1965, 33p. (NASA SP-75)

**2571**
Kiely, John R., ed.
  Sampler: Stanford Advanced Mars Project for Life Detection, Exploration and Research. Stanford, Calif., School of Engineering, Stanford University, June 1965, 543p.

**2572**
Stanford University. Design Div.
  Multivator Design Studies. (NASA-CR-70723). Alexandria, Va., NTIS, Sept. 1965, 115p. ($4.00)

**2573**
Heintze, Carl
  Search Among the Stars. Princeton, Van Nostrand, 1966, 175p.

**2574**
Holmes, David C.
  The Search for Life on Other Worlds. New York, Sterling Pub. Co., 1966, 240p.

**2575**
Horowitz, Norman H.
  The Biological Significance of the Search for Extraterrestrial Life. (NASA-CR-77550; JPL-TR-32-1000). Presented at the American Astronautical Society, Anaheim, Calif., 23 May, 1966. Pasadena, Jet Propulsion Lab, California Institute of Technology, 1966, 15p. (Available from NTIS: $1.00).

NOTES:

**2576**
Melpar, Inc.
   Research and Development of an Instrument for Detection of Extraterrestrial Life by Optical Rotary Dispersion. Final Technical Report. (NASA-CR-423). Washington, NASA, 1966, 99p. ($1.45) (Available from NTIS)

**2577**
Tzonis, Konstantin
   Exobiological Studies of Interplanetary Space and Upper Atmospheric Layers. (Tr. into English of a Russian paper presented at the International Astronautical Federation, 16th International Astronautical Congress, Athens, 13-18 September, 1965. (NASA-TT-F-10055). Alexandria, Va., NTIS, 1966, 9p. ($1.00)

**2578**
U.S. National Academy of Sciences-National Research Council
   Planetary and Lunar Exploration. (In its Space Research. Directions for the Future. (NAS-NRC-1403). Washington, 1966, pp. 1-144).

**2579**
Usdin, Earl, & George R. Perez
   Life Detection Subsystem. Progress Report no. 1. (NASA-CR-77032). Springfield, Va., NTIS, 1966, 55p. (50¢)

**2580**
Imshenetskii, A.A.
   A Comparative Evaluation of the Methods for Search for Life on Mars; Paper. COSPAR Plenary Meeting, 11th, Tokyo, Japan, May 9-21, 1968. Paris, COSPAR, 14p.

**2581**
Reynolds, Orr E.
   Biosatellite II Mission; Paper. COSPAR Plenary Meeting, 11th, Tokyo, Japan, May 9-21, 1968. Paris, COSPAR, 19p.

**2582**
Freundlich, M. M., & Bernard M. Wagner, eds.
   Exobiology: the Search for Extraterrestrial Life. American Astronautical Society and American Association for the Advancement of Science Symposium, New York, N.Y., Dec. 30, 1967. Proceedings. Tarzana, Calif., American Astronautical Society (AAS Science and Technology Series, Vol. 19), 1969, 183p. ($9.75)

**2583**
Martin Marietta Corp. Research Institute for Advanced Studies
   Extraterrestrial Life Detection by Enzymatically Induced Exchange of O-18. Annual Report, 15 May, 1968-15 May 1969. Baltimore, 1969, 10p. (NASA-CR-106454). Available from NTIS.

**2584**
Merek, E.L., & V.I. Oyama
   Integration of Experiments for the Detection of Biological Activity in Extraterrestrial Exploration. Paper. Paris, COSPAR, 1969, 16p. (Delivered at COSPAR Plenary Meeting. 12th, Prague, May 11-24, 1969).

**2585**
U.S. National Academy of Sciences-National Research Council. Space Sciences Board
   The Outer Solar System: a Program for Exploration. Sponsored by NASA. (NASA-Cr-107324). Washington, June 1969, 90p. Available from NTIS.

**2586**
Illinois Institute of Technology. Research Institute. Astro Sciences Center
   Orbital Imagery for Planetary Exploration. Vol. 2: Definitions of Scientific Objectives, ed. by D.A. Klopp. (NASA-Cr-73451). Chicago, Aug. 1969, 175p. Available on microfilm from NTIS.

**2587**
Bendix Aviation Corp. Aerospace Systems Div.
   An Analysis of Carbon 14 Radiation Detection Systems, by K. Wainio. (NASA-Cr-73384; BSR-2781). Ann Arbor, Mich., Oct. 1969, 92p. (Available on microfilm from NTIS).

NOTES:

# THE SEARCH FOR LIFE

**2588**
Perkins-Elmer Corp. Aerospace Systems
  Design and Construction of Heated Ion Source for Organic Analysis and Design and Construction of Double Focusing Mass Spectrometer Analyzer. Final Report. (NASA-CR-110150). Prepared for Jet Propulsion Lab. Pomona, Calif., Dec. 1969, 20p. Available from NTIS on microfilm.

**2589**
California Institute of Technology. Jet Propulsion Laboratory
  Earth-Based Research on the Outer Planets During the Period 1970-1985, by R.L. Newburn, Jr., et al. (NASA-CR-109250; JPL-Tr-32-1456). Pasadena, March 15, 1970, 32p. Available from NTIS on microfilm.

**2590**
California Institute of Technology. Jet Propulsion Laboratory
  Scientific Objectives for Imaging Experiments at the Outer Planets, by T.H. Reilly. (NASA-CR-109871; JPL-TM-33-454). Pasadena, June 15, 1970, 13p. Available on microfilm from NTIS.

**2591**
Archer, J. L., & A. J. O'Donnell
  The Scientific Exploration of Near Stellar Systems. Paper Presented at American Astronautical Society, 17th Annual Meeting, Seattle, Wash., June 28-30, 1971; Paper AAS-71-166. Tarzana, Calif., 1971, 35p.

**2592**
Kozlov, B.L.
  Future Prospects for use of TV Microscopy to Detect Extraterrestrial Life. NASA-TT-F-13733. (Tr. into English from Russian Report Pr-12, 1969). Washington, NASA, 1971, 8p. (Available on microfilm from NTIS).

**2593**
Oliver, Bernard M.
  Project Cyclops: a Design Study of a System for Detecting Extraterrestrial Intelligent Life. NASA-TM-X-68634. Moffett Field, Calif., NASA-Ames Research Center, 1971, 252p. ($14.75) (Available from NTIS).

**2594**
U.S. National Aeronautics & Space Administration
  Planetary Atmosphere Experiments Test (PAET). Washington, 1971, 23p. (NASA News Release-71-99. Available from NASA Scientific & Technical Information Facility, Box 33, College Park, Md. 20740).

**2595**
Biospherics, Inc.
  Automated Microbial Metabolism Laboratory. Annual Report 1972. Rockville, Md., 1972, 170p. (NASA-CR-129508). Available from NTIS: $10.50.

**2596**
Imshenetsky, A.A., ed.
  Extraterrestrial Life and its Detection Methods. NASA-TT-F-710. Washington, NASA, 1972, 262p. ($3.00) (Available from NTIS).

**2597**
Mariner Mars 1971 Project. V. 2: Preliminary Space Results. NASA-Cr-125548; JPL-TR-32-1550-Vol. 2. 1972, 73p. (Available from NTIS).

**2598**
Project Cyclops: a Design Study of a System for Detecting Extraterrestrial Intelligent Life. (Sponsored by NASA at Stanford University). NASA-CR-114445. Alexandria, Va., 1972, 253p. ($14.75). Available from NTIS.

**2599**
A Survey of Life-Detection Experiments for Mars. (NASA-TM-X-54946). Alexandria, Va., NTIS, Aug 1973, 77p, ($3.00).

NOTES:

THE SEARCH FOR LIFE

## PERIODICAL ARTICLES

2600
"The Theory of Life on Mars is Dealt a Blow by Spectra." Science Newsletter 34, July 16, 1938, p. 36.

2601
"U.S. Air Force Ends Probe, Discounts Reports." New York Times, Dec. 28, 1949, p. 8, col. 4.

2602
Fontes, Olavo T.
"Project Argus and the 'Anonymous' Satellite." Flying Saucers, Oct. 1959, pp. 8-12.

2603
"Dr. Lederberg Reports Device Being Developed at Stanford University Medical School to Determine if Life Exists by Microscopic Analysis of Surface Matter Radioed Back to Earth." New York Times, Jan. 14, 1960, p. 6, col. 4.

2604
Lederberg, Joshua
"The Search for Life Beyond the Earth." London, New Scientist (7: 170), Feb. 18, 1960, p. 386.

2605
"J. B. Edson Article on Exploring Possible Existence of Earthlike Planets by Means of Radio Astronomy." New York Times, March 13, 1960, sec. VI, p. 31.

2606
Price, George R.
"U.S. Begins Search for Beings in Other Worlds." Popular Science 176, April 1960, pp. 66-69, 209.

2607
"Project Ozma, Designed to try to Pick up Intelligible Signals That Might Have Been Sent From Other Worlds, set for 1 Month of Observations With 85-Foot Instrument, National Radio Astronomy Observatory, Green Bank, W. Va." New York Times, April 3, 1960, p. 30, col. 1; April 5, 1960, p. 13, col. 2.

2608
"Project Ozma." Time 75, April 18, 1960, p. 53.

2609
Dyson, F. J.
"The Search for Artificial Stellar Sources of Infrared Radiation." Science 131, June 3, 1960, p. 1667.

2610
Bradbury, Ray
"A Serious Search for Weird Worlds." Life 49, Oct. 1960, pp. 116-118, 120, 123-124, 126-128, 130.

2611
Briggs, Michael H.
"New Evidence on Martian Life." London, Spaceflight (2:8), Oct. 1960, pp. 237-238, 259.

2612
Bradbury, R.
"The Serious Search for Weird Worlds; Project Ozma." Life 49, Oct. 24, 1960, pp. 116-118.

2613
"Ozmology." Scientific American 203, Nov. 1960, p. 97.

2614
Parks, R. J.
"The U.S. Planetary Exploration Program." Astronautics (ARS Journal) 6, Jan.-June 1961, pp. 22-24, 92-96.

NOTES:

# THE SEARCH FOR LIFE

**2615**
Aho, Arthur C.
"Magnetic UFO Detector." Flying Saucers From Other Worlds, Feb. 1961, pp. 25-26.

**2616**
Drake, Frank D.
"Project Ozma." Physics Today (14:4), April 1961, pp. 40-46.

**2617**
"Project Ozma off." Senior Scholastic, Teacher's Edition 78, April 26, 1961, p. 20.

**2618**
"Exploring Venus by Radar." London, Spaceflight 3, Sept. 1961, pp. 180-182.

**2619**
"Project Ozma and Other Efforts to Determine if Life Exists on Other Worlds Discussed." New York Times, Oct. 8, 1961, sec. XII, p. 9, col. 1.

**2620**
Sharp, Peter F.
"The Search for Life Beyond the Earth." London, Flying Saucer Review 7, Nov.-Dec. 1961, pp. 12-15.

**2621**
"Search for Civilizations." Science Newsletter 80, Dec. 23, 1961, p. 414.

**2622**
Brewer, Fred
"What's up There?" E. Hartford, Bee-Hive 37, Jan. 1962, pp. 2-7.

**2623**
Pettengill, G. H., et al
"A Radar Investigation of Venus." Astronomical Journal 67, May 1962, pp. 181-190.

**2624**
Levin, Gilbert V., et al
"Gulliver--a Quest for Life on Mars." Science 138, Oct. 1962, pp. 114-121.

**2625**
Beller, W.
"Mariner B to Test for Life on Mars." Missiles and Rockets (11:15), Oct. 8, 1962, pp. 24-25.

**2626**
"Tiny Instrument to see if Life on Mars." Science Newsletter 82, Oct. 27, 1962, p. 272.

**2627**
Berrill, N. J.
"Our Gamble in Space: the Search for Life." Atlantic (212:2), 1963, pp. 135-150.

**2628**
Drake, F. D.
"Project Ozma." (In Cameron, A. G. W., ed. Interstellar Communication. New York, Benjamin, 1963, pp. 176-177). ($8.50)

**2629**
Jastrow, R., & H. Newell
"Why Land on the Moon?" Atlantic (212:2), 1963, pp. 41-45.

**2630**
"Device (Biotelescanner) Unveiled to Relay Data on Life in Space." Air Force Times 23, Feb. 16, 1963, p. 3.

**2631**
Moore, P.
"Hopes of Finding Life on Venus Fade." London, New Scientist 17, Feb. 28, 1963, p. 479.

NOTES:

**2632**
Mayer, Cornell H.
"Radio Observations of the Moon and Planets." (In International Union of Geodesy and Geophysics. 13th General Assembly, U.S. National Report, 1960-1963. American Geophysical Union: Transactions, Vol. 44. Washington, June 1963, pp. 457-461).

**2633**
Pursglove, S. D.
"Out of This World Life Detectors: Gulliver Astrorobot." Popular Mechanics 119, June 1963, pp. 72-75.

**2634**
"Harvard University Scientist, C. Z. Sagan, Says Astronomer on Mars Would Probably be Unable to Prove Existence of Life on Earth Using Most Advanced Equipment Known Today." New York Times, June 10, 1963, p. 13, col. 2.

**2635**
Briggs, Michael H.
"Automated Life-Detection Devices." London, Spaceflight 5, July 1963, pp. 128-133.

**2636**
"NASA to Send Sampling Device to Mars in 1966 to Find out if Life Exists There." New York Times, Aug. 11, 1963, sec. IV, p. 9, col. 6.

**2637**
"H. E. Newell Discusses NASA Plans to Probe Mars for Signs of Life." New York Times, Aug. 13, 1963, p. 11, col. 1.

**2638**
"Life Detector: Multivator." Time 82, Aug. 30, 1963, p. 52.

**2639**
Meinschein, W. G.
"Detecting Extraterrestrial Life." Industrial Research (5:8), Sept. 1963, pp. 20-24.

**2640**
"Life Beyond Earth Sought: Project Ozma." Science Newsletter 84, Sept. 14, 1963, p. 166.

**2641**
"Life Detectors: Wolftrap and Multivator." Newsweek 62, Sept. 30, 1963, p. 56.

**2642**
Tello, Luis R.
"Necesidad Ecológica de la Exploración del Espacio." Buenos Aires, Revista Nacional Aeronáutica y Espacial 23, Oct. 1963, pp. 20-21.

**2643**
Tufty, B.
"8 Planet-probe Robots." Science Newsletter 84, Oct. 12, 1963, p. 227.

**2644**
Imshenetskii, A. A.
The Possiblity of Existence and Methods of Detecting Extraterrestrial Life. (In Problemy Kosmicheskoy Biologii 1, Moscow, Akademia Nauk SSR, 1962, pp. 153-160). English Trans. by National Aeronautics & Space Administration, Washington, Nov. 1963: NASA-TT-F-164; N64-11665.

**2645**
Brown, A. H.
The Search for Extraterrestrial Life. (In Kaufman, William C., ed. Bioastronautics: Fundamental and Practical Problems: Proceedings ...the American Association for the Advancement of Science, Annual Meeting, 130th, Cleveland, Ohio, December 30, 1963. Advances in Astronautical Sciences: 17. North Hollywood, Western Periodicals Co., 1964, pp. 41-60).

**2646**
Quimby, Freeman H.
"Some Criteria of Living Systems Useful in the Search for Extraterrestrial Life." Developments in Industrial Microbiology 5, 1964, pp. 224-234.

NOTES:

2647
"Search for Life in Space." Science Newsletter 85, Jan. 18, 1964, p. 46.

2648
"Philco Labs Developing Protein Detector to Probe for Life on Mars." Missiles and Rockets 14, March 2, 1964, pp. 26-27.

2649
Le Croissette, Dennis H.
"Systems Constraints on the Search for Extraterrestrial Life." (In U.S. Air Force Academy. Proceedings of the First Annual Rocky Mountain Bioengineering Symposium, Held at the Academy, 4-5 May, 1964 (AD-450818). Colorado Springs, Colo., U.S. Air Force Academy, 1964, pp. 84-87.

2650
Stuart, Jerry L.
"Instrumentation Required for Life Detection Systems." (In Annual Rocky Mountain Bioengineering Symposium. First, U.S. Air Force Academy, Colorado Springs, Colorado, May 4-5, 1964. Proceedings. Ed. by Grover J. D. Schock. Colorado Springs, U.S. Air Force Academy, 1964, pp. 88-97).

2651
Weston, Charles R.
"Principles of Optical Measurements Applied to Biological Growth in the Wolf-trap." (In U.S. Air Force Academy. Proceedings of the First Annual Rocky Mountain Bioengineering Symposium, Held at the Academy, 4-5 May, 1964 (AD-450818). Colorado Springs, Colo., U.S. Air Force Academy, 1964, pp. 99-109.)

2652
"American Biological Sciences Institute Publication, Bioscience, Urges NASA Land Life-Detecting Instruments on Mars by 1969." New York Times, June 11, 1964, p. 2, col. 3.

2653
"Looking for Martians (Design for an Automated Biological Laboratory, ABL)." Flight International 85, June 18, 1964, p. 1042.

2654
"Aeronutronic Studying Automated bio lab: Search for Life on Mars." Aviation Week and Space Technology 81, Sept. 28, 1964, p. 83.

2655
Boton, E. A.
"An Instrumented Search for Extraterrestrial Life." Dordrecht, Holland, Space Science Reviews 3, Dec. 1964, p. 715.

2656
Blei, Ira, & J. W. Liskowitz
"Review of Concepts and Investigations for the use of Optical Rotation as a Means of Detecting Extraterrestrial Life." (In Life Sciences and Space Research, Vol. 3. International Space Science Symposium. 5th, Florence, Italy, May 12-16, 1964. Ed. by Marcel Florkin. New York, Wiley, 1965, pp. 86-94).

2657
Sall, T.
"Some Considerations Concerning Detection of Extraterrestrial Life." Annals of the New York Academy of Science 134, Part 1, 1965, pp. 452-453.

2658
Botan, E. A., et al
"Study of an Instrumented Analytical System for Extraterrestrial Study of Atmospheres, Possible Life Forms and Soils." Aerospace Medicine 36, Jan. 1965, pp. 21-35.

2659
Corliss, William R.
"Detecting Life in Space." International Science & Technology, Jan. 1965, pp. 28-34.

NOTES:

**2660**
Pay, R.
 "Boosters Pace Mars Life Detection." Missiles and Rockets 16, Jan. 18, 1965, p. 32.

**2661**
"Dr. Newell Defends Search for Life on Mars, House of Representatives Committee Hearing." New York Times, Feb. 20, 1965, p. 10, col. 6.

**2662**
"Dr. Newell Says Unmanned Shots to Look for Life on Mars are Planned for 1971 and 1973, Senate Committee." New York Times, March 24, 1965, p. 21, col. 8.

**2663**
"Mariner IV Could not see Life Forms on Mars." Science Newsletter 87, April 17, 1965, p. 245.

**2664**
"Panel Appointed by National Sciences Academy at Government Request Urges Effort to Land Automated Biological Lab on Mars by 1971 to Test for Presence of Life...Finds it Reasonable That Mars is Inhabited With Living Organisms..." New York Times, April 27, 1965, p. 1, col. 4.

**2665**
"Search for Martian Life." Time 85, May 7, 1965, p. 78.

**2666**
Spencer, S. M.
 "Is Anybody out There? Mars-bound Mariner IV." Saturday Evening Post 238, June 19, 1965, pp. 44-46.

**2667**
Neumann, Temple W.
 "The Automated Biological Laboratory, its Scientific and Engineering Objectives." Space World B-17, 21, July 1965, pp. 16-21.

**2668**
Lederberg, Joshua
 "Signs of Life: Criterion System of Exobiology." London, Nature 207, July 3, 1965, pp. 9-13.

**2669**
"W. Sullivan Article on Past and Planned Efforts to Find Life on Mars." New York Times, July 11, 1965, sec. VI, pp. 1, 12.

**2670**
Oyama, Vance I.
 "A Future Automated Biological Laboratory" (In VPI Proceedings of the Conference on the Exploration of Mars and Venus). Alexandria, Va., NTIS, Aug. 1965, 14p. ($7.00)

**2671**
Lovelock, J. E.
 "A Physical Basis for Life Detection Experiments." London, Nature 207, Aug. 7, 1965, pp. 568-570.

**2672**
Dmitriyev, A.L.
 "The Universe Under the Microscope." (Tr. into English from Vokrut Sveta (USSR), #9, 1963, pp. 37-39. (FTD-TT-65-1341/1 & 2 & 4; AD-627119). Alexandria, Va., NTIS, 3 Jan. 1966, 14p. ($1.00)

**2673**
Lafonta, Paul
 "Détecteurs Magnétiques." Paris, Phénomènes Spatiaux, March 1966, pp. 12-16.

**2674**
Thomson, J. R., & K. W. Dockter
 "Finding Life in Outer Space." Today's Health 44, April 1966, pp. 42-45.

**2675**
"Study Mars With Photos." Science Newsletter 89, April 2, 1966, p. 215.

NOTES:

## THE SEARCH FOR LIFE

2676
"Dr. N. Horowitz of Jet Propulsion Lab Says Increasing Evidence of Hostile Environment on Mars Should not Deter Search for Life. . .Symposium on Search for Extraterrestrial Life, Sponsored by American Astronautical Society." New York Times, May 25, 1966, p. 22, col. 1.

2677
Wilks, Willard, & Rex Pay
"Quest for Martian Life Re-emphasized." Technology Week 18, June 1966, pp. 26-28.

2678
"New Method may Detect Martian Life." Science Newsletter 89, June 4, 1966, p. 435.

2679
Goddard, Jimmy
"What Mariner IV saw." Flying Saucers, Aug. 1966, pp. 12-17.

2680
Pryor, H.
"Search for Extraterrestrial Life: Report on Symposium, Anaheim, California." Science Digest 60, Aug. 1966, pp. 28-36.

2681
Normyle, W. J.
"NASA Report to Congress Details Planetary Alternatives for Future: Emphasis on Venus." Aviation Week and Space Technology 85, Sept. 19, 1966, p. 91.

1001
Graf, E. R., et al
"A new Criterion in the Quest for Life in our Solar System." (In Saturn V-Apollo and Beyond: National Symposium, Huntsville, Ala., June 11-14, 1967. Transactions 4, Ed. by S. S. Hu, Tarzana, Calif., American Astronautical Society, 1967, 6 pp.).

2683
Kok, Bessel, & Joseph E. Varner
"Extraterrestrial Life Detection by Means of Isotopic Oxygen Exchange." (In Life Sciences and Space Research V. International Space Science Symposium. 7th, Vienna, May 10-18, 1966. Papers. Ed. by A. H. Brown and F. G. Favorite. Amsterdam, North-Holland Pub., 1967, pp. 200-216).

2684
Rea, D. G.
"The Role of Infrared Spectroscopy in the Biological Exploration of Mars." (In International Symposium on Basic Environmental Problems of Man in Space. 2d, Paris, France, June 14-18, 1965. Proceedings. Ed. by H. Bjurstedt. Vienna, Springer, 1967, pp. 506-531).

2685
Smith, W. B.
"Search for Extraterrestrial Life...Address by N. H. Horowitz, Oct. 13, 1965, Reply." Science 155, Feb. 17, 1967, p. 852.

2686
Hitchcock, Dian R., et al
"Detecting Planetary Life From Earth." London, Science Journal 3, April 1967, pp. 56-67.

2687
"NASA Tests Boost Jupiter Life Possibility." Technology Week 20, April 24, 1967, p. 28.

2688
Lamb, Peter
"Pictures From Outer Space." London, Spacelink 4, Summer 1967, pp. 13-17.

2689
Tompkins, Daniel N.
"Mars Biological Laboratory." (In Instrument Society of America. National Aerospace Instrumentation Symposium. 13th, San Diego, Calif., June 13-16, 1967. Proceedings. Pittsburgh, 1967, pp. 297-308).

NOTES:

**2690**
Cortright, Edgar M.
"The Voyager Program." London, *Spaceflight* 9, July 1967, pp. 222-227.

**2691**
Sloan, Richard K.
"Life Detection by Visual Imaging." *Journal of the Astronautical Sciences* 14, Sept.-Oct. 1967, pp. 218-224.

**2692**
Hitchcock, Dian R., & James E. Lovelock
"Life Detection by Atmospheric Analysis." *Icarus* 7, Sept. 1967, pp. 149-159.

**2693**
"Lust for Life: Search for Life on Other Worlds." *Science News* 92, Sept. 30, 1967, p. 320.

**2694**
"Better Life Spies for Space." *Science Digest* 62, Dec. 1967, p. 37.

**2695**
Sponsler, G. C.
"The Automated Biological Laboratory, a Proposal." (In International Symposium on Space Technology and Science, 7th, Tokyo, Japan, May 15-22, 1967. *Proceedings*. Ed. by Y. Kuroda. Tokyo, Agne Pub., 1968, pp. 811-818).

**2696**
Hill, Arthur
"Methods to Detect Planet Life Detailed at AAS Meeting; Automated Microbian Metabolism Laboratory." *Aerospace Technology* 21, Jan. 15, 1968, pp. 45-46.

**2697**
Lloyd, Dan
"UFO Detector Network in the United Kingdom." London, *Flying Saucer Review* 14, March-April 1968, pp. 27-28.

**2698**
Merek, Edward L., & Vance I. Oyama
"Analysis of Methods for Growth Detection in the Search for Extraterrestrial Life." *Applied Microbiology* 16, May 1968, pp. 724-731.

**2699**
Gazenko, O. G.
"Space Research and the Biological Sciences." (In *Soviet Report Delivered at the UN Space Conference, New York, June 1968*. Washington, Oct. 9, 1968, pp. 1-12. Tr. from Russian by Joint Publications Research Service. (JPRS-46630: $3.00).

**2700**
Imshenetsky, A.A.
"Problems in Detection of Extraterrestrial Life" (In *Bioastronautics and the Exploration of Space*. Brooks Air Force Base, Texas, School of Aerospace Medicine, Nov. 1968, pp. 563-568). (AD-687893). Available from NTIS.

**2701**
Getmantsev, G.G. et al
"On Extra Atmospheric Radioastronomical Investigations." Tr. into English from *Vestn. Akad. Nauk. SSR* (Moscow), V. 4, 1969, pp. 55-60. (NASA-CR-103444; ST-RA-SP-IGA-10842). Greenbelt, Md, NASA Goddard Space Flight Center, 1969, 10p. (Available from NTIS).

**2702**
Klein, Harold P.
"Lookout for Life." *Space World* F-2-62, Feb. 1969, pp. 42-43.

**2703**
Watson, Charles E.
"An Analytic Evaluation of Interstellar Exploration." *Flying Saucers* 62, Feb. 1969, pp. 28-30.

**2704**
"What's up There in Space?" *Space World* F-2-62, Feb. 1969, pp. 40-43.

NOTES:

**2705**
"Flyby Hints Life." *Christian Science Monitor*, March 19, 1969, p. 8, col. 1.

**2706**
Larkin, L. S.
"Three UFO Detectors That Really Work." *Flying Saucers* 63, April 1969, pp. 13-15.

**2707**
Halpern, B.
"Optical Activity for Exobiology and the Exploration of Mars." *Applied Optics* 8, July 1969, pp. 1349-1353.

**2708**
Lederberg, Joshua
"Mars Through a Crystal Ball." *Applied Optics* 8, July 1969, pp. 1269-1270.

**2709**
Rea, D. G.
"Exobiology and the Exploration of Mars." *Applied Optics* 8, July 1969, pp. 1267-1269.

**2710**
Young, R. S.
"Extraterrestrial Life Detection." *Applied Optics* 8, July 1969, pp. 1355-1360.

**2711**
"Jet Propulsion Lab Reports Preliminary Analysis of Mariner 6 Data has Shown Mars to be Unlike Earth and Less Hospitable to Life Than Previously Believed." *New York Times*, Aug. 3, 1969, p. 1, col. 1.

**2712**
"The Venus Probes, Venera-5 and Venera-6." *Space World* F-10-70, Oct. 1969, pp. 44-49.

**2713**
Leighton, Robert B., et al
"Mariner 6 and 7 Television Pictures: Preliminary Analysis." *Science* 166, Oct. 3, 1969, pp. 49-67.

**2714**
Lucero, Daniel P., et al
"A Hydrogen Flame Ionization Detector for Martian Lunar Life Detection Experiments." (In Instrument Society of America. International Aerospace Instrumentation Symposium. 16th, Seattle, Wash., May 11-13, 1970. *Instrumentation in the Aerospace Industry, v. 16. Proceedings*. Ed. by B. Washburn. Pittsburgh, 1970, pp. 176-186).

**2715**
Troitskii, V.
"The Search for Extraterrestrial Life." Tr. into English from *Aviat. Kosmonaut.*, USSR, #8, 1968, pp. 77-80. (AD-713920; FSTC-HT-23-009-70). Washington, 1970, 11p. (Available on microfilm from NTIS).

**2716**
Hord, C. W., et al
"Ultraviolet Spectroscopy Experiment for Mariner Mars 1971." *Icarus* 12, Jan. 1970, pp. 63-77.

**2717**
"Dr. R. Jastrow Article Holds...Exploration of Mars in Next 30 Years May Help Scientists Solve Mystery of Origin of Life and Ascertain Whether Extraterrestrial Life Could Exist..." *New York Times*, May 10, 1970, sec. VI, p. 30.

**2718**
Laprade, Armand
"Did NASA Probe the Moons of Mars?" *Flying Saucers* 69, June 1970, pp. 23-25.

---

NOTES:

**2719**
California Institute of Technology. Jet Propulsion Laboratory
"Mariner Mars 1971 Project." (In its Space Programs Summary #37-64. Vol. 1, May-June 1970. Pasadena, July 31, 1970, pp. 1-29). (NASA-CR-113720; JPL-SPS-37-64-Vol. 1). Available on microfilm from NTIS.

**2720**
Ponnamperuma, Cyril, & Harold P. Klein
"The Coming Search for Life on Mars." Quarterly Review of Biology 45, Sept. 1970, pp. 235-258.

**2721**
Imshenetskii, A. A., et al
"Methods of Searching for Extraterrestrial Life." (In Life Sciences and Space Research IX. COSPAR Plenary Meeting. 13th, Open Meeting of Working Group 5, Leningrad, USSR, May 20-29, 1970. Proceedings. Ed. by Wolf Vishniac. Berlin, Akademie-Verlag, 1971, pp. 147-151).

**2722**
Schurmeier, M. H.
"The 1969 Mariner View of Mars." (In Astronautical Research 1970. Proceedings of the Internautical Federation. 21st, Konstanz, West Germany, Oct. 4-10, 1970. Amsterdam, North-Holland Pub., 1971, pp. 193-210).

**2723**
"U.S. Federal Government Proposes Plan to Build Mammoth Radio Telescope, Which Would be World's Largest Scientific Instrumentation Plains of San Agustin, Near Datil, New Mexico...Scientists Hope Observatory Will help Measure Physical Processes...and Perhaps Discover Presence of Life Elsewhere in Universe." New York Times, March 16, 1971, p. 24, col. 3.

**2724**
"J. N. Wilford Reports USSR Plans to Land a Robot-Like Detection Lab on Mars, Possibly by Late '73...Designed to Collect and Analyze Samples of Martian Soil and Water...There is Possibility That Life Exists on Mars." New York Times, April 16, 1971, sec. IV, p. 12, col. 2.

**2725**
"...Unmanned U.S. Mariner 9 Flight to Mars May Have Been Most Fruitful Scientific Experiment Ever Conducted in Space... Asserts There is Possibility That Life Exists on Mars." New York Times, April 16, 1971, sec. IV, p. 12, col. 2.

**2726**
Thomsen, D. E.
"Toward a Universal Biology: the Search for Life on Mars." Science News 100, July 24, 1971, pp.64-65.

**2727**
"Upper Atmosphere and Space Research." (In Soviet Bloc Research in Geophysics, Astronomy & Space #257, Washington, July 30, 1971, pp. 26-56. (Available on microfilm from NTIS).

**2728**
Holden, C.
"Soviet-American Conference Urges Search for Other Worlds." Science 174, Oct. 8, 1971, pp. 130-131.

**2729**
Radmer, Richard, & Bessel Kok
"A Unified Procedure for the Detection of Life on Mars." Science 174, Oct. 15, 1971, pp. 233-239.

**2730**
Horowitz, Norman H.
"The Search for Life on Mars: Where we Stand Today." Bulletin of the Atomic Scientists 27, Nov. 1971, pp. 13-17.

**2731**
Ciardi, John
"Is Anyone There?" Saturday Review 54, Nov. 20, 1971, p. 27.

NOTES:

# THE SEARCH FOR LIFE

**2732**
DeMarcus, W. D.
"The Outer Planets: Fly-by Missions" (In NASA. Marshall Space Flight Center. <u>Space for Mankind's Benefit</u>. Lexington, Ky., 1972, pp. 307-311). NASA-SP-313. ($14.50). Available From NTIS.

**2733**
Ponnamperuma, Cyril
"A la Recherche de la vie Extraterrestre." Paris, <u>Sciences</u> 12, Jan.-Feb. 1972, pp. 48-55.

**2734**
Klein, Harold P., et al
"Biological Experiments: the Viking Mars." <u>Icarus</u> 16, Feb. 1972, pp. 139-146.

**2735**
Soffen, G. A., & A. T. Young
"The Viking Mission to Mars." <u>Icarus</u> 16, Feb. 1972, pp. 1-16.

**2736**
Sagan, Carl, et al
"Message from Earth; Plaque on Pioneer 10." <u>Science</u> 175, Feb. 25, 1972, pp. 881-884.

**2737**
Agnew, Irene
"Mars Probes Spur Research on Extraterrestrial Life." <u>Science Digest</u> 71, April 1972, pp. 70-71.

**2738**
Dooling, D.
"Project Viking." London, <u>Spaceflight</u> 14, May 1972, pp. 162-166.

**2739**
Haviland, R. P.
"On the Search for Extra-solar Intelligence." London, <u>Spaceflight</u> 14, June 1972, pp. 217-219.

**2740**
U.S. National Academy of Sciences Survey Committee...Calls for Special National Project and new Instruments for Detecting Inhabited Worlds Beyond Solar System." <u>New York Times</u>, June 4, 1972, p. 36, col. 1.

**2741**
Lear, John
"Search for Man's Relatives Among the Stars." <u>Saturday Review</u> 55, June 10, 1972, pp. 30-37.

**2742**
Trofimov, Alexei, et al
"Will Robots Find Life on Mars?" <u>Space World</u> I-7-103, July 1972, pp. 40-43.

**2743**
Wiley, John P., Jr.
"Sky Reporter; Communication Plaque on Pioneer 10." <u>Natural History</u> 81, April 1972, pp. 44-45; Aug. 1972, pp. 72-73.

**2744**
Dick, William, & Henry Gris
"The Search for Life in Outer Space." <u>National Enquirer</u>, Aug. 20, 1972, back page.

**2745**
Fawcett, George D.
"Science Must Investigate." <u>Flying Saucers</u> 78, Sept. 1972, pp. 10-11.

**2746**
Martin, Anthony R.
"Mission to Jupiter." London, <u>Spaceflight</u> (part 1: 14), Aug. 1972, pp. 294-299; (part 2: 14), Sept. 1972, pp. 325-332.

**2747**
Pokrovskii, Giorgii I.
"Where Should Space Neighbors be Looked for?" <u>Space World</u> I-7-103, July 1972, pp. 46-47; I-10-106, Oct. 1972, pp. 15-16.

NOTES:

**2748**
"Chances of Contacting Extraterrestrial Civ-
ilizations Seem Poor." Science News 103,
Feb. 24, 1973, p. 118.

**2749**
Miller, James R.
"Speeded-up Research for Life in Space."
Readers Digest 102, May 1973, pp. 255-256.

**2750**
Sagan, Carl
"Of Mars, Martians and Mariner 9; Excerpt
from Mars and the Mind of Man." Horizon 15,
Summer 1973, pp. 26-37.

**2751**
Milton, S.
"Is Anyone out There?" London, New Scientist
59, Aug. 16, 1973, pp. 380-382.

**2752**
"Project Viking." London, Spaceflight 15,
Sept. 1973, pp. 322-331.

**2753**
Cathcart, John M.
"Top Space Scientist: U. S. Space
Probe Will Find Life on Mars in 1976."
National Enquirer, Sept. 8, 1974, p.7.

NOTES:

# SPACE FLIGHT

## BOOKS

**2754**
Russen, D.
   Iter Lunare: or, a Voyage to the Moon. Containing Some Considerations on the Nature of That Planet. The Possibility of Getting Thither. With Other Pleasant Conceits About the Inhabitants, Their Manners and Customs. London, Printed for J. Nutt, 1703, 139p.

**2755**
McDermot, M., pseud.
   A Trip to the Moon...Containing Some Observations and Reflections Made by him During his Stay in That Planet, Upon the Manners of the Inhabitants...Printed at Dublin; and Reprinted at London for J. Roberts, 1728, pp. 6, 90.

**2756**
Nicolson, Marjorie H.
   Voyages to the Moon. New York, Macmillan, 1948, 297p.

**2757**
Marbarger, J. P., editor
   Space Medicine; the Human Factor in Flights Beyond the Earth. Urbana, University of Illinois Press, 1951, 83p. ($3.00)

**2758**
Space, Gravity, and the Flying Saucer. Introd. by Desmond Leslie. London, T. Werner Laurie, Ltd., 1954, 182p.

**2759**
Muller, W. D.
   Du Wirst die Erde Sehn als Stern; Probleme der Weltraumfahrt. Stuttgart, Deutsche Verlags-Anstalt, 1955, 315p. (14.80 marks)

**2760**
Plantier, J.
   La Propulsion des Soucoupes Volantes par Action Directe sur l'Atome. Tours, Mame, 1955, 123p.

**2761**
Lau, G.
   Die Welt Wird Grosser; das Technische Wunder der Raumfahrt, Fliegende Untertassen, Raketen, Kunstliche Satelliten. Lahr, Schwarzwald, Astro Pubs., 1956, 94p.

**2762**
Sanger, E.
   The Attainability of the Fixed Stars. Trans. From the German by R. Schamberg. Report #T-69. Santa Monica, Rand Corporation, 1956, 16p.

**2763**
Avenal, A.
   View From Orbit II. London, W. Laurie, 1957, 167p. (18 shillings)

NOTES:

# SPACE FLIGHT

**2764**
Muller, W. D.
  Man Among the Stars. New York, Criterion Books, 1957, 307p. ($4.95)

**2765**
Scamehorn, Howard
  Balloons to Jets. Chicago, Regnery, 1957, 271p.

**2766**
Fritts, C. A.
  We are in a Race to Conquer Outer Space. New York, Vantage, 1958, 106p. ($2.95)

**2767**
Bucheim, R. W., et al
  Space Handbook: Astronautics and its Applications. New York, Random House, 1959, pp. 18-26. ($3.95)

**2768**
Gallagher, B. J.
  God, man and the Atomic Age; Modern Technology and its Impact on the Future. New York, Exposition, 1959, 106p. ($2.50)

**2769**
Lang, D.
  From Hiroshima to the Moon; Chronicles of Life in the Atomic Age. New York, Simon & Schuster, 1959, 496p.

**2770**
Paluzie Borrell, A.
  Como se Realizaron los Viajes Interplanetarios. Barcelona, Ed. Rauter, 1959.

**2771**
Brookings Institute, Washington, D. C.
  Proposed Studies on the Implications of Peaceful Space Activities for Human Affairs, Prepared for NASA...Report of the Committee on Science and Astronautics, U.S. House of Representatives, 87th Congress, first session, March 24, 1961. Washington, Government Printing Office, 1961, 272p.

**2772**
Pirie, N. W., ed.
  Biology of Space Travel. London, Institute of Biology, 1961, 120p. (25 shillings)

**2773**
Lovell, A. C. B.
  The Exploration of Outer Space. New York, Harper & Row, 1962, pp. 70-81. ($3.50)

**2774**
Ordway, Frederick I., et al
  Basic Astronautics. Englewood Cliffs, Prentice-Hall, 1962, 587p. (see Chapter 6).

**2775**
Cramp, Leonard G.
  Piece for a Jigsaw. E. Cowes, Isle of Wight, Eng., Somerton Pub. Co., 1966, 388p.

**2776**
Alceoli
  La Luna, Stazione del Traffico Cosmico. Catania, Edizioni di Studi e Ricerche, 1967, 23p. (400 lira)

**2777**
Chase, Frank M.
  Document 96; a Rationale for Flying Saucers. Clarksburg, W. Va., Saucerian Pubs., 1968, 128p.

**2778**
Harder, James A.
  The UFO Propulsion Problem. Berkeley, Calif., 1968, 10p.

**2779**
Macvey, John W.
  How we will Reach the Stars. New York, Collier Books, 1969, 244p. ($1.25)

NOTES:

# SPACE FLIGHT

2780
California Institute of Technology. Jet Propulsion Laboratory
Publications of the Jet Propulsion Laboratory, July 1968 Through June 1969. (NASA-CR-106958. JPL-Bibl-39-10). Pasadena, Calif., Oct. 1969, 75p. Available from NTIS on microfilm.

2781
Gibbs-Smith, Charles H.
Aviation: an Historical Survey ...
London, HMSO, 1970, 315p.

2782
Pallmann, Ludwig F.
Cancer Planet Mission. London, Foster P., 1970, 216p. (30 shillings)

2783
Rynin, Nikolai A.
Interplanetary Flight and Communication. Tr. from Russian by R. Lavoott. Ed. by M. Lowej. Jerusalem, Israel Program for Scientific Translations, 1970. (Available from NTIS).

2784
Maluquer Wahl, J.J.
Data for a History of Astronautics in Spain up to 1939. (NASA-TT-F-13168). Tr. into English. Washington, Aug. 1970, 35p. Available on microfilm from NTIS.

2785
Afshar, H. K.
The Innovative Consequences of Space Technology and the Problems of the Developing Countries. Teheran, Teheran University Press, 1971, 419p.

2786
Brock, Rudolf
Planetenreise. Modelle u. Konstruktionen f. d. Interplanetäre Raumfahrt. Düsseldorf, Schwann, 1971, 110p. (12.80 marks)

2787
Deerwester, Jerry M., & Susan M. Norman
Reference System Characteristics for Manned Stopover Missions to Mars and Venus. Washington, National Aeronautics and Space Administration; for sale by the National Technical Information Service, Springfield, Va., 1971, 72p. ($3.00) (Nasa technical note, NASA TN D-6226)

2788
Fishbach, Laurence H., et al
Approximate Trajectory Data for Missions to the Major Planets. Washington, National Aeronautics and Space Administration; for sale by the National Technical Information Service, Springfield, Va., 1971, 213p. ($3.00) (NASA technical note, NASA TN D-6141)

## PERIODICAL ARTICLES

2789
Stapledon, O.
"Interplanetary Man?" London, Journal of the British Interplanetary Society (7: 6), Nov. 1948, p. 213.

2790
Kuiper, Gerard P.
"The Law of Planetary and Satellite Distances." Astrophysical Journal (109: 2), March 1949, p. 308.

2791
"Populating Other Planets." Science Digest 26, July 1949, pp. 74-75.

NOTES:

# SPACE FLIGHT

**2792**
"Look! It's Flying Discs Again." Popular Mechanics 96, Aug. 1951, pp. 120-121.

**2793**
Spitzer, L.
"Interplanetary Travel Between Satellite Orbits." ARS Journal 22, March-April 1952, p. 93.

**2794**
Keyhoe, Donald E.
"How the Saucers fly." Fate 7, Nov. 1954, pp. 27-43.

**2795**
"Flying Saucers and Science." American Mercury 85, July 1957, pp. 121-125.

**2796**
Korcsmaros, Jesse
"Radar--Clue to UFO Propulsion?" Fate 10, Aug. 1957, pp. 64-69.

**2797**
Krafft, Carl F.
"The Atom and the UFO." Fate 10, Oct. 1957, pp. 51-53.

**2798**
Strughold, Hubertus
"The Possibilities of an Inhabitable Extra-terrestrial Environment Reachable From the Earth." The Journal of Aviation Medicine 28, Oct. 1957, pp. 507-512.

**2799**
Holden, A. R.
"Flying Saucer Propulsion." London, Flying Saucer Review 4, Jan.-Feb. 1958, pp. 18-21.

**2800**
"On to the Planets." Senior Scholastic. Teacher's Edition 72, March 28, 1958, pp. 24-26.

**2801**
Ogden, Richard C.
"The Creation of the Solar System." London, Flying Saucer Review 4, July-Aug. 1958, pp. 14-18.

**2802**
Burridge, Gaston
"Townsend Brown and his Anti-gravity Discs." Fate 11, Nov. 1958, pp. 40-48.

**2803**
Castells Adriaensens, Francisco de
"Posibilidades del Viaje Lunar." Madrid, Ejército 227, Dec. 1958, pp. 49-59.

**2804**
Pierce, J. R.
"Relativity and Space Travel." Institute of Radio Engineers, Proceedings, #47, June 1959, pp. 1053-1061.

**2805**
Calvin, Melvin
"Round Trip From Space." Evolution 3, Sept. 1959, pp. 362-377.

**2806**
Krafft, Carl F.
"Atomic Structure in Relation to Spaceship Propulsion." London, Flying Saucer Review 5, Sept.-Oct. 1959, pp. 21-22.

**2807**
Korcsmaros, Jesse
"Flying Discs, Clouds, and Falling ice." Flying Saucers, Dec. 1959, pp. 30-33.

**2808**
Cole, Dandridge M.
"Extraterrestrial Colonies." Navigation, Journal of the Institute of Navigation 7, Summer-Autumn 1960, pp. 82-98.

**2809**
Lovell, A. C. B.
"The Exploration of Outer Space." London, Spaceflight (2:7), July 1960, pp. 194-203.

NOTES:

SPACE FLIGHT

2810
Berkner, L. V.
"Space Flight and Science." *Astronautics (ARS Journal)* (6:10), 1961, pp. 46-47, 138-150.

2811
Moore, P., & S. W. Greenwood
*Venus as an Astronautical Objective.* (In Ordway, F. I., ed. *Advances in Space Science and Technology* 3. New York, Academic Press, 1961, pp. 113-147).

2812
Quiroga, R. A.
"La Conquista de los Grandes Planetas," Buenos Aires, *Revista Nacional de Aeronáutica Espacial* 21, Jan. 1961, pp. 8-9.

2813
Finch, B. F.
"The Saucer--a Flying Plasma." London, *Flying Saucer Review* 7, July-Aug. 1961, pp. 13-16.

2814
Margaria, Rodolfo
"La Conquista di Pianeti." Rome, *Missili e Spazio* 4, Aug. 1961, pp. 49-52.

2815
"Cautious Mars Landing." *Science Newsletter* 81, April 21, 1962, p. 243.

2816
Gradecak, Vjekoslav
"Electricity for Space Exploration." London, *Flying Saucer Review* 8, May-June 1962, pp. 16-19.

2817
Von Hoerner, Sebastian
"The General Limits of Space Travel." *Science* 137, July 1962, pp. 18-23.

2818
Von Hoerner, S.
"The General Limits of Space Travel." *Science* (137: 3523), July 6, 1962, p. 18.

2819
Kopyev, V. Y.
"Biology and Flights to Outer Space." Moscow, *Nauka I. Zhizn.* 9, Sept. 1962, pp. 13-22. Trans. by Joint Publications Research Service, Washington, JPRS-16766.

2820
"The Next 25 Years on Mars: Colonies on the Moon, and First men on Mars." *U.S. News and World Report* 53, Oct. 1, 1962, p. 78.

2821
Cramp, Leonard G.
"A Challenge to the Technical Press." London, *Flying Saucer Review* 9, Jan.-Feb. 1963, pp. 6-10, 111.

2822
Lafonta, Paul
"Delendus est Clypeus." Paris, *Phénomènes Spatiaux*, Nov. 1964, pp. 8-12.

2823
Woolen, R. W., & E. J. Merz
"Mars-Voyager Systems." (In *AIAA Unmanned Spacecraft Meeting, Los Angeles, Calif., March 1-4, 1965.* Publication CP-12. New York, American Institute of Aeronautics and Astronautics, 1965, pp. 395-412).

2824
Norman, Paul
"Gravity Powered Objects?" London, *Flying Saucer Review* 11, March-April 1965, pp. 18-20.

2825
Eugster, Jack
"What Does Space Flight Teach Us?" *Aerospace Medicine* 36, April 1965, pp. 345-350.

NOTES:

# SPACE FLIGHT

**2826**
Keyhoe, Donald E.
  "I Know the Secret of the Flying Saucers." True 47, Jan. 1966, pp. 34-36, 94-95.

**2827**
Jones, Bob
  "Magnetic Space Propulsion." Flying Saucers, Aug. 1966, p. 25.

**2828**
Lauritzen, Hans
  "Disclosure of the Motive Power Systems of the Flying Saucers." Amsterdam, Het Interplanetaire Nieuwsbulletin 1, Sept. 1966, pp. 10-12.

**2829**
Lauritzen, Hans
  "The Motive Power of the Flying Saucers." Saucer News 13, Winter 1966-1967, pp. 7-9.

**2830**
Spencer, Dwain F.
  "Fusion Propulsion for Interstellar Missions." New York Academy of Science, Annals 140, Dec. 16, 1966, pp. 407-418.

**2831**
Lauritzen, Hans
  "Flying Saucers--Superconducting Whirls of Plasma." Flying Saucers, March 1967, pp. 10-11.

**2832**
Lauritzen, Hans
  "Magnetic Motors: Power Systems of the Flying Saucers?" Flying Saucers, March 1967, pp. 13-21.

**2833**
Goupil, Jean
  "L'hypothèse du Champ Magnétique Canalisé: Tentative D'explication de Quelques Phénomènes Etranges." Paris, Phénomènes Spatiaux, June 1967, pp. 2-4.

**2834**
Goupil, Jean
  "Une Conséquence Curieuse de L'hypothèse du Champ Repulsif: la Forme des 'O.V.N.I.'" Paris, Phénomènes Spatiaux, June 1967, pp. 9-11.

**2835**
Evans, Gordon H.
  "UFO: Theories of Flight." Science and Mechanics 38, Aug. 1967, pp. 48-51, 72-74.

**2836**
Beach, David
  "Some UFO Thoughts From Britain." Flying Saucers, June 1968, pp. 27-28.

**2837**
Silver, Brent W.
  "Grand Tour of the Jovian Planets." Journal of Spacecraft & Rockets 5, June 1968, pp. 633-637.

**2838**
Slater, Robert M.
  "Solving the Secret of UFO Propulsion." Flying Saucers, June 1968, pp. 15-17.

**2839**
Clarke, Arthur C.
  "When Earthmen Become the 'UFO' of Other Planets." Flying Saucers, Aug. 1968, pp. 8-17.

**2840**
Clarke, Arthur C.
  "Next: the Planets." (In Bioastronautics and the Exploration of Space. Brooks Air Force Base, Texas, School of Aerospace Medicine, Nov. 1968, p. 511). (AD-687893) (Available from NTIS).

**2841**
California Institute of Technology. Jet Propulsion Laboratory
  "Viking Project Orbiter System and Project Support." (In its Space Programs Summary #37-61, Vol. 1, Jan. 31, 1970, p. 21). Available on microfilm from NTIS.

NOTES:

## SPACE FLIGHT

**2842**
"Astronautics, Development and Goals." (In
Goals and Means in the Conquest of Space,
by R. G. Perelman. Tr. into English from
Russian. Washington, NASA, May 1970, pp.
1-46). (NASA-TT-F-595). Available on
microfilm from NTIS.

**2843**
Dickinson, Terence
"The Zeta Reticuli Incident." Astronomy
(2:12), Dec. 1974, pp. 4-18.

NOTES:

## TELEPORTATIONS

**BOOKS**

2844
Weiss, Sara
  Journeys to the Planet Mars; or Our Mission to Ento (Mars). Rochester, N.Y., Bradford Press, 1903, 548p.

2845
Bethurum, T.
  Aboard a Flying Saucer. Non-fiction; a True Story of Personal Experience. Los Angeles, DeVorss & Co., 1954, 192p. ($3.00)

2846
Adamski, George
  Inside the Space Ships. New York, Abelard-Schuman, 1955, 256p. ($3.50)

2847
Michael, Cecil
  Roundtrip to Hell in a Flying Saucer. New York, Vantage Press, 1955, 61p.

2848
Nelson, Buck
  My Trip to Mars, the Moon, and Venus. Mountain View, Mo., 1956, 44p.

2849
Rember, Winthrop A.
  Eighteen Visits to Mars. New York, Vantage Press, 1956, 439p.

2850
Gibbons, G.
  They Rode in Space Ships. New York, Citadel Press, 1957, 217p. ($3.50)

2851
Gibson, E. H.
  A.D. 2018; Recollections of the Chaplain of a Space Ship. New York, Greenwich, 1958, 62p. ($2.50)

2852
Martin, D. M.
  Seven Hours Aboard a Space Ship. Detroit, 1959? 29p.

2853
Menger, H.
  From Outer Space to You; Autobiography. Clarksburg, Saucerian Books, 1959, 256p. ($4.50)

2854
Zinsstag, Lou
  On George Adamski. Basle, Switzerland, 1959, 5p.

NOTES:

## TELEPORTATIONS

2855
Ferguson, William
  My Trip to Mars. Clarksburg, W. Va., Saucerian Books, 196-?, 14p.

2856
Fry, D. W.
  The White Sands Incident, and to Men on Earth; a Technician Talks With a Spacemen and Rides in a Flying Saucer. New Combined Edition. Merlin, Oregon, Understanding, 1964? 66, 41p.

2857
Macvey, John W.
  Journey to Alpha Centauri. New York, Macmillan, 1965, 256p.

2858
Fuller, J. G.
  The Interrupted Journey: Two Lost Hours Aboard a Flying Saucer. New York, Dial Press, 1966, 301p. ($5.95)

2859
Nelson, Buck
  My Trip to Mars, the Moon and Venus. Mountain View, Mo., 1956, 44p.

2860
Rampa, T. L.
  My Visit to Venus. Clarksburg, W. Va., Saucerian Books, 1966, 42p.

2861
  Zemkla, Interplanetary Avatar. Azusa, Calif., Galaxy Press, 1966.

2862
Martin, Dan
  Seven Hours Aboard a Space Ship. Clarksburg, W. Va., Saucerian Pubs., 1968, 17p. ($1.00)

2863
  UFO Flight. Visit to Planet Selo. Azusa, Calif., Galaxy Press, 1968, 80p.

2864
McDermot, M., pseud.
  A Trip to the Moon (1728) by Murtagh McDermot. A Trip to the Moon (1764-5) by Sir Humphrey Lunatic, Pseudonym of Francis Gentleman. Ed. by Jeanne K. Welcher and George E. Bush, Jr. Gainesville, Fla., Scholars' Facsimiles & Reprints, 1970, 204p.

2865
Dohmen, J. G.
  A Identifier. Le cas Adamski. Que Penser des Soucoupes Volantes? Biarritz, Travox, 1972, 240p.

## PERIODICAL ARTICLES

2866
Gelatt, R.
  "In a Saucer From Venus." Review of Book, Behind the Flying Saucers, by F. Scully. Saturday Review of Literature, Sept. 23, 1950, pp. 20-21.

2867
Wells, Leslie E.
  "They Disappeared into the Unknown." Fate 9, July 1956, pp. 63-67.

2868
Miller, Max B.
  "Report on the UFO." Fate 9, Dec. 1956, pp. 31-34.

2869
Nelson, Buck
  "I Visited Mars, Venus, and the Moon." Search, Dec. 1956, pp. 6-20.

NOTES:

# TELEPORTATIONS

**2870**
Hall, Richard
 "Is There a Veil of Secrecy Around the Flying Saucers?" Flying Saucers From Other Worlds, June 1957, pp. 51-56.

**2871**
 "Mr. Cooke Goes to Zomdic." London, Flying Saucer Review 4, July-Aug. 1958, pp. 26-27.

**2872**
Keyhoe, Donald F.
 "Flying Saucers, Menace or Myth?" Argosy 350, June 1960, pp. 17, 80-83.

**2873**
Comella, Tom
 "Have UFO's 'Swallowed' our Aircraft?" Fate 14, May 1961, pp. 32-37.

**2874**
 "The Brazilian Abduction." London, Flying Saucer Review 8, Nov.-Dec. 1962, pp. 10-12.

**2875**
Van den Berg, Basil
 "My Discovery Will Prove Adamski's Claim." London, Flying Saucer Review 8, Nov.-Dec. 1962, pp. 3-5.

**2876**
Maney, Charles A.
 "Why the Air Force Can't Investigate UFO's." Fate 16, May 1963, pp. 26-28.

**2877**
Lorenzen, Coral
 "The Disappearance of Rivalino da Silva, Kidnapped by a UFO?" Fate 16, June 1963, pp. 26-33.

**2878**
Creighton, Gordon
 "The Most Amazing Case of all, Part 1--a Brazilian Farmer's Story." London, Flying Saucer Review 11, Jan.-Feb. 1965, pp. 13-17.

**2879**
Creighton, Gordon
 "The Most Amazing Case of all. Part II--Analysis of the Brazilian Farmer's Story." London, Flying Saucer Review 11, March-April 1965, pp. 5-8.

**2880**
Creighton, Gordon
 "Teleportations." London, Flying Saucer Review, March-April 1965, pp. 14-16.

**2881**
 "Important Discoveries." London, Flying Saucer Review 12, Nov.-Dec. 1966, p. 17.

**2882**
Greenwald, H.
 "Interrupted Journey," by J. G. Fuller (Review), Saturday Review 49, Dec. 31, 1966, pp. 22-23.

**2883**
Spraggett, Allen
 "Kidnapped by a UFO." Fate 20, Jan. 1967, pp. 34-41.

**2884**
Keel, John A.
 "The UFO Kidnappers." Saga 33, Feb. 1967, pp. 10-14, 50, 52-54, 56-60, 62.

**2885**
Creighton, Gordon
 "Attempted Abduction by UFO Entity?" London, Flying Saucer Review 13, March-April 1967, pp. 23-24.

NOTES:

## TELEPORTATIONS

**2886**
Creighton, Gordon
"Even More Amazing." London, <u>Flying Saucer Review</u> 12, July-Aug. 1966, pp. 23-27; Sept.-Oct. 1966, pp. 22-25; Nov.-Dec. 1966, pp. 14-16; v.13, Jan.-Feb. 1967, pp. 25-27; v.14, Jan.-Feb. 1968, pp. 18-20.

**2887**
Noonan, Allen
"The Man who saw Venus; Interview." Ed. by Lloyd Mallan. <u>Mechanics Illustrated</u> 64, May 1968, pp. 58-60.

**2888**
Dendle, Brian J.
"Romantic Voyage to Saturn: Tirso Aguimana de Veca's <u>Una Temporada en el más Bello de los Planetas</u>." <u>Studies in Romanticism</u> 7, Summer 1968, pp. 243-247.

**2889**
Bryant, Larry W.
"The UFO Cover-up at Langley Air Force Base." <u>Flying Saucers From Other Worlds</u>, June 1968, pp. 11-14.

**2890**
Galíndez, Oscar A.
"Teleportation from Chascomusto, Mexico." London, <u>Flying Saucer Review</u>, Sept.-Oct. 1968, pp. 3-4.

**2891**
Creighton, Gordon
"More Teleportations." London, <u>Flying Saucer Review</u> (16: 5), Sept.-Oct. 1970, pp. 11-13, 32.

**2892**
Slate, B. A.
"The Great UFO Ride." <u>Fate</u> 254, May 1971, pp. 38-50.

**2893**
Creighton, Gordon
"Another Teleportation and its Sequel." London, <u>Flying Saucer Review</u> (17:5), Sept.-Oct. 1971, pp. 15-17+.

**2894**
Creighton, Gordon
"'Forty eight Hours in a Flying Saucer.'" London, <u>Flying Saucer Review</u> (17:6), Nov.-Dec. 1971, pp. 15-17.

**2895**
Creighton, Gordon
"Uproar in Brazil." London, <u>Flying Saucer Review</u>, Nov.-Dec. 1971. pp. 24-29.

**2896**
Galíndez, Oscar A.
"L'Incident Brunelli-Porchietto; une Téléportation?" Paris, <u>Phénomènes Spatiaux</u> 35, March 1973, pp. 21-31.

**2897**
Galíndez, Oscar A.
"A New Teleportation Near Córdoba: Strange luminous object by the roadside, 81 kilometers missing and a petrol tank half-full." London, <u>Flying Saucer Review</u> (19:3), May-June 1973, pp. 6-12.

NOTES:

## UNIDENTIFIED FLYING OBJECTS, INCLUDING FLYING SAUCERS

**BOOKS**

**2898**
U.S. Air Force
 Estimate of the Situation. Washington, G.P.O., Sept. 1948, 4p.

**2899**
Heline, Corinne
 America's Invisible Guidance. Los Angeles, New Age Press, 1949, 175p.

**2900**
U.S. Air Force
 Unidentified Flying Objects, Project Grudge. Dayton, Wright-Patterson Air Force Base, 1949, 366p. (Its Technical Report #102-AC 49/15100).

**2901**
U. S. Air Force. Air Material Command
 Unidentified Aerial Objects; Project Sign. Dayton, Wright-Patterson Air Force Base, 1949, 35p. (Technical Report No. F-TR-2274-IA).

**2902**
Arnold, Kenneth A.
 The Flying Saucer as I Saw it. Boise, Idaho, 1950, 16p.

**2903**
Johnson, DeW. B.
 Flying Saucers, Fact or Fiction? Los Angeles, 1950, 339p. (Master of Arts Thesis in Journalism at UCLA).

**2904**
Keyhoe, Donald E.
 The Flying Saucers are Real. Greenwich, Fawcett Pubs., 1950, 175p.

**2905**
Layne, Meade
 Flying Discs--the Ether Ship Mystery and its Solution. San Diego, Borderland Sciences Research Associates, 1950, 38p.

**2906**
Sanctillean
 Flying Saucers: Portents of These 'Last Days.' Santa Barbara, I.F. Rowny Press, 1950, 39p.

**2907**
Scully, Frank
 Behind the Flying Saucers. New York, Holt & Co., 1950, 230p. ($2.75)

**2908**
Canada. Deputy Minister of Transport for Air Services
 Project MAGNET and Project SECOND STORY. Ottawa, Queen's Printer, Dec. 1950, 47p.

---

NOTES:

UNIDENTIFIED FLYING OBJECTS INCLUDING FLYING SAUCERS

2909
Vliegende Schotels. Fantasie of Werkelijkheid? The Hague, Succes (Prinsevinkenpark 2), 1951, 63p. (0.50 florins)

2910
Arnold, Kenneth A., & Ray Palmer
  The Coming of the Saucers. Boise, Idaho, 1952, 192p.

2911
Gardner, M.
  Flying Saucers (In the Name of Science).
New York, G. P. Putnam's Sons, 1952, pp. 56-58.

2912
Leslie, Desmond, & George Adamski
  Flying Saucers Have Landed. New York, British Book Centre, 1953, 232p. ($3.50)

2913
Menzel, Donald H.
  Flying Saucers. Cambridge, Harvard University Press, 1953, 319p. ($4.75)

2914
The Project A Report. Ada, Ohio, Ohio Northern University, 1953, 9p.

2915
Zeñabí, J.
  El Misterio de los Discos Voladores.
Santiago de Chile, 1953, 77p.

2916
Calvillo Madrigal, Salvador
  Plativología, Ensayo Nesciente. México, 1954, 21p.

2917
Digard-Comet, S.
  Les Soucoupes: Leur Provenance, Leur But, par Isnomimis, Pseud. Paris, Aryana, 1954, 43p.

2918
Duclout, J. A.
  Orígen, Estructura y Técnica de Los Platos Voladores. Buenos Aires, America Técnica, Editorial, 1954, 180p. (80 pesos, Argentine)

2919
Feryer, Rich (pseud.)
  Fliegende Untertassen: UFO's. Greifen Ausserirdische Mächte in Unsere Verhältnisse ein? Woher Kommen sie...Boniswil/Ag, Switzerland, Schaefer, 1954, 32p. (1.50 Swiss francs)

2920
Kutsakes, Demetrios D.
  Hyparchei zōē eis Tous Allous Kosmous; Mia Ereuna Metaxu Plantetōn, Aplanōn kai Galaxiōn. Athens, 1954, 74p.

2921
Pauquet, P. P.
  Schöpfer, Weltall, "Untertassen" Grübelein um Bewohnte Sternenwelten. "Fliegende Untertassen" und Schöpferabsichten. Cologne, Bachem Pubs., 1954, 61p. (1.80 German Marks)

2922
Pedrajo, Manuel
  Los Platillos Volantes y la Evidencia.
(n.p., n.p.) 1954.

2923
Wilkins, Harold T.
  Flying Saucers From the Moon. London, P. Owen, Ltd., 1954, 320p.

2924
Wilkins, Harold T.
  Flying Saucers on the Attack. New York, Citadel Press, 1954, 329p. ($3.50)

2925
Allingham, Cedric
  Flying Saucer From Mars. New York, British Book Centre, 1955, 153p.

NOTES:

## UNIDENTIFIED FLYING OBJECTS INCLUDING FLYING SAUCERS

**2926**
Angelucci, Orfeo M.
  The Secret of the Saucers. Amherst, Wis., Amherst Press, 1955, 167p.

**2927**
Busson, Le Roy
  Les Derniers Secrets de la Terre. Paris, Eds. de la Table Ronde, 1955.

**2928**
Girvan, Ian W.
  Flying Saucers and Common Sense. London, F. Muller, 1955, 160p.

**2929**
Jessup, Morris K.
  The Case for the UFO (Unidentified Flying Objects). New York, Citadel Press, 1955, 239p. ($3.50)

**2930**
Keyhoe, Donald E.
  The Flying Saucer Conspiracy. New York, H. Holt Co., 1955, 315p. ($3.50)

**2931**
Miller, R. D.
  You do Take it With you. New York, Citadel Press, 1955, 238p.

**2932**
Osorio, Luis E.
  Llega la Era Interplanetaria. Bogotá, Eds. de la Idea, 1955? 96p.

**2933**
Schopfer, Siegfried
  Fliegende Untertassen: ja Oder Nein? Stuttgart, Walter Hädecke Verlag, 1955, 32p.

**2934**
Sievers, F.
  Flying Saucer Uber Sudafrika. Zur Frage d. Besuche aus d. Wellenraum. Pretoria, Sagittarius Pub. Co., 1955, 415p. (26 shillings)

**2935**
U.S. Air Force. Air Technical Intelligence Center
  Special Report no. 14. (Analysis of Reports of Unidentified Aerial Objects.) Dayton, Wright-Patterson Air Force Base, 1955, 308p. Project no. 10073.

**2936**
Barker, Gray, ed.
  The Saucerian Review. Clarksburg, W. Va., 1956, 98p.

**2937**
Busson, Bernard
  Los Ultimos Secretos de la Ciencia. Tr. into Spanish by J. Garriga Pujol. Barcelona, Ed. Vergara, 1956, 220p.

**2938**
Fry, Daniel W.
  Steps to the Stars. Lakemont, Ga., CSA Pubs., 1956, 83p.

**2939**
Garreau, C.
  Alerte Dans le Ciel! Documents Officiels sur les Objets Volants non Identifiés. Paris, Editions du Grand Damier, 1956, 257p.

**2940**
Girvan, W.
  Flying Saucers and Common Sense. New York, Citadel Press, 1956, 156p. ($3.50)

**2941**
Guieu, Jimmy
  Flying Saucers Come From Another World. Trans. by Charles Ashleigh. London, Hutchinson, 1956, 248p. (12 shillings, 6 pence)

**2942**
Jessup, Morris K., ed.
  UFO Annual, 1956. New York, Arco Pub. Co., 1956, 380p. ($4.00)

NOTES:

UNIDENTIFIED FLYING OBJECTS INCLUDING FLYING SAUCERS

2943
Michel, Aimé
  The Truth About Flying Saucers. Tran. From French by Paul Selver. New York, Criterion Books, 1956, 225p. ($3.95)

2944
Vogt, Cristián
  El Misterio de los Platos Voladores. Buenos Aires, Editorial 'La Mandrágora,' 1956, 190p.

2945
Barton, Michael X.
  Educational and Inspirational Courses of Study. 13v. Los Angeles, Futura Press, 1957-64.

2946
Constance, Arthur
  The Inexplicable Sky. New York, Citadel Press, 1957, 287p. ($3.95)

2947
Jessup, Morris K.
  The Expanding Case for the UFO. New York, Citadel Press, 1957, 253p.

2948
Miller, Max B.
  Flying Saucers: Fact or Fiction? Twelve Years Research of UFO's in Our Skies Revealed by the Top Scientists, Astronomers, Air Force Personnel, and Technical Observers. Los Angeles, Trend Books, 1957, 128p. ($0.75)

2949
Norkin, Israel
  Saucer Diary. New York, Pageant Press, 1957, 137p.

2050
Reeve, Bryant, & Helen Reeve
  Flying Saucer Pilgrimage. Amherst, Wis., Amherst Press, 1957, 304p.

2951
Speer, Herbert V.
  Nicht von Dieser Erde. Tatsachenbericht über die Interplanetar. Fliegenden Schreiben und über den Oberbefehlshaber der Raumschiff-Flotte Ashrar-Sheran. 3v. Heiden, Schönenberger, 1957. (3.35 Swiss francs)

2952
Stringfield, Leonard H.
  Inside Saucer Post 3-0 Blue. Cincinnati, Moeller Printing Co., 1957 (available from author: 4412 Grove Ave., Cincinnati, Ohio 45227).

2953
Sylvester, J.
  Flying Saucer. London, Ward, Lock & Co., Ltd., 1957, 101p. (3 shillings, 6 pence)

2954
Van Tassel, George
  The Council of Seven Lights. Los Angeles, De Vorss, 1957.

2955
Fox, C. V. J.
  The Pentagon Case. New York, Freedom Press, 1958.

2956
  Flying Saucer Review; World Roundup of UFO Sightings and Events. New York, Citadel Press, 1958, 224p. ($3.75)

2957
Gibbons, G.
  The Coming of the Space Ships. New York, Citadel Press, 1958, 188p. ($3.50)

2958
Girvin, Calvin C.
  The Night has a Thousand Saucers. El Monte, Calif., Understanding Pub. Co., 1958, 168p.

NOTES:

## UNIDENTIFIED FLYING OBJECTS INCLUDING FLYING SAUCERS

**2959**
MacDougall, C. D.
  Hoaxes. New York, Ace Books, 1958.

**2960**
Michel, Aimé
  Flying Saucers and the Straight-line Mystery. Pref. by L. M. Chassin. New York, Criterion Books, 1958, 285p.

**2961**
Michel, Aimé
  Mystérieux Objets Célestes. Paris, Arthaud, 1958, 393p.

**2962**
Perego, Alberto
  Sono Extraterrestri! Il Piano Operativo dell'Aviazione Elettromagnetica 1944-1958. Rome, Ediz. Alper, 1958, 64p. (800 lira)

**2963**
Stanford, Ray, & Rex Stanford
  Look Up. Corpus Christi, Tex., Essene Press, 1958, 66p.

**2964**
Williamson, George H.
  UFO's Confidential!; the Meaning Behind the Most Closely Guarded Secret of all Time. Corpus Christi, Essene Press, 1958, 100p.

**2965**
Allen, W. G.
  Spacecraft From Beyond Three Dimensions. New York, Exposition, 1959, 202p. ($3.50)

**2966**
Jung, Carl G.
  Flying Saucers; a Modern Myth of Things Seen in the Skies. Trans. From the German by R. F. C. Hull. New York, Harcourt, Brace, 1959, 186p.

**2967**
Miller, Max B.
  Flygende Tefat, Fantasi Eller Verklighet? Tr. into Swedish by A. W. Edstrom. Halsingborg, Parthenon, 1959, 259p.

**2968**
Simões, Auriphebo Berrance
  Os Discos Voadores; Fantasia e Realidade. São Paulo, Edart, 1959, 390p.

**2969**
Stranges, Frank E.
  Flying Saucerama. New York, Vantage Press, Inc., 1959, 115p. ($3.00)

**2970**
Barton, Michael X.
  We Want you. Los Angeles, Futura Press, 1960, 34p.

**2971**
Keyhoe, Donald E.
  Flying Saucers: Top Secret. New York, G. P. Putnam's Sons, 1960, 283p. ($3.95)

**2972**
Morison, R.
  Driving Whirlwinds. London, Lew Singer, 1960, 16p. (2 shillings, 6 pence)

**2973**
Stranges, Frank E.
  Danger From the Stars; a Warning From Frank E. Stranges. Van Nuys, Calif., I.E.C., Inc., 196-,($1.00)

**2974**
Tacker, L. J.
  Flying Saucers and the U.S. Air Force. Princeton, D. Van Nostrand Co., 1960, 164p. ($3.50)

**2975**
Gonzalez Ganteaume, H.
  Platillos Voladores Sobre Venezuela; los Hechos Presentados. Caracas, Tipografía Olimpia, 1961, 250p.

NOTES:

## UNIDENTIFIED FLYING OBJECTS INCLUDING FLYING SAUCERS

2976
Littman, Arnold
   Peter hat Pech! Die Jag d. Nach d. Fliegenden Untertasse. Munich, Hueber, 1961, 71p. (1.90 marks)

2977
Maney, Charles A., & Richard Hall
   The Challenge of Unidentified Flying Objects. Washington, 1961, 208p.

2978
Qvarnström, G.
   Dikten och den nya Vetenskapen; det Astronautiska Motivet. Lund, C. W. K. Gleerup, 1961, 304p.

2979
Ribera Jorda, Antonio
   Objetos Desconocídos en el Cielo. Barcelona, Librería Editorial Argos, 1961, 289p.

2980
Alemán Velasco, Miguel
   Los Secretos y las Leyes del Espacio. México, 1962, 258p.

2981
Herrmann, Joachim
   Das Falsche Weltbild. Stuttgart, Kosmos Verlag, Franckh'sche Verlagshandlung, 1962, 162p.

2982
Lorenzen, Coral
   The Great Flying Saucer Hoax; the UFO Facts and Their Interpretation. New York, William-Frederick Press, for the Aerial Phenomena Research Organization of Tucson, 1962, 257p.

2983
Manas, J. H.
   Flying Saucers and Space Men; a Scientific and Metaphysical Dissertation in Interplanetary Traveling. New York, Pythagorean Society (152 W. 42nd St., New York, N. Y. 10036), 1962, 124p. ($2.00)

2984
Terblanche, Le Roux
   Die Vlieënde Piering. Johannesburg, Pronkboeke, 1962, 109p.

2985
Fragner, Wolfram
   Physik der Uraniden. Munich, UFO-Forschungsgruppe Munchen-Nuernberg, 1963, 41p.

2986
Menzel, Donald H., & L. G. Boyd
   The World of Flying Saucers; A Scientific Examination of a Major Myth of the Space Age. New York, Doubleday & Co., 1963, 302p. ($4.50)

2987
Veit, Karl L.
   Erforschung Ausserirdischer Weltraumschiffe--ein Wissenschaftliches Anliegen des 20. Jahrhunderts. Wiesbaden-Schierstein, Ventla-Verlag, 1963, 95p.

2988
Beer, Lionel
   An Introduction to Flying Saucers. London, 1964, 44p.

2989
Bull, F. M.
   UFO Handbook 2. London, British UFO Research Association, 1964, 31p.

2990
Díez Gomez, J. M.
   Los Platillos Voladores. Barcelona, Editorial Molino, 1964. (3.00 Spanish Reales)

2991
Hall, Richard, ed.
   The UFO Evidence. Kensington, Md. (National Investigations Committee on Aerial Phenomena, 3535 University Blvd., W.), 1964.

NOTES:

## UNIDENTIFIED FLYING OBJECTS INCLUDING FLYING SAUCERS

2992
Jung, Carl G.
　Civilization in Transition. (In his Collected Works. 2d ed. v. 10. Ed. by Gerhard Adler, et al. Tr. by R. F. Hull. Princeton, Princeton University Press, 1964, pp. 314-433).

2993
Jung, Carl G.
　Ein Moderner Mythus. Von Dingen, die am Himmel Gesehen Werden. 2d ed. Zürich, Rascher, 1964, 143p. (8.20 Swiss francs)

2994
King, George
　The Flying Saucers; a Report on the Flying Saucers, Their Crews, and Their Mission to Earth. Los Angeles, Aetherius Society, 1964, 16p. ($0.50)

2995
Layne, Meade
　The Coming of the Guardians: an Interpretation of the "Flying Saucers" as Given From the Other Side of Life. Vista, Calif., Borderland Sciences Research Associates Foundation, 1964, 72p. ($3.00)

2996
Miller, W.
　Looking for the General. New York, McGraw Hill, 1964.

2997
National Investigations Committee on Aerial Phenomena
　The UFO Evidence (Unidentified Flying Objects) Ed. by Richard H. Hall. Washington, National Investigations Committee on Aerial Phenomena, 1964, 184p. ($5.00)

2998
Platillos Volantes. Barcelona, Plaza and Janés, 1964.

2999
Rand Corporation
　Rand Report R-414-PR. Santa Monica, Rand Corp., 1964, 151p.

3000
Ribera, Antonio
　Objetos Desconocídos en el Cielo. Barcelona, Libería Editorial Argos, 1964, 350p. (180 pesetas)

3001
Barker, Gray
　Gray Barker's Book of Saucers. Clarksburg, W. Va., Saucerian Books, 1965, 77p.

3002
Crum, W. L.
　Lunar Lunacy and Other Commentaries. Philadelphia, 1965, 284p.

3003
Gibney, F., & G. J. Feldman
　The Reluctant Space-farers; a Study in the Politics of Discovery. New York, New American Library, 1965, 174p.

3004
Ribera, Antonio
　El Gran Enigma de los Platillos Volantes. Barcelona, Editorial Pomaire, 1965, 435p.

3005
Uriondo, Oscar A.
　Objetos Aéreos no Identificados, un Enigma Actual. Buenos Aires, 1965, 155p.

3006
Vallée, Jacques
　Anatomy of a Phenomenon: Unidentified Objects in Space--A Scientific Phenomenon. Chicago, Henry Regnery & Co., 1965, 210p. ($4.95)

NOTES:

## UNIDENTIFIED FLYING OBJECTS INCLUDING FLYING SAUCERS

3007
Babcock, Edward J., & Timothy Green, eds.
 UFO's Around the World. (n. p.) Interplanetary News Service, 1966, 64p.

3008
Bordeleau, Henri
 J'ai vu des Soucoupes Volantes. Montreal, Editions du Jour, 1966, 124p. ($1.00 Canadian currency)

3009
Carter, Joan F.
 14 Footsteps from Outer Space. Dallas, Royal Pub., 1966, 168p.

3010
Chartrand, Robert L., & William F. Brown
 Facts About Unidentified Flying Objects. Washington, Library of Congress, Legislative Reference Service, 1966, 29p.

3011
Coelho Netto, P.
 A Realidade dos Discos Voadores. Rio de Janeiro, Editôra Minerva, 1966, 59p.

3012
Cramp, Leonard G.
 Piece for a Jig-Saw. (Somerton Pub. Co., Newport Rd., Cowes, Isle of Wight, England), 1966, 388p. (27 shillings, 6 pence)

3013
Davidson, Leon
 Flying Saucers: an Analysis of the Air Force Project Blue Book Special Report no. 14. 3d ed. Ramsey, N. J., Ramsey-Wallace Corp., 1966, 84p.

3014
Knaggs, Oliver
 Let the People Know. Cape Town, So. Africa, Howard B. Timmins (Pty.) Ltd., 1966, 113p. (1.15 rands)

3015
Lindsay, Gordon
 The Riddle of the Flying Saucers. Dallas, The Voice of Healing Pub. Co., 1966, 31p.

3016
 The Mel Noel Story, the Inside Story on the U.S. Air Force Secrecy on UFO's. Inglewood, Calif., 1966, 26p.

3017
Michel, Aimé
 Mystérieux Objets Célestes. New ed. Paris, Editions Planète, 1966, 303p. (18.50 francs)

3018
Misraki, Paul
 Flying Saucers Through the Ages. Tr. from the French by Gavin Gibbons. London, Spearman, 1966, 192p. (21 shillings)

3019
Pereira, Flavio A.
 O Livro Vermelho dos Discos Voadores. São Paulo, Ediçōes Florença Ltda., 1966, 486p.

3020
Rehn, K. G.
 De Flygande Tefaten. Dokument och Teori. Gothenburg, Zinderman, 1966, 174p. (24.50 kroner)

3021
Stanton, L. J.
 Flying Saucers, Hoax or Reality? New York, Belmont Books, 1966, 157p.

3022
Steiger, Brad
 Strangers From the Skies. New York, Award Books, 1966, 158p.

NOTES:

**3023**
Stranges, Frank E.
  New Flying Saucerama. 4th ed. Glendale, N. Y., International Evangelism Crusades, 1966, 117p.

**3024**
Strentz, Herbert
  A Survey of Press Coverage of UFO's, 1947-1966. Ph. D. Dissertation, Journalism, Northwestern University. Evanston, 1966.

**3025**
Thomas, P.
  Flying Saucers Through the Ages. Tran. from French by G. Gibbons. London, Neville, Spearman, 1966, 192p. (21 shillings)

**3026**
Twitchell, Cleve
  The UFO Saga. Lakemont, Ga., CSA Press, 1966, 94p.

**3027**
U.S. Air Force
  Research and Development. No. AFR 80-17. Washington, G.P.O., 1966, 18p.

**3028**
U. S. Congress. House. Committee on Armed Services.
  Unidentified Flying Objects. Hearing, Eighty-ninth Congress, second session, April 5, 1966. Washington, U. S. Govt. Print. Office, 1966, pp. 5991-6075.

**3029**
U.S. Scientific Advisory Panel on Unidentified Flying Objects. Ad hoc Committee to Review Project Blue Book
  Special Report. Washington, 1966, 10p.

**3030**
Vallée, Jacques, & Janine Vallée
  Challenge to Science; the UFO Enigma. Chicago, Regnery, 1966, 268p.

**3031**
Weor, Samael, A.
  Los Platillos Voladores. San Salvador, Editorial Lea, 1966?, 28p.

**3032**
Adler, Bill
  The Flying Saucer Reader. New York, New American Library, 1967, 244p. ($5.95)

**3033**
Adler, Bill, comp.
  Letters to the Air Force on UFOs. New York, Dell Pub. Co., 1967, 157p.

**3034**
Barker, Gray
  Book of Space Ships and Their Relationships With the Earth. Clarksburg, Saucerian Books, 1967, 127p. ($3.95)

**3035**
Binder, Otto
  What we Really Know About Flying Saucers. Greenwich, Conn., Fawcett Publications, 1967, 224p. ($0.75)

**3036**
Bloecher, Ted
  Report on the UFO Wave of 1947. Washington, 1967 (Available from California UFO Research Institute, Box 941, Lawndale, California 90260).

**3037**
Bray, Arthur R.
  Science, the Public and the UFO. Ottawa, Canada, Bray Book Service (Box 5051, Station F), 1967, 193p. ($2.75 Canadian currency)

**3038**
Buckle, Eileen
  The Scoriton Mystery. London, Neville Spearman, 1967, 303p. (30 shillings)

NOTES:

## UNIDENTIFIED FLYING OBJECTS INCLUDING FLYING SAUCERS

3039
Chambers, Howard V.
  UFO's for the Millions. Los Angeles, Sherbourne Press, Inc., 1967, 158p.

3040
Coelho Netto, Paulo
  A Realidade dos Discos Voadores. 2d ed., Ampliada. Rio de Janeiro, Editôra Minerva, 1967, 155p.

3041
Cox, Donald W., ed.
  America's Explorers of Space, Including a Special Report on UFO's. Maplewood, N.J., Hammond, 1967, 93p.

3042
Cremaschi, Inisero, & Giuseppe Pederiali
  Dischi Volanti, Benvenuti. Bologna, Carroccio, 1967, 157p.

3043
Danyans, Eugenio
  Platillos Volantes en la Antiguedad. Barcelona, Editorial Pomaire, 1967, 262p.

3044
David, Jay, ed.
  The Flying Saucer Reader. New York, New American Library, 1967, 244p. ($5.95)

3045
Dinotos, Sábado
  A Antiguedade dos Discos Voadores. São Paulo, 1967, 173p.

3046
Earley, George W.
  Unidentified Flying Objects, an Historical Perspective. Bloomfield, Conn., 1967, 14p.

3047
Edwards, Frank
  Flying Saucers; here and now! New York, Lyle Stuart, 1967, 261p. ($5.95)

3048
The Flying Saucer Menace. New York and London, Universal Pub. and Distributing Corp., 1967, 64p.

3049
Flying Saucers Illustrated. Compiled by the eds. of Real Magazine. Studio City, Calif., Kling House Ltd., 1967, 79p.

3050
Flying Saucers Pictorial. Compiled by the eds. of Real magazine. Tucson, Ariz., Arizill Realty and Pub. Co., 1967, 73p.

3051
Flying Saucers, UFO Reports #1-4. New York, Dell Pub., 1967.

3052
Gaddis, Vincent H.
  Mysterious Fires and Lights. New York, McKay, 1967, 280p.

3053
Green, Gabriel, & Warren Smith
  Let's Face the Facts About Flying Saucers. New York, Popular Library, 1967, 127p.

3054
Greenfield, Irving A.
  The UFO Report. New York, Lancer Books, 1967, 141p.

3055
Kon, Alejandro
  La Verdad Revelada Sobre los Platos Voladores y lo que Vendrá. Buenos Aires, N. R. Pinto, 1967 (unpaged).

NOTES:

**3056**
Kronos
  Essai de Méditations Immaterielles...3 v.
Macon, Impr. Protat Frères, 1967.

**3057**
Lleget Colomer, Mario
  Mito y Realidad de los Platillos Volantes.
Barcelona, Ediciones Telstar, 1967, 191p.

**3058**
McDonald, James E.
  Unidentified Flying Objects: Greatest
Scientific Problem of our Times. Pittsburgh,
Pittsburgh Subcommittee, National Investigations Committee on Aerial Phenomena, 1967,
36p.

**3059**
The McGraw-Hill Encyclopedia of Space. New
  York, McGraw-Hill, 1967, p. 469.

**3060**
Mallan, Lloyd
  The Official Guide to UFOs. New York,
Science & Mechanics Pub. Co., 1967, 96p.
($0.75)

**3061**
Menzel, Donald H.
  UFO: Fact or Fiction? Cambridge, 1967, 15p.

**3062**
Michell, John F.
  The Flying Saucer Vision; the Holy Grail
Restored. London, Sidgwick & Jackson, Ltd.,
1967, 176p. (25 shillings)

**3063**
Moseley, James W., comp.
  Jim Moseley's Book of Saucer News. Clarksburg, W. Va., Saucerian Press, 1967, 118p.
($5.00)

**3064**
The new Report on Flying Saucers. Greenwich,
  Conn., Fawcett Pubs., Inc., 1967, 80p.

**3065**
Olsen, Thomas M.
  The Reference for Outstanding UFO Sighting
Reports. Riderwood, Md., UFO Information
Retrieval Center, Inc., 1967 -

**3066**
Pereira, Flávio A.
  O Livro Vermelho dos Discos Voadores.
São Paulo, Edições Florença Ltda., 1967, 486p.

**3067**
Rehn, K. G.
  The Question of Proof--a new Slant. Bromma,
Sweden, 1967, 5p.

**3068**
Sherwood, John C.
  Flying Saucers are Watching you. Clarksburg, W. Va., Saucerian Books, 1967, 78p.

**3069**
Shuttlewood, A.
  The Warminster Mystery. London, Neville,
Spearman, Ltd., 1967, 265p. (25 shillings)

**3070**
Soule, Gardner
  UFOs and IFOs; a Factual Report on Flying
Saucers. New York, Putnam, 1967, 189p.

**3071**
Sprinkle, R. L.
  Patterns of UFO Reports. Laramie, Wyoming,
1967, 15p.

**3072**
Steiger, Brad, & Joan Whritenour
  Flying Saucers are Hostile. New York,
Award Books, 1967, 160p.

NOTES:

UNIDENTIFIED FLYING OBJECTS INCLUDING FLYING SAUCERS

3073
The True Report on Flying Saucers, by leading UFO authorities, including Donald E. Keyhoe. Exclusive Project Blue Book sighting, photos from U. S. Air Force files. Compiled by the editors of True. Greenwich, Conn., Fawcett Publications, 1967, 96p.

3074
United Press International
Flying Saucers; a Look Special. New York, 1967, 65p.

3075
U.S. Kennedy Space Center
NASA Management Instruction: Processing Reports of Space Vehicle Fragments. Cape Canaveral, Fla., June 28, 1967, 5p. (Item #D16)

3076
Young, Mort
UFO: Top Secret. New York, Essandess Special Editions, 1967, 156p.

3077
Anglada Font, Luis
La Realidad de los OVNI a Través de los Siglos. Buenos Aires, Editorial Kier, 1968, 380p. (6.75 pesos)

3078
Baker, Robert M. L.
Investigations of Anomalistic Observational Phenomena. El Segundo, Calif., 1968, 23p.

3079
Benham, Wilfrid E.
Biometric Analysis of the 'Flying Saucer' Photographs. Archers' Court, Hastings (Sussex), Metaphysical Research Group, 1968, 26p. (7 shillings, 6 pence)

3080
Bourquin, Gilbert A.
L'invisible Nous Fait Signe. Moutier, Ed. Robert, 1968, 199p. (17 Swiss francs)

3081
Campione, Michael J.
UFO's: 20th Century's Greatest Mystery. (Michael J. Campione, Cinnaminson, New Jersey, 08077), 1968, 129p. ($3.95)

3082
Carrión, Felipe Machado
Discos Voadores: Imprevisiveis e Conturbadores. Pôrto Alegre, Escola Graf. Educandario São Luiz, 1968, 179p. (9 cruzeiros)

3083
Chase, Frank M.
Document 96: a Rationale for Flying Saucers. Clarksburg, W. Va., Saucerian Books, 1968, 123p. ($5.00)

3084
Coelho Netto, Paulo
Discos Voadores e Misterios da Aviação. Rio de Janeiro, Ed. Minerva, 1968, 125p. (4 cruzeiros)

3085
Colorado. University
Final Report of the Scientific Study of Unidentified Flying Objects Conducted by the University of Colorado. Edward U. Condon, scientific director. Daniel S. Gillmor, ed. 3 v. Boulder, Colo., 1968.

3086
Cremaschi, Inisero, & Giuseppe Pederiali
Dischi Volanti: Benvenuti. Sessantotto Illustrazioni. Bologna, Carroccio, 1968, 157p. (1000 lira)

3087
Farris, Joseph
UFO-ho ho! Cartoons for Flying Saucer Lovers. New York, Popular Library, 1968, 96p. ($0.60)

NOTES:

## UNIDENTIFIED FLYING OBJECTS INCLUDING FLYING SAUCERS

**3088**
Fernandez Luna, Walter
  La Ufología en el Uruguay. Montevideo, 1968, 14p.

**3089**
Flying Saucers and UFO's 1968.
  New York, K.M.R. Pubs., 1968, 73p.

**3090**
Flying Saucers; Twenty-one Years of UFO's, the Great Mystery of our Time, by the editors of Cowles and UPI. New York, Cowles Education Corp., 1968, 137p. (95¢)

**3091**
Fuller, John G., ed.
  Aliens in the Skies; the Scientific Rebuttal to the Condon Committee Report. Testimony by 6 Leading Scientists on Science and Astronautics, before the House Committee, July 29, 1968. New York, Putnam, 1969, 217p. ($5.95)

**3092**
Galindez, Oscar A.
  Informe Sobre los Objetos Voladores no Identificados. Córdoba, Argentina, Z. Mariani, 1968, 89p. (4 Argentine pesos)

**3093**
Hood, Joseph F.
  The Story of Airships; When Monsters Roamed the Skies. London, Barker, 1968, 141p.

**3094**
Hynek, J. Allen
  The Scientific Problem Posed by Unidentified Flying Objects. Evanston, Ill., 1968, 15p.

**3095**
Klass, Philip J.
  UFOs—Identified. New York, Random House, 1968, 290p.

**3096**
Kozel, Carlos
  Woher Kommen die Fliegenden Untertassen..? São Paulo, Agência Brasileira de Pubs., 1968? 355p.

**3097**
Loftin, Robert
  Identified Flying Saucers. New York, McKay, 1968, 245p. ($5.95)

**3098**
Lore, Gordin I. R., & Harold H. Deneault
  Mysteries of the Skies: UFOs in Perspective. Englewood Cliffs, N. J., Prentice-Hall, 1968, 237p.

**3099**
McDonald, James E.
  Are UFO's Extraterrestrial Surveillance Craft? Tucson, 1968, 4p.

**3100**
McDonald, James E.
  Science, Technology, and UFO's. Tucson, 1968, 14p.

**3101**
McDonald, James E.
  Statement on Unidentified Flying Objects. Tucson, 1968, 39p.

**3102**
McDonald, James E.
  UFO's--an International Scientific Problem. Tucson, 1968, 40p.

**3103**
Miller, Robert W., & Rick R. Hilberg
  The Saucer Enigma. Cleveland, UFO Magazine Pubs., 1968, 23p.

**3104**
The Official Guide to UFO's, a Special Science & Mechanics News Book. New York, Science & Mechanics Pub. Co., 1968, 96p.

NOTES:

## UNIDENTIFIED FLYING OBJECTS INCLUDING FLYING SAUCERS

**3105**
Rand Corp.
 UFO: What to do? Santa Monica, Calif., Nov. 27, 1968, 43p.

**3106**
Saunders, David R., & R. R. Harkins
 UFO's Yes. Where the Condon Committee Went Wrong. New York, Signet Books, 1968, 191p.

**3107**
Santesson, Hans S., ed.
 Flying Saucers in Fact and Fiction. New York, Lancer Books, 1968, 224p.

**3108**
Sprinkle, R. L.
 Personal and Scientific Attitudes, a Survey of Persons Interested in UFO Reports. Laramie, Wyoming, 1968, 11p.

**3109**
Stanway, Roger H., & Anthony R. Pace
 Flying Saucer Report: UFOs Unidentified, Undeniable. Stoke-on-Trent, Newchapel Observatory, 1968, 87p. (12 shillings, 6 pence)

**3110**
Steiger, Brad, & Joan Whritenour
 The Allende Letters. New York, Award Books, 1968, 155p.

**3111**
Tyler, Steven
 Are the Invaders Coming? New York, Tower Pubs., 1968, 139p.

**3112**
UFO: Flying Saucers. Poughkeepsie, N.Y., Western Pub. Co., 1968, 64p.

**3113**
UFO Terminology. Cleveland, Flying Saucer Digest, 1968, 6p.

**3114**
U. S. Air Force
 Aids to Identification of Flying Objects. Washington, for sale by the Supt. of Docs., U. S. Govt. Print. Office, 1968, 35p.

**3115**
U.S. Congress. House. Committee on Science & Astronautics
 Symposium on Unidentified Flying Objects. Hearings Before the Committee on Science and Astronautics, 90th Congress, Second Session, #7, July 29, 1968. Washington, U.S. Government Printing Office, 1968, 254p.

**3116**
Uriondo, Oscar A.
 El Problema Científico de los OVNI. Buenos Aires, Plus Ultra, 1968, 174p. (480 pesos)

**3117**
Vesco, Renato
 Intercettateli Senza Sparare! La Vera Storia dei Dischi Volanti. Milan, U. Mursia, 1968, 351p. (2600 lira)

**3118**
Westphal, Peter G.
 UFO UFO. Das Buch von den Fliegenden Untertassen. Stuttgart, Deutsche Verlagsanstalt, 1968, 103p. (9.80 marks)

**3119**
Wright, T.
 The Intelligent Man's Guide to Flying Saucers. South Brunswick, N. J., A. S. Barnes, 1968, 279p. ($5.95)

**3120**
Bowen, Mollie
 Flying Saucers and Outer Space. Ed. by Dan Lloyd. London, Tyndall Mitchell, 1969, 45p. (children's book, 7 shillings, 6 pence)

**3121**
Chapman, Robert
 Unidentified Flying Objects. London, Arthur Barker, Ltd., 1969, 168p. (30 shillings)

NOTES:

## UNIDENTIFIED FLYING OBJECTS INCLUDING FLYING SAUCERS

**3122**
Dean, John W.
  Flying Saucers Closeup. Clarksburg, W. Va., G. Barker, 1969, 224p.

**3123**
Fleissig, Josef
  Zahada Naseho Stoleti. Letajici Talire. Nazory Americkych, Sovetskych a Ceskoslovenskych Vedcu. Liberec, Severočes. Nakl., T. Liberecke Tisk., 1969, 191p. (23.00 korunas)

**3124**
Hervey, Michael
  UFO's in the Southern Hemisphere. Sydney, Australia, Horwitz Publications, 1969, 192p. (80¢ Australian)

**3125**
Hirano, Imao
  Aporo to Soratobu Emban. Tokyo, 1969, 208p.

**3126**
Le Poer Trench, Brinsley
  Operation Earth. London, Neville Spearman, 1969, 128p. (30 shillings)

**3127**
Lehr, Georges
  Contre les Soucoupes Volantes...Paris, Berger Levrault, 1969, 73 & 75pp. (6 francs)

**3128**
Lore, Gordon I. R.
  Strange Effects From UFO's. Kensington, Md. (National Investigations Committee on Aerial Phenomena, 3535 University Blvd., W.), 1969.

**3129**
Lore, Gordon I. R., & Harold H. Deneault
  Mysteries of the Skies; UFO's in Perspective. London, R. Hale, 1969, 237p. (36 shillings)

**3130**
Macaluso, Giuseppe
  La Verità sui Dischi Volanti e sui Loro Piloti...Rome, Pensiero e Azione, (Nicoletti & Terenzi), 1969, 417p. (1800 lira)

**3131**
McDonald, James E.
  Objets Volants non Identifiés, le plus Grand Problème Scientifique de Notre Temps?... Textes Traduits par René Fouéré...Science, Technologie et UFO's, Etude Présenté a un Seminaire Général des United Aircraft Research Laboratories, East Hartford, Conn., 26 Janvier, 1968. Paris, Phénomènes Spatiaux, 1969, 89p. (7.50 francs)

**3132**
Michel, Aimé
  Pour les Soucoupes Volantes, par Aimé Michel. Contre les Soucoupes Volantes, par Georges Lehr. Nancy, Berger-Levrault, 1969, 80p. (6 francs)

**3133**
National Investigations Committee on Aerial Phenomena
  UFO's:: a new look; a special report. Ed. by Donald E. Keyhoe and Gordon I. R. Lore, Jr. Washington, 1969, 46p.

**3134**
Patrovsky, Venceslav
  Zahady Letajicich Taliru. Prague, NV, 1969, 218p. (15.00 korunas)

**3135**
Pedersen, Frank, & Iver O. Kjems
  UFO-Orientering. 2d ed. Copenhagen, UFO-NYT, Peder Lykkes Vej 55, 1969, 120p. (8.00 krones)

**3136**
Rehn, K. G.
  UFO! Nya Fakta om de Flygende Tefaten. Gothenburg, Zinderman, 1969, 203p.

NOTES:

**3137**
Ribera Jorda, Antonio
  Platillos Volantes en Iberoamérica y España.
Barcelona, Editorial Pomaire, 1969, 429p.

**3138**
Romaniuk, Pedro
  Naves Extraterrestres: sus Incursiones a la Tierra. Buenos Aires, Editorial Merlin, 1969, 108p. (2.25 pesos)

**3139**
Sanderson, Ivan T.
  Uninvited Visitors: a Biologist Looks at UFO's. London, Spearman, 1969, 245p. (30 shillings).

**3140**
Saunders, David R., & R. R. Harkins
  UFO's? Yes! Where the Condon Committee Went Wrong. Cleveland, The World Pub. Co., 1969, 256p. ($5.95)

**3141**
Steckling, Fred
  Why are They Here? Spaceships from Other Worlds. New York, Vantage Press, 1969, 148p. ($3.95)

**3142**
Tarade, Guy
  Soucoupes Volantes et Civilisations d'Outre-space. Paris, Eds. J'ai lu, 1969, 319p. (4 francs)

**3143**
Tucci, Eduardo A., & Alberto Giordano
  Los Platos Voladores y sus Tripulantes. Buenos Aires, Editorial Glem, 1969, 228p. ($2.75)

**3144**
U. S. Congress. House. Committee on Science and Astronautics.
  Aliens in the Skies; the Scientific Rebuttal to the Condon Committee Report. Testimony by six leading scientists before the House Committee on Science and Astronautics, July 29, 1968. Edited and with an introd. and commentary by John G. Fuller. New York, Putnam, 1969, 217p. ($5.95)

**3145**
U.S. National Academy of Sciences-National Research Council. National Academy of Sciences Panel
  Review of the University of Colorado Report on Unidentified Flying Objects: Special Report, by Gerald M. Clemence, et al. (AD-688541; AFOSR-69-1276TR). Washington, Jan. 1969, 10p. (Available from NTIS).

**3146**
Vallée, Jacques
  Passport to Magonia: From Folklore to Flying Saucers. Chicago, H. Regnery Co., 1969, 372p. ($6.95)

**3147**
Vesco, Renato
  I Velivoli del Mistero. I Segreti Tecnici del Dische Volanti. Milano, U. Mursia, 1969, 423p.

**3148**
Adler, Bill
  Flying Saucers Have Arrived! Ed. by Jay David. New York, World Pub. Co., 1970, 352p. ($6.95)

**3149**
Bemelmans, Hans
  "Reports From Ibiuna." London, Flying Saucer Review (16:1), Jan.-Feb. 1970, pp. 15-19.

**3150**
Bordeleau, Henri
  J'ai Percé le Mystère des Soucoupes Volantes. Montreal, Société Nefer, 1970, 323p.

**3151**
Burt, Eugene H.
  UFO's and Diamagnetism; Correlations of UFO and Scientific Observations. New York, Exposition Press, 1970, 134p. ($5.00)

---

NOTES:

## UNIDENTIFIED FLYING OBJECTS INCLUDING FLYING SAUCERS

**3152**
Campione, Michael J.
  UFO Manual. (M. J. Campione, Cinnaminson, New Jersey 08077), 1970 (folio) ($2.00)

**3153**
Chapman, Robert
  Unidentified Flying Objects. London, Mayflower, 1970, 160p. (5 shillings).

**3154**
Dello Strologo, Saulla
  Quello che i Goveri ci Nascondono sui Dischi Volanti. Milan, G. De Vecchi, 1970, 190p. (3000 lira)

**3155**
Durrant, Henry
  Le Livre Noir des Soucoupes Volantes. Paris, Laffont, 1970, 301p. (18 francs)

**3156**
Dutta, Rex
  Flying Saucer Viewpoint. London, Pelham Books (31 Corsica St., Highbury Corner), 1970, 115p. (30 shillings)

**3157**
'Flying Saucer Review' Case Histories. London, Flying Saucer Service (49a Kings Grove, London S.E. 15), Oct. 1970- (supp. #1) (4 shillings)

**3158**
Glemser, Kurt
  UFO Report, 1969. Kitchener, Canada, 1970, 44p.

**3159**
Gurney, Gene, & Clare Gurney
  Unidentified Flying Objects. New York, Abelard-Schuman, 1970, 144p. ($4.50)

**3160**
Keel, John A.
  UFO's, Operation Trojan Horse. New York, Putnam, 1970, 320p. ($6.95)

**3161**
Kolosimo, Peter
  Des Ombres sur les Etoiles. Paris, Michel, 1970, 384p. (25 francs)

**3162**
Kuningas, Tapani
  UFOja Suomen Taivaala. Helsinki, Kirjayhtyma, 1970, 175p.

**3163**
Larson, Kenneth
  The Topstone. Los Angeles, Calif. 90057 (415-1/2 S. Coronado St.), 1970.

**3164**
Leslie, Desmond
  Flying Saucers Have Landed. London, Spearman, 1970, 281p. (2.10 pounds)

**3165**
Perego, Alberto
  Gli Extraterrestri Sono Tornati. Il Misterio dell' Apollo 13. Il Rapporto sull'-Aviazione di Altri Pianeti. Rome, Ed. del Centro Italiano Studi Aviaziones Elettromagnética, 1970, 221p. (3,000 lira)

**3166**
Philipp, Franz
  Deutscher Raumflug ab 1934; ein Unbequemes Buch. 2d ed. Berlin (Spandau), F. Philipp, 1970, 104p.

**3167**
Santos, Maurice
  Les Soucoupes Volantes aux Frontières de L'Impossible. Monte Carlo, Regain, 1970, 189p. (17 francs)

NOTES:

UNIDENTIFIED FLYING OBJECTS INCLUDING FLYING SAUCERS

3168
UFO'er-Flyvende Cigarer...Ed. by Leif Eckhoff Pedersen. Copenhagen, Hirtshals Dansk IGAP, Peder Rimmensgade 21, 1970, 21p.

3169
Atmananda
Ajñātācā Śodha va Bodha. Poona, India, Gayatri Prakashana, 1971, 407p.

3170
Bandini, Franco
Il Mistero dei Dischi Volanti. Florence, Centro Internazionale del Libro, 1971. 115p.

3171
Bernard, Raymond
La Terre Creuse; la Plus Grande Découverte Géographique de l'Histoire Humaine. Tr. by Robert Genin. Paris, A. Michel, 1971, 236p. (19.50 francs)

3172
Bordeleau, Henri
J'ai Chassé les Pilotes des Soucoupes Volantes. Montreal, Société Nefer, 1971, 204p.

3173
Cathie, Bruce L., & P. N. Temm
Harmonic 695; the UFO and Anti-Gravity. Wellington, Reed, 1971, 201p. ($4.95 Australian)

3174
Codr, Milan
Volání Dálných Světů, Usti nad Lahem, Czechoslovakia, Severočes, nakl., t. SG o1, Liverec, 1971, 210p. (24.00 koruna)

3175
Davidson, Leon
UFO's: an Analysis of Project Blue Book, Special Report 14. 4th ed. rev., 1971 (available from Stanton Friedman, UFO Research Institute, P.O. Box 941, Lawndale, California) 90260

3176
Eastern UFO Symposium, Baltimore, 1971
Proceedings. Ed. by Coral E. Lorenzen. Tucson, Aerial Phenomena Research Organization, 1971, 40p.

3177
Flammande, Paris
The Age of Flying Saucers; Notes on a Projected History of Unidentified Flying Objects. New York, Hawthorn Books, 1971, 288p. ($8.95)

3178
Friedman, Stanton T.
UFOs—Myth & Mystery; 1971 Midwest UFO Conference, sponsored by Midwest UFO Network (MUFON), featuring Stanton T. Friedman. (n. p.), 1971, unpaged.

3179
Garreau, Charles
Soucoupes Volantes; Vingt ans d'enquêtes. Paris, Mame, 1971, 213p. (18 francs)

3180
Gifford, Dennis
Science Fiction Film. New York, Dutton, 1971, 160p.

3181
Hobana, Ion
OZN, o Sfidare Pentru Ratiunea Umana. Bucharest, Editura Enciclopedica Romana, 1971, 199p. (6.25 leu)

3182
Kuningas, Tapani
Ufojen Jäljillä Uusimmat UFO-Havainnot Suomesta ja Muualta. Helsinki, Kirjayhtyma, 1971, 263p. (19 markkow)

NOTES:

## UNIDENTIFIED FLYING OBJECTS INCLUDING FLYING SAUCERS

3183
Midwest UFO Conference, 2d, St. Louis, 1971
UFO's—Defiance to Science; Conference Proceedings. Participating: Walter H. Andrus, Jr., et al. Moderator: John F. Schuessler. O'Fallon, Mo., UFO Study Group of Greater St. Louis, 1971, 115p. ($3.00)

3184
Moseley, James W.
The Wright Field Story. Clarksburg, W. Va., Saucerian Books, 1971, 80p.

3185
Proceedings of the Eastern UFO Symposium, Jan. 23, 1971, Baltimore, Md. Sponsored by Aerial Phenomena Research Organization, 3910 E. Kleindale Rd., Tucson, Arizona 85712. Ed. by Coral E. Lorenzen. Tucson, Aerial Phenomena Research Organization, 1971, 40p.

3186
Rodríguez, Vicente C.
OVNI: Estudios Sobre Naves Interplanetarias. Bahía Blanca, Arg., Gráfica del Sur, 1971, 192p.

3187
Shuttlewood, Arthur
UFO's, Key to the New Age. London, Regency Press, 1971, 216p. (1.80 pounds)

3188
Tarade, Guy
Les Dossiers de l'Etrange. Paris, R. Laffont, 1971, 318p.

3189
Uriondo, Oscar A.
Los Supuestos de la Actitud Científica Hacia los OVNI. Comunicación Presentada en el 2do. Simposio Nacional de Investigaciones Sobre OVNI, Reunido en Buenos Aires, los Días 29, 30 y 31 de Octubre de 1970. Buenos Aires, Centro de Estudios de Fenómenos Aéreos Inusuales, 1971, 16p.

3190
Vesco, Renato
Intercept—but Don't Shoot; the True Story of the Flying Saucers. Tr by D. D. Paige. New York, Grove Press, 1971, 338p. ($8.50).

3191
Weverbergh, Louis J., & Ion Hobana
UFO's in Oost en West. Deventer, Holland, N. Kluwer, 1971. (17.90 florins)

3192
Araújo, Hernani Ebecken de
Os Discos Voadores e a Teoria da Relatividade do Dr. Einstein, 1957-1969. Flying Saucers and the Relativity Theory from Dr. Einstein, 1957-1969. 2d ed. Rio de Janeiro, 1972? 69p.

3193
Bergier, Jacques, et le Groupe INFO
Le Livre de l'Inexplicable. Adapted by Georges H. Gallet. Paris, Michel, 1972, 244p. (24 francs)

3194
Bergrun, Norman R.
Tomorrow's Technology Today. Campbell, California, Academy Press, 1972, 65p.

3195
British Unidentified Flying Object Research Association
A Guide to the UFO Phenomenon. London, BUFORA, 1972, 176p.

3196
Casault, Jean
Manifeste Pour l'Avenir. Quebec, Eds. AFFA, 1972, 200p.

3197
Daniel (pseud.)
Luci All'Orizzonte. Macchine Extraumane da Almeno 10,000 Anni, la Natura della Giostra della Mano del Signore, la Metapsichica dell'Anima. Rome, Tip. Grafikarte, 1972? 145p.

NOTES:

## UNIDENTIFIED FLYING OBJECTS INCLUDING FLYING SAUCERS

**3198**
Dohmen, J. G.
  A Identifier (suivi de) Le Cas Adamski. Que Penser des Soucoupes Volantes? Biarritz, Travox (26 Ave. de l'Impératrice), 1972, 240p. (39 francs)

**3199**
Dutta, Rex
  Flying Saucer Message. London, Pelham Books, (31 Corsica St., Highbury Corner), 1972, 117p. (1.75 pounds).

**3200**
Ferguson, Jean
  Tout sur les Soucoupes Volantes. Montreal, Leméac, 1972, 258p.

**3201**
  Forschung in Fessein: Elektro-Gravitation UFO-Phänomen, das Rätsel der Elektro-Gravitation. Wiesbaden-Schierstein, Ventla-Verlag, 1972, 272p.

**3202**
Gallup, George H.
  The Gallup Poll: Public Opinion 1935-1972. 3v. New York, Random House, 1972.

**3203**
Glemser, Kurt
  UFO's: Menace from the Skies. Kitchener, Canada, Galaxy Press, 1972, 36p.

**3204**
Guieu, Jimmy
  Black out sur les Soucoupes Volantes. Paris, Omnium Littéraire, 1972, 292p.

**3205**
Guieu, Jimmy
  Les Soucoupes Volantes Viennent d'un Autre Monde. Paris, Omnium Littéraire, 1972, 292p.

**3206**
Hynek, J. Allen
  The UFO Experience; a Scientific Inquiry. Chicago, Regnery, 1972, 276p. ($6.95)

**3207**
Kettelkamp, L.
  Investigating UFO's. London, Ronald Stacy, Ltd., (56 Doughty St.), 1972, 96p. (1.25 pounds)

**3208**
Lax, Soini
  Pudasjärven Ufot. Uraania vai Utopiaa? Helsinki, Kirjayhtymä, 1972, 117p.

**3209**
Lindsay, Gordon
  The Riddle of the Flying Saucers. Kitchener, Canada, Galaxy Press, 1972, 31p.

**3210**
Lob, Jacques, & R. Gigi
  Le Dossier des Soucoupes Volantes. Neuilly, Dargaud, 1972, 72p.

**3211**
Midwest UFO Conference. 3d, Quincy, Ill., 1972
  MUFON '72 Conference Proceedings, June 17, 1972, Quincy, Ill. Quincy, Midwest UFO Network, 1972, 145p.

**3212**
Pace, A. R., & R. H. Stanway
  UFO's Unidentified Undeniable. Newchapel Observatory, Stoke-on-Trent, Staffordshire, England, 1972.

**3213**
Schwartz, A. W., ed.
  Theory and Experiment in Exobiology, v. 2. Groningen, Wolters-Noordhoff, 1972, 147p.

**3214**
  UFO's; a Scientific Debate. Ed. by Carl Sagan & Thornton Page. Ithaca, Cornell University Press, 1972, 310p. ($12.50)

NOTES:

## UNIDENTIFIED FLYING OBJECTS INCLUDING FLYING SAUCERS

3215
Vendégek a Világűrből? Miről vall a "Repülő" Csészealjak" es a Paleoasztronautika Irodalma. Bucharest, 1972, 336p.

3216
Vesco, Renato
Operazione Plenilunio.I Voli Spaziali dei Dischi Volanti. Milan, U. Mursia, 1972, 553p.

3217
Wat zijn UFO's? A Talk by Netty de Bruyn Kops, et al, on television, under Direction of Jack van Belle. Hilversum, Holland, Nederlandse Omroep Stichting (Postbus 40,000), 1972, 45p.

3218
Weverbergh, Julian, & Ion Hobana
UFO's in Oust en West. Deventer, Holland, Kluwer, 1972, 336p. (19.50 florins)

3219
Aréjula, Francisco
Hacia una Física de los OVNIS. Barcelona, Distribuidora CEDEL, 1973, 117p.

3220
Durrant, Henry
Les Dossiers des OVNI. Paris, Laffont, 1973, 309p.

3221
Eden, Jerome
Planet in Trouble; the UFO Assault on Earth. New York, Exposition Press, 1973, 214p.

3222
Edwards, Frank
Strange World. New York, Bantam Books, 1973, 237p.

3223
Gaston, Patrice
Disparitions Mystérieuses, Le Cosmos Nous Observe. Paris, R. Laffont, 1973, 336p.

3224
Gheorgita, Florin
O. Z. N., o Problema Moderna. Jassy, Rumania, Junimea, 1973, 168p.

3225
Guasp, Miguel
Teoría de Procesos de los OVNI. Valencia, Spain, 1973, unpaged.

3226
Hirano, Imao
Emban ni tsuite no Majime na Hanashi. Tokyo, 1973, 289p.

3227
Holiday, Frederick W.
The Dragon and the Disc: an Investigation into the Totally Fantastic. London, Sidgwick & Jackson, 1973, 247 + 12 pp.

3228
Jacobs, David
The Controversy Over Unidentified Flying Objects in America, 1896-1973. Madison, 1973, 372p.

(Doctoral dissertation, University of Wisconsin-Madison)

3229
Kampen, Hans van
Vliegende Schotels. Waan of Wetenschap? Baarn, Holland, H. Meulenhoff, 1973, 180p.

3230
Keyhoe, Donald E.
Aliens From Space. Garden City, Doubleday, 1973. ($7.95)

NOTES:

## UNIDENTIFIED FLYING OBJECTS INCLUDING FLYING SAUCERS

**3231**
Lagarde, Fernand, et al
  *Mystérieuses Soucoupes Volantes.*
Paris, Eds. Albatros, 1973, 318p.

**3232**
McCampbell, James M.
  *Ufology: new Insights From Science and Common Sense.* Belmont, Calif., IUFOB Jaymac Co. (12 Bryce Court), 1973.

**3233**
Netto, Paulo Coelho
  *Astrônomos e Discos Voadores.*
Rio de Janeiro, Livraria São José, 1973, 28p.

**3234**
Tomas, Andrew
  *We are not the First. Riddles of Ancient Science.* New York, Bantam Books, 1973, 180p.

**3235**
Tomorrow Show, NBC-TV Network, Oct. 23, 1973, 12-1 A.M., channel 4-Milwaukee, Wis. (Program dedicated to UFO's).

**3236**
Translations From Priroda #6, 1973. (Trans. into English From Priroda, Moscow, Oct. 11, 1973). Arlington, Va., (Joint Pubs. Research Service), 1973, 18p.

**3237**
Uriondo, Oscar A.
  *OVNIS en la Argentina: Suplemento del Catálogo de Avistamientos Tipo-I.*
Buenos Aires, Centro de Estudios de Fenómenos Aéreos Inusuales, 1973, 26p.

**3238**
Zungri, Giuseppe
  *L'Enigma dei Cieli.* Turin, MEB, 1973, 181p.

**3239**
Bourret, Jean C.
  *La Nouvelle Vague des Soucoupes Volantes.*
Paris, Eds. France-Empire, 1974, 297p.

**3240**
Brovetto, P., & V. Maxia
  *On the Airglow Phenomenon Usually Referred to as UFO.* Cagliari, Italy, Instituto di Fisica, Università di Cagliari, 1974, 24p.

**3241**
Emenegger, Robert
  *UFO's, Past, Present and Future.* New York, Ballantine, 1974.

**3242**
Fowler, Raymond E.
  *UFO's: Interplanetary Visitors; a UFO Investigator Reports on the Facts, Fables and Fantasies of the Flying Saucer Conspiracy.* Jericho, N. Y., Exposition Press, 1974, 365p.

**3243**
Gösta Rehn, K.
  *UFO's Here and now.* New York, Abelard-Schuman, 1974, 192p.

**3244**
Klass, Philip J.
  *UFO's Explained.* New York, Random House, 1974, 369p.

**3245**
Pinotti, Roberto
  *UFO: la Congiura del Silenzio.* Milan, Armenia Ed., 1974, 245p.

**3246**
Ribera Jorda, Antonio
  *El Gran Enigma de los Platillos Volantes desde la Prehistoria hasta la Epoca Actual.* Esplugas de Llobregat, Spain, Plaza & Janés, 1974, 563p.

NOTES:

UNIDENTIFIED FLYING OBJECTS INCLUDING FLYING SAUCERS

**3247**
Salisbury, Frank B.
  The Utah UFO Display: a Biologist's Report. New York, Devin-Adair, 1974, 227p.

**3248**
"Seen any Flying Saucers Lately? The UFO Controversy." The David Suskind Show, Broadcast Over Channel 10 Television, Milwaukee, Wisconsin, Friday, February 15, 1974, 9:30 P.M. (Public Broadcasting System).

**3249**
Tomorrow Show (Devoted to UFO's). NBC-TV (Channel 4, Milwaukee, Wis.). 12 Midnight-1 A.M., Oct. 30, 1974. Host: Tom Snyder. Discussion with Herb Schirmer, et al.

**3250**
UFO, Past, Present and Future.
  Producer: Allan F. Sandler.
  Directors: Ramon Rivas and Robert Emenegger.
  16 & 35 mm. film. Running time: 80 minutes. Hollywood, Sandler Films, 1974.

**3251**
UFO's: do you Believe? Reported by Jim Hartz. Directed and Produced by Craig Leake. Television Program, NBC-TV Network, Dec. 15, 1974, 10PM, Eastern Standard Time.

**3252**
Gris, Henry
  "Secret Evidence Shows UFO's Come From Other Worlds." National Enquirer July 15, 1975, p. 4.

## PERIODICAL ARTICLES

**3253**
Douglas, A. V.
  "Other Little Ships." Atlantic Monthly 136, Aug. 1925, pp. 169-174.

**3254**
Simons, Rodger L.
  "The Space Ship Hokum." The Catholic World 140, Nov. 1934, pp. 164-170.

**3255**
"Flying Saucers: the Somethings." Time 50, July 14, 1947, p. 18.

**3256**
"Remember the Flying Saucers?" Science Digest 22, Oct. 1947, pp. 69-71.

**3257**
"Dr. C. C. Wylie on Mass Hysteria; Urges sky Patrol to Guard Against Repetition." New York Times, Dec. 27, 1947, p. 28, col. 1.

**3258**
"Editorial on Dr. C. C. Wylie, Dec. 8, on Mass Hysteria." New York Times, Jan. 1, 1948, p. 22, col. 3.

**3259**
Stanley, Neil, & Chester S. Geier
  "The Flying Saucer Jigsaw Puzzle." Fate 1, Fall 1948, pp. 101-104.

**3260**
Shalett, Sidney
  "What you can Believe About Flying Saucers." Part I. Saturday Evening Post 221, April 30, 1949, pp. 20-21, 136-139.

NOTES:

UNIDENTIFIED FLYING OBJECTS INCLUDING FLYING SAUCERS

**3261**
"Things That go Whiz; Flying Saucers." Time 53, May 9, 1949, p. 98.

**3262**
"What the Air Force Believes About Flying Saucers." Fate 2, Nov. 1949, pp. 69-83.

**3263**
"Quarter's Polls on Flying Saucers." Public Opinion Quarterly (14:3), 1950, pp. 597-598.

**3264**
"Psychoanalyzing the Flying Saucers." Air Force 33, Feb. 1950, pp. 15-19.

**3265**
Klemin, Alexander
"The Flying Saucer." Aero Digest 32, March 1950, pp. 129-130.

**3266**
"The Flying Saucer Mystery." Science News Letter 57, March 25, 1950, p. 188.

**3267**
Rougeron, Camille
"Soucoupistes et Antisoucoupistes." Paris, Illustration 6, April 1950, p. 422.

**3268**
Whittington, George
"What's True About the Disc?" The Journal of Spaceflight and the Rocket News Letter, April 1950, pp. 2-5.

**3269**
"U.S. News & World Report Claims Saucers are Revolutionary U.S. Navy Craft...H. J. Taylor Reports Saucers are Secret U.S. Military Inventions." New York Times, April 4, 1950, p. 10, col. 2.

**3270**
Waithman, Robert
"These Flying Saucers." London, The Spectator 184, April 14, 1950, pp. 489-490.

**3271**
"Flying Saucers Again." Newsweek 35, April 17, 1950, p. 29.

**3272**
"Saucer-Eyed Dragons." Time 55, April 17, 1950, pp. 52-54.

**3273**
"Psychoanalyzing the Flying Saucers." Science Digest 27, May 1950, pp. 29-34.

**3274**
Bowman, Norman J.
"A Scientific Analysis of the Flying Disk Reports." The Rocket News Letter 3, June 1950, pp. 2-7.

**3275**
Fuller, Curtis
"Flying Saucers, Fact or Fiction?" Flying 47, July 1950, pp. 16-17.

**3276**
Wilkins, Harold T.
"Flying Saucers." London, The Contemporary Review 78, July 1950, pp. 49-53.

**3277**
"Flying Saucers." Public Opinion Quarterly 14, Fall 1950, pp. 597-598.

**3278**
"More Flying Saucers." Review of Book, Behind the Flying Saucers, by F. Scully. Science News Letter 58, Sept. 16, 1950, p. 181.

NOTES:

**3279**
Day, Langston
"Flying Saucers: Fact or Fiction?" London, Chambers' Journal, Nov. 1950, pp. 668-670.

**3280**
Jones, Harold S.
"The Flying Saucer Myth." London, The Spectator, Dec. 15, 1950, pp. 686-687.

**3281**
Palmer, Ray
"New Report on the Flying Saucers." Fate 4, Jan. 1951, pp. 63-81.

**3282**
Wood, R. H.
"Saucers, Secrecy and Security." Aviation Week 54, Feb. 19, 1951, p. 50.

**3283**
"Flying Saucers: Ballooney, not Baloney." Aviation Week 54, Feb. 19, 1951, p. 13.

**3284**
"Flying Saucers Again." Senior Scholastic, Teacher's Edition 58, Feb. 21, 1951, p. 12.

**3285**
"Belated Explanation on Flying Saucers." Time 57, Feb. 26, 1951, p. 22.

**3286**
"Saucers? No, Skyhooks." Newsweek 37, Feb. 26, 1951, p. 17.

**3287**
Wilson, Richard
"A Nuclear Physicist Exposes Flying Saucers." Look 15, Feb. 27, 1951, pp. 60-62, 64.

**3288**
"Flying Saucers (Editorial)." London, Engineer, March 30, 1951, pp. 416-417.

**3289**
Ley, Willy
"More About out There." Review of The Flying Saucers, by C. Heard. Saturday Review of Literature 34, April 28, 1951, pp. 20-21.

**3290**
Webster, Robert N.
"The Saucers Aren't Balloons." Fate 4, May-June 1951, pp. 4-7.

**3291**
Vinther, L. W.
"Another Saucer Mystery." Flying 48, June 1951, p. 23.

**3292**
Wood, R. H.
"Where are the Flying Saucers?" Aviation Week 54, June 25, 1951, p. 74.

**3293**
"What Were the Flying Saucers?" Popular Science 159, Aug. 1951, pp. 74-75.

**3294**
"U.S. Air Force Warns Weather Balloons Might be Mistaken for Saucer." New York Times, Oct. 12, 1951, p. 39, col. 4.

**3295**
Kaempffert, Waldemar
"Expert Sees Flying Object--Saucer or Balloon." Science Digest 31, Feb. 1952, p. 74.

**3296**
"Editorial on Life's Account of Saucers." New York Times, April 12, 1952, p. 10, col. 2.

**3297**
"Flying Saucers are old Stuff." Popular Science 160, May 1952, pp. 145-147.

NOTES:

UNIDENTIFIED FLYING OBJECTS INCLUDING FLYING SAUCERS

3298
Lambert, R. S.
"Flying Saucers--Their Lurid Past." Toronto, Canada, Saturday Night (67:9), May 17, 1952, p. 18.

3299
Ginna, R. E.
"Saucer Reactions." Life 32, June 9, 1952, p. 20.

3300
Wylie, C. C.
"Saucers Elude Astronomers." Science News Letter 61, June 14, 1952, p. 375.

3301
Lee, B. S.
"AF vs. Saucers." Aviation Week 56, June 23, 1952, p. 16.

3302
Aswell, J. R.
"Flying Saucers, New in Name Only." Reader's Digest 61, July 1952, pp. 7-9.

3303
"U.S. Air Force Cites 24-Hour Readiness to Challenge Unidentified Objects; Explains 2-Hour lag in Chasing Objects Over Washington..." New York Times, July 29, 1952, p. 23, col. 6.

3304
"U.S. Air Force Calls Objects Natural Phenomena...Major General Samford's Views Summarized...New York State Air Raid Filter Center Reports Rise in Sightings." New York Times, July 30, 1952, p. 1, col. 2.

3305
Fuller, B. A. G.
"Flying Saucers." The Journal of Philosophy 49, Aug. 1952, pp. 545-559.

3306
"Study Radar 'Ghosts'." Science Newsletter 62, Aug. 1952, p. 99.

3307
"Major General Ramey Reports None of 1500 Reports to U.S. Air Force Since 1947 Shows Evidence of Material Objects; Says no Pattern Established." New York Times, Aug. 4, 1952, p. 3, col. 4.

3308
"Cause of Flying Saucers." Science News Letter 62, Aug. 9, 1952, p. 82.

3309
"Saucers Under Glass." Newsweek 40, Aug. 18, 1952, p. 49.

3310
"U.S. Air Force Releases Details of Talk Between Jet Pilot and Airport Control Tower Just Before he Crashed While Chasing Unidentified Object Near Godman Base, Ky., Jan. 7, 1948." New York Times, Aug. 21, 1952, p. 3, col. 5.

3311
"Dr. O. Struve's Speech on Life Possibilities on Other Planets and Solar Systems as Debunking Belief That Saucers are Space Ships From Another World." New York Times, Aug. 24, 1952, sec. IV, p. 8, col. 3.

3312
"No Visitors From Space." Science News Letter 62, Aug. 30, 1952, p. 140.

3313
"Flying Saucers." London, Journal of the British Interplanetary Society 11, Sept. 1952, pp. 224-226.

NOTES:

## UNIDENTIFIED FLYING OBJECTS INCLUDING FLYING SAUCERS

**3314**
Levitt, I. M.
"Scientist Diagnoses the Flying Saucer." Popular Mechanics 98, Sept. 1952, p. 95.

**3315**
Menzel, Donald H.
"New Theory of the Flying Saucers." Science Digest 32, Sept. 1952, pp. 11-16.

**3316**
Mulholland, J.
"Magicians Scoff at Flying Saucers." Popular Science 161, Sept. 1952, pp. 96-98.

**3317**
"Scientists at International Congress on Astronautics, Stuttgart, not Interested in Saucers; do not Believe They Originate in USSR or on Another Planet." New York Times, Sept. 5, 1952, p. 18, col. 6.

**3318**
Lang, Daniel
"A Reporter at Large, Something in the sky." The New Yorker 28, Sept. 6, 1952, pp. 68-89.

**3319**
Black, V.
"Flying Saucer Hoax." American Mercury 75, Oct. 1952, pp. 61-66.

**3320**
"Apparaissaient des Soucoupes Volantes." Paris, France Illustration 8, Oct. 4, 1952, pp. 363-372.

**3321**
"How to fly a Saucer." Collier's 130, Oct. 4, 1952, pp. 50-51.

**3322**
Campbell, John W.
"How do Saucers fly?" Pic 23, Nov. 1952, pp. 16-17, 73-74.

**3323**
Elliott, L.
"Flying Saucers: Myth or Menace? Picture Story." Coronet 33, Nov. 1952, pp. 47-54.

**3324**
"Hollywood Builds Flying Saucers." Popular Science 161, Nov. 1952, pp. 132-134.

**3325**
"Deputy Premier Pervukhin (USSR) Links Saucers to U.S. war Jitters." New York Times, Nov. 7, 1952, p. 6, col. 3.

**3326**
Fuller, Curtis
"Let's get it Straight About the Saucers." Fate 5, Dec. 1952, pp. 20-31.

**3327**
Thomson, M.
"Saucers?" Science 116, Dec. 5, 1952, p. 640.

**3328**
Gelatt, R.
"Flying Saucer Hoax." Saturday Review 35, Dec. 6, 1952, p. 31.

**3329**
"Civil Aeronautics Authority Reports Saucers are Reflections Caused by Temperature Inversions; Says Cold air Breaking out From Layers of hot air Cause Radar Beams and Ground Lights to Rebound." New York Times, Dec. 11, 1952, p. 35, col. 8.

**3330**
"Feasible Flying Saucers." London, Flight 63, Jan. 9, 1953, p. 46.

NOTES:

## UNIDENTIFIED FLYING OBJECTS INCLUDING FLYING SAUCERS

**3331**
Fuller, Curtis
"Saucers and Ionization." *Fate* 6, Feb. 1953, pp. 8-10.

**3332**
Perrett, J.
"Barbu, Marc de Café, Hallebardes, la Question des Soucoupes." Paris, *Mercure de France* 317, March 1953, pp. 414-418.

**3333**
Hynek, J. Allen
"Unusual Aerial Phenomena." *Journal of the Optical Society of America* 43, April 1953, pp. 311-314.

**3334**
"Man-Made Flying Saucer." London, *RAF Flying Review* 8, April 1953, pp. 11-12.

**3335**
May, J. A.
"Flying Saucers are Hard to Wash up." Ottawa, *Roundel* 5, April 1953, p. 47.

**3336**
"Canadian 'Saucer' may be Delta Fighter." London, *Air Pictorial & Air Reserve Gazette* 15, May 1953, p. 153.

**3337**
Clarke, Arthur C.
"Flying Saucers." London, *Journal of the British Interplanetary Society* 12, May 1953, pp. 97-100.

**3338**
Wylie, C. C.
"Those Flying Saucers." *Science* 118, July 31, 1953, pp. 124-126.

**3339**
Hull, John
"Obituary of the Flying Saucers." *Air Line Pilot* 22, Sept. 1953, pp. 13-14.

**3340**
Keyhoe, Donald E.
"Flying Saucers, Fact or Fancy?" *Air Line Pilot* 22, Oct. 1953, pp. 9-10.

**3341**
Webster, Robert N.
"Saucers: Material or Immaterial?" *Fate* 6, Oct. 1953, pp. 4-5.

**3342**
"ATIC Begins Study of Saucer Reports." *Aviation Week* 59, Oct. 19, 1953, p. 18.

**3343**
Keyhoe, Donald E.
"Flying Saucers From Outer Space." *Look* 17, Oct. 20, 1953, p. 114.

**3344**
Straight, M.
"To our Readers." *New Republic* 129, Oct. 26, 1953, p. 22.

**3345**
Bowman, Norman J.
"The Need for Critical Analysis of Flying Disc Reports." London, *Journal of Spaceflight* 5, Nov. 1953, p. 11.

**3346**
"Canada Building Lab to try to Prove or Disprove Existence." *New York Times*, Nov. 12, 1953, p. 8, col. 5.

NOTES:

**3347**
"Flying Saucer Controversy; Meteorological Balloons, and Weather Conditions Which May Provide Explanations of the Phenomenon." London, Illustrated London News 223, Dec. 5, 1953, pp. 936-937.

**3348**
Long, A.
"Air Force Looks at Saucers." Science Digest 35, Jan. 1954, pp. 9-10.

**3349**
"On the Flying Saucer Trail." The American Magazine 157, April 1954, p. 56.

**3350**
Ross, John C.
"Canada Hunts for Saucers." Fate 7, May 1954, pp. 12-15.

**3351**
Ruppelt, Edward J.
"What our Air Force Found out About Flying Saucers." True, May 1954, pp. 19-20, 22, 24, 26, 30, 124-134.

**3352**
Keyhoe, Donald E.
"Flying Saucers: are They Interplanetary?" London, RAF Flying Review 9, Aug. 1954, pp. 11-12.

**3353**
"No new Theories, Evidence, Saucers, Says Pentagon in Lifting UFO lid." Air Force Times 15, Aug. 28, 1954, p. 9.

**3354**
Crum, Norman J.
"Flying Saucers and Book Selection." Library Journal 79, Oct. 1, 1954, pp. 1719-1722.

**3355**
Genet
"Letter From Paris." New Yorker 30, Oct. 23, 1954, p. 159.

**3356**
Dowding, Hugh C.
"I Believe in Flying Saucers." Fate 7, Nov. 1954, pp. 24-26.

**3357**
Lang, Daniel
"Ils Livrent Combat aux Soucoupes Volantes." Paris, Constellation 79, Nov. 1954, pp. 133-154.

**3358**
Gibbs-Smith, Charles H.
"Flying Saucers." London, The Queen 202, Nov. 17, 1954, p. 64.

**3359**
"Pres. Eisenhower Reports U.S. Air Force has Assured him Saucers are not Invading Earth...Aide Reports only 10% Cannot be Evaluated." New York Times, Dec. 16, 1954, p. 1, col. 8.

**3360**
"Finds Saucers Exist Solely in Imagination." Science Digest 37, Jan. 1955, p. 24.

**3361**
"Saucer Sorcery: Cartoons." New York Times Magazine, Jan. 9, 1955, p. 25.

**3362**
"Flying Saucers." London, Practical Mechanics, Feb. 1955, pp. 217-218.

**3363**
Flick, David
"Tripe for the Public." Library Journal 80, Feb. 1, 1955, p. 202, 204.

NOTES:

## UNIDENTIFIED FLYING OBJECTS INCLUDING FLYING SAUCERS

**3364**
"Confusion in the sky." Fate 8, March 1955, pp. 24-29.

**3365**
Kocivar, B.
"Is This the Real Flying Saucer?" Look 19, June 14, 1955, pp. 44-46.

**3366**
Espigador
"Now You See 'em, Now You Don't." Americas 7, Aug. 1955, p. 36.

**3367**
Fuller, Curtis
"The Saucers are Flying." Fate 8, Aug. 1955, pp. 6-16.

**3368**
Mandel, Siegfried
"The Great Saucer Hunt." The Saturday Review 38, Aug. 6, 1955, pp. 28-29.

**3369**
Wassilko-Serecki, Zoe
"Startling Theory on Flying Saucers." American Astrology 23, Sept. 1955, pp. 2-5.

**3370**
"Dr. Menzel Maintains Saucers are Natural Phenomena." New York Times, Sept. 2, 1955, p. 3, col. 1.

**3371**
"Couzinet's Saucer." London, RAF Flying Review 11, Oct. 1955, pp. 23-24.

**3372**
"U.S. Air Force, Announcing Development of Flying-Saucer Type Aircraft and Releasing Sketch of Such Craft Being Built by Avro of Canada for RCAF and U.S. Air Force, Again Insists Previous Sightings of Saucers Were Either Illusions or Observations of Conventional Phenomena...Issues Report on 8-Year Survey of 'Unknown' Objects." New York Times, Oct. 26, 1955, p. 1, col. 2.

**3373**
"D. E. Keyhoe Charges U.S. Air Force Conceals Facts." New York Times, Oct. 28, 1955, p. 12, col. 3.

**3374**
"Air Force has 'Jet Saucer.'" Army-Navy-Air Force Register 76, Oct. 29, 1955, p. 1.

**3375**
"Saucers Downed (Air Force Report on)." Air Force Times 16, Oct. 29, 1955, p. 1.

**3376**
"Flying Saucers may Provide Key to Planes of Future." Air Force Times 16, Nov. 5, 1955, p. 11.

**3377**
"No Flying Saucers yet...but." American Aviation 19, Nov. 7, 1955, p. 32.

**3378**
"Saucer Blue Book." Time 66, Nov. 7, 1955, p. 52.

**3379**
Comella, Thomas M.
"Why the Real Saucer is Interplanetary." Fate 8, Dec. 1955, pp. 17-23.

---

NOTES:

# UNIDENTIFIED FLYING OBJECTS INCLUDING FLYING SAUCERS

**3380**
Geiger, E. E.
"Flying Saucers...in Brazil." Intervia 10, Dec. 1955, p. 941.

**3381**
"Major Keyhoe's Book, The Flying Saucer Conspiracy, and E. Ruppelt's The Report on Unidentified Flying Objects Reviewed." New York Times, Jan. 22, 1956, sec. VII, p. 25, col. 1.

**3382**
Dorsey, Tom
"Balloons may Increase Flying Saucer Reports." Air Force Times 16, Jan. 28, 1956, p. 4.

**3383**
"Baldwin Article Discusses use of Jet Stream Lift Principle in Wingless Craft." New York Times, Feb. 12, 1956, p. 32, col. 1.

**3384**
"Latest on the Flying Saucer." Saturday Review 39, Feb. 25, 1956, p. 23.

**3385**
Ruppelt, Edward J.
"Inside Story of the Saucers." Science Digest 39, April 1956, pp. 35-41.

**3386**
"Unidentified Flying Objects, Motion Picture Documentary on Flying Saucers, Previewed, Hollywood." New York Times, April 24, 1956, p. 26, col. 1.

**3387**
"Unidentified Objects?" Sky & Telescope 15, Aug. 1956, p. 444.

**3388**
Keyhoe, Donald E.
"Flying Saucer Conspiracy." (Review by E. K. Roosevelt). American Mercury 83, Sept. 1956, pp. 153-156.

**3389**
Miller, Max B.
"Report on the UFO." Fate 9, Dec. 1956, pp. 31-34.

**3390**
"Form Saucers Investigation Group." Science Digest 41, Feb. 1957, p. 39.

**3391**
Goble, H. C.
"Did Jones Chart an Unknown World?" Fate 10, April 1957, pp. 68-70.

**3392**
Ruppelt, Edward J.
"Report on Unidentified Flying Objects." Fate 10, April 1957, pp. 27-43.

**3393**
Edwards, Frank
"Frank Edwards' Report, To see or not to see Flying Saucers." Fate 10, May 1957, pp. 17-32.

**3394**
"Ambassador H. J. Taylor Warns Senate Foreign Relations Committee not to Dismiss Saucer Reports." New York Times, May 5, 1957, p. 42, col. 1.

**3395**
De Tastelero, Mira
"Flying Saucers in the Movies." Flying Saucers From Other Worlds, June 1957, pp. 40-43.

NOTES:

## UNIDENTIFIED FLYING OBJECTS INCLUDING FLYING SAUCERS

3396
Edwards, Frank
 "Frank Edwards' Report, The Plot to Silence me." Fate 10, June 1957, pp. 17-23.

3397
Edwards, Frank
 "The Plot to Silence me." Fate 10, June 1957, pp. 17-23.

3398
Hall, Richard
 "Is There a Veil of Secrecy Around the Flying Saucers?" Flying Saucers From Other Worlds, June 1957, pp. 51-56.

3399
Slaboda, Emil
 "He Collected on a Flying Saucer." Fate 10, June 1957, pp. 66-69.

3400
Edwards, Frank
 "Frank Edwards' Report." Fate 10, July 1957, pp. 27-31.

3401
"Flying Saucers and Science." American Mercury 85, July 1957, pp. 121-125.

3402
Norris, Geoffrey
 "Something in the sky." London, RAF Flying Review 12, July 1957, pp. 14-16.

3403
"Is the U.S. Government Expecting Invaders From Space?" Flying Saucers From Other Worlds, Aug. 1957, pp. 16-21.

3404
Abel, R. C.
 "Arcoids (Unidentified Flying Objects and Their Propulsion)." London, Aeronautics 37, Sept. 1957, pp. 82-85.

3405
"Science, Cups or Saucers?" Time 70, Sept. 9, 1957, p. 67.

3406
Edwards, Frank
 "Frank Edwards' Report, ten Years of UFO's." Fate 10, Oct. 1957, pp. 65-69.

3407
"Znaniye-Sila (publication) Reports USSR Develops Saucer." New York Times, Oct. 2, 1957, p. 18, col. 6.

3408
"Bell Aircraft Engineer Says Objects Come From Outer Space; Harvard Observatory Director Menzel Calls Them Mirages." New York Times, Nov. 6, 1957, p. 12, cols. 2, 3.

3409
"No Evidence for Saucers." Science News Letter 72, Nov. 16, 1957, p. 307.

3410
"Dinner Time." Time 70, Nov. 18, 1957, p. 28.

3411
Edwards, Frank
 "Frank Edwards' Report, Science Heads for Outer Space." Fate 11, Jan. 1958, pp. 33-40.

NOTES:

3412
"Official Air Force Statements on Unidentified Flying Objects." The UFO Investigator 1, Jan. 1958, pp. 25-26.

3413
"Serious Flaws in AF Special Report, Revealed by NICAP Analysis." The UFO Investigator 1, Jan. 1958, pp. 16-18.

3414
"D. E. Keyhoe cut off air During TV Discussion of Flying Saucers When he Digresses From Script." New York Times, Jan. 23, 1958, p. 55, col. 1.

3415
"D. E. Keyhoe Says Public Congressional Hearing on Information Supplied by Aerial Phenomenon Investigating Committee Would Have Proved Reality of Flying Saucers." New York Times, Jan. 24, 1958, p. 14, col. 1.

3416
Schroeder, W.
"Only 4 Hours to Reach the Moon." London, Flying Saucer Review 4, Jan.-Feb. 1958, pp. 12-14.

3417
Edwards, Frank
"Frank Edwards' Report, Scientists and Satellites." Fate 11, Feb. 1958, pp. 69-76.

3418
Everson, Vincent
"The Fowls That fly in Mid-heaven." Fate 11, Feb. 1958, pp. 62-68.

3419
Edwards, Frank
"Frank Edwards' Report, why Shoot the Moon?" Fate 11, May 1958, pp. 41-47.

3420
Edwards, Frank
"Frank Edwards' Report, is the UFO 'Curtain' Lifting?" Fate 11, June 1958, pp. 47-54.

3421
Edwards, Frank
"Frank Edwards' Report, Some Answers to the Saucer." Fate 11, July 1958, pp. 45-51.

3422
"Dr. Jung Says Flying Saucers are not Quirks of Imagination; Says Something has Been Seen." New York Times, July 30, 1958, p. 13, col. 2.

3423
Edwards, Frank
"Frank Edwards' Report, an Open Forum on UFO's." Fate 11, Aug. 1958, pp. 75-81.

3424
Maney, Charles A.
"Scientific Aspects of UFO Research." London, Flying Saucer Review 4, Sept.-Oct. 1958, pp. 10-12, 30.

3425
Edwards, Frank
"Frank Edwards' Report: you can Break the UFO Barrier." Fate 11, Nov. 1958, pp. 77-83.

3426
"The Flying Saucer Myth." Armed Forces Chemical Journal 12, Nov.-Dec. 1958, p. 11.

3427
Barker, Gray
"Hostile Spacecraft." Flying Saucers, Dec. 1958, pp. 12-23, 78.

NOTES:

## UNIDENTIFIED FLYING OBJECTS INCLUDING FLYING SAUCERS

3428
"The Truth About the Book 'The Report on Unidentified Flying Objects' by Edward J. Ruppelt." Flying Saucers From Other Worlds, Dec. 1958, pp. 35-42, 56.

3429
"Space Ships--and the People They Visit." Flying Saucers From Other Worlds, Feb. 1959, pp. 6-7, 15.

3430
Webster, Robert N., & John C. Ross
"Air Force Report on UFO's." Fate 12, Feb. 1959, pp. 54-66.

3431
"Crew and Passengers on American Airlines Plane Report 3 Objects Accompany Plane for 45 Minutes, Newark-Detroit Flight." New York Times, Feb. 26, 1959, p. 2, col. 5.

3432
Clarke, Arthur C.
"What's up There?" Holiday 25, March 1959, pp. 32, 34-37, 39-40.

3433
Edwards, Frank
"An Astronomer Reports on UFO's." Fate 12, March 1959, pp. 34-41.

3434
"Admiral Dufek Sees 'Flying Saucers' Possible." New York Times, March 12, 1959, p. 10, col. 8.

3435
Edwards, Frank
"Arthur C. Clarke Looks at the Universe." Fate 12, May 1959, pp. 68-75.

3436
Lorenzen, Coral
"The Fitzgerald Investigation, What it Means." Flying Saucers, May 1959, pp. 47-50.

3437
Pack, Warren E.
"Interview With Donald Keyhoe." Fate 12, Aug. 1959, pp. 86-90.

3438
Edwards, Frank
"Censorship and UFO's." Fate 12, Sept. 1959, pp. 47-52.

3439
Robey, D. H.
"Theory About Flying Saucers." Saturday Review 42, Sept. 5, 1959, pp. 51-55.

3440
"Service Holds Ground on Flying Saucers." Air Force Times 20, Sept. 26, 1959, p. 42.

3441
Robey, Donald H.
"A Theory About Flying Saucers." Air Force & Space Digest, Nov. 1959, pp. 82-84.

3442
Cort, D.
"Saucery and Flying Saucer." Nation 189, Nov. 7, 1959, pp. 331-332.

3443
Edwards, Frank
"Frank Edwards' Report: Have UFO's Learned to Outwit Radar?" Fate 12, Dec. 1959, pp. 50-55.

3444
Kor, Peter
"UFO's From the Critic's Corner, the Myth of the Flying Saucer Mystery." Flying Saucers From Other Worlds, Dec. 1959, pp. 48-53.

Notes:

## UNIDENTIFIED FLYING OBJECTS INCLUDING FLYING SAUCERS

**3445**
Nollet, A. R.
"Flying Saucers...a Hard Look." *Marine Corps Gazette* 43, Dec. 1959, pp. 20-25.

**3446**
Ogden, Richard C.
"The Case for the R. E. Straith Letter." *Flying Saucers*, Dec. 1959, pp. 34-39, 47.

**3447**
Dagenis, Arleigh J.
"Do you Believe in Flying Saucers?" *Michigan Technic*, Jan. 1960, pp. 16-17, 50-51.

**3448**
"Air Force Demands Respect for 'UFO Report'." *Air Force Times* 20, Jan. 9, 1960, p. 7.

**3449**
Davidson, Leon
"Why I Believe Adamski." London, *Flying Saucer Review* 6, Jan.-Feb. 1960, pp. 3-8.

**3450**
Hiestand, Edgar W.
"Senators Want Saucer Truth." *Flying Saucers*, Feb. 1960, pp. 25-26.

**3451**
"It's Official: UFO's not Flying Saucers; Most Reports Traced to Aircraft, Stars." *Air Force Times* 20, Feb. 13, 1960, p. 38.

**3452**
Davidson, Leon
"ECM & CIA = UFO." London, *Flying Saucer Review* 6, March-April 1960, pp. 9-12.

**3453**
"Saucers Explained." *Science Newsletter* 77, April 30, 1960, p. 279.

**3454**
Edwards, Frank
Frank Edwards' Report, Do you Still Believe in Flying Saucers?" *Fate* 13, June 1960, pp. 27-34.

**3455**
Keyhoe, Donald F.
"Flying Saucers, Menace or Myth?" *Argosy* 350, June 1960, pp. 17, 80-83.

**3456**
Edwards, Frank
"Frank Edwards' Report: Air Force Warns Flying Saucers no Joke." *Fate* 13, July 1960, pp. 44-52.

**3457**
"Flying Saucers Ruled out by Latest UFO Fact Sheet." *Air Force Times* 20, July 30, 1960, p. 17.

**3458**
Fuller, Curtis
"Lyndon B. Johnson Calls for UFO Alert." *Fate* 13, Nov. 1960, pp. 27-29.

**3459**
Drake, Walter R.
"Man on the Threshold of Space." London, *Flying Saucer Review* 6, Nov.-Dec. 1960, pp. 22-24.

**3460**
Oberth, Hermann
"Warum Ufoforschung?" (In *Mitteilungen der Gesellschaft für Interplanetarik*. Vienna, Europäischer-Verlag, 1961, pp. 1-7).

**3461**
Sharp, Peter F.
"An Appraisal of the Present UFO Position." London, *Flying Saucer Review* 7, 1961, pp. 19-22.

NOTES:

UNIDENTIFIED FLYING OBJECTS INCLUDING FLYING SAUCERS

**3462**
Edwards, Frank
 "Frank Edwards' Report, Authorities who Believe in Flying Saucers." Fate 14, Jan. 1961, pp. 29-35.

**3463**
Tacker, Lawrence J.
 "...This is our Position (the Official Air Force Position on Flying Saucers, From Flying Saucers and the U.S. Air Force)." Airman, Official Journal of the Air Force 5, Jan. 1961, pp. 2-5.

**3464**
"1961: This may be the Year." Vogue 137, Jan. 1, 1961, p. 137.

**3465**
Fawcett, George D.
 "The Flying Saucers are Hostile." Flying Saucers, Feb. 1961, pp. 12-21, 58.

**3466**
Binder, Otto O.
 "Flying Saucer Phenomena." Space World 1, March 1961, pp. 40-41.

**3467**
"Book, Assault on the Unknown, Reviewed." New York Times, April 6, 1961, Op. 31, col. 1.

**3468**
Daniel-Rops, Henri
 "Une Enigme Sous nos Yeux." Paris, France Illustration 7, May 5, 1951, p. 490.

**3469**
"Danger From the Stars: a Warning From the Space Administration." London, Flying Saucer Review 7, May-June 1961, p. 7.

**3470**
Foght, Paul
 "Inside the Flying Saucers, Pancakes." Fate 14, Aug. 1961, pp. 32-36.

**3471**
Smith, E. R.
 "UFO's and Artificial Satellites." London, Flying Saucer Review 7, Sept.-Oct. 1961, pp. 6-11.

**3472**
"Saucers, Pancakes and Such." Science 135, Feb. 16, 1962, p. 518.

**3473**
"Fifteen-Year Air Force Flying Saucer Verdict: They Exist Only in Domestic Spats." Air Force Times 22, Feb. 17, 1962, p. 20.

**3474**
Davidson, Leon
 "An Open Letter to Saucer Researchers." Flying Saucers From Other Worlds, no. FS-24, March 1962, pp. 36-51.

**3475**
Wellman, Wade
 "Extra-solar UFO's." London, Flying Saucer Review 8, March-April 1962, pp. 8-10.

**3476**
"W. L. Laurence on House of Representatives Committee Hearings." New York Times, April 1, 1962, sec. IV, p. 9, col. 1.

**3477**
"Taking no Chances." TIG Brief 14 (U.S. Air Force), April 13, 1962, p. 18.

**3478**
Oberth, Hermann
 "Dr. Hermann Oberth Discusses UFO's." Fate 15, May 1962, pp. 36-43.

NOTES:

3479
"The U.S. Air Force News Release." London, Flying Saucer Review 8, July-Aug. 1962, pp. 11-13.

3480
Moseley, James W.
"UFO's, the Universe, and Mr. John M. Cage." Fate 15, Sept. 1962, pp. 78-84.

3481
Sykes, Egerton
"Flying Saucers and Negative Matter." Atlantis 5, Sept. 1962, pp. 49-51.

3482
Ellerby, Christopher
"Saucers From Mars?" London, Flying Saucer Review 8, Sept.-Oct. 1962, pp. 16-18.

3483
"Professor Hermann Oberth Defends the Flying Saucer." London, Flying Saucer Review 8, Sept.-Oct. 1962, pp. 15-16.

3484
Vallée, Jacques, & Janine Vallée
"Mars and the Flying Saucers." London, Flying Saucer Review 8, Sept.-Oct. 1962, pp. 5-11.

3485
Bray, Derek R. M.
"Flying Saucers: a Startling Theory." Haarlem, Netherlands, Panorama 1, Nov. 1962, pp. 4-7.

3486
"U.S. Space Plans Offer Clue to UFO Problem." The UFO Investigator 2, Jan.-Feb. 1963, pp. 6-7.

3487
Drake, Walter R.
"Did UFO's Stop a war?" London, Flying Saucer Review 9, March-April 1963, pp. 13-14.

3488
"The Italian Scene, Part 2." London, Flying Saucer Review 9, March-April 1963, pp. 3-6.

3489
Edwards, Frank
"Are our Satellites Hunting the Saucers?" Fate 16, May 1963, pp. 29-35.

3490
Creighton, Gordon W.
"The Italian Scene--Part 4." London, Flying Saucer Review 9, July-Aug. 1963, pp. 10-12.

3491
Zinsstag, Lou
"Conversations With Dr. Jung." London, Flying Saucer Review 9, July-Aug. 1963, pp. 14-16.

3492
"Ufology: New Report Debunks Belief That Unidentified Flying Objects are Buzzing the Earth." Newsweek 62, Aug. 5, 1963, p. 44.

3493
Menzel, Donald H., & L. G. Boyd
"The World of Flying Saucers" (Review by D. Cohen). Science Digest 54, Sept. 1963, p. 64.

3494
"A Speech by Wilbert B. Smith." London, Flying Saucer Review 9, Sept.-Oct. 1963, pp. 13-16.

3495
Tufty, Barbara
"8 Planet-probe Robots." Science News Letter 84, Oct. 1963, p. 227.

3496
Evans, Gordon H.
"What you Don't Know About Space." Flying Saucers, Nov. 1963, pp. 16-32.

NOTES:

## UNIDENTIFIED FLYING OBJECTS INCLUDING FLYING SAUCERS

3497
"Another Speech by Wilbert B. Smith." London, Flying Saucer Review 9, Nov.-Dec. 1963, pp. 11-14.

3498
Sharp, Peter F.
"The Truth: Some Suggestions for the Investigator." London, Flying Saucer Review 9, Nov.-Dec. 1963, pp. 7-9.

3499
Menzel, Donald H.
"The World of Flying Saucers." Bologna, Scientia (99:8), 1964, pp. 158-165.

3500
Bryant, Larry W.
"A Hard Look at UFO News Management." Fate 17, Feb. 1964, pp. 41-44.

3501
Cox, Adrian R.
"A Question of Time." London, Flying Saucer Review 10, March-April 1964, pp. 18-21, 34; July-Aug. 1964, pp. 7-9.

3502
Michel, Aimé
"Where Dr. Menzel has Gone Wrong." London, Flying Saucer Review 10, March-April 1964, pp. 8-10.

3503
"Reporting Unidentified Flying Objects." TIG Brief (U.S. Air Force), May 22, 1964, p. 17.

3504
Cadman, A. G.
"A Layman's Time and Space." London, Flying Saucer Review 10, Nov.-Dec. 1964, pp. 19-21.

3505
Fitch, C. W.
"UFO's and Governmental Secrecy." Saucer News 11, Dec. 1964, pp. 16-18.

3506
Le Poer Trench, Brinsley
"The Three W's." Saucer News 11, Dec. 1964, pp. 7-10.

3507
"J. W. Macvey's Book, Alone in the Universe, Reviewed." New York Times, Dec. 8, 1964, sec. VII, p. 34.

3508
"W. Sullivan's Book, We are not Alone: The Search for Intelligent Life on Other Worlds, Reviewed." New York Times, Dec. 20, 1964, sec. VII, p. 10.

3509
Hynek, J. Allen
"Unidentified Flying Object (UFO)." (In Encyclopaedia Britannica. v.22, Chicago, 1965, pp. 696-697).

3510
Fouéré, René
"Météors à Hublots." Paris, Phénomènes Spatiaux, Feb. 1965, pp. 7-9.

3511
Vogt, Christian
"Non-Dramatique Incident." Paris, Phénomènes Spatiaux, Feb. 1965, pp. 5-6.

3512
Gearhart, Livingston
"Bombed by Meteors." Fate 18, March 1965, pp. 80-82.

NOTES:

3513
"U.S. Air Force...Reports (no UFO's) Indicated Threat to National Security." New York Times, March 6, 1965, p. 21, col. 5.

3514
Creighton, Gordon
"Astronauts Forced Down by UFO's?" London, Flying Saucer Review 11, May-June 1965, pp. 15-18.

3515
Michel, Aimé
"Reflections of an Honest Liar." London, Flying Saucer Review 11, May-June 1965, pp. 11-14.

3516
Cleary-Baker, John
"A UFO Stops the Clocks." London, BUFORA Journal and Bulletin 1, Summer 1965, p. 19.

3517
Maney, Charles A.
"Scientific Measurement of UFO's." Fate 18, June 1965, pp. 31-39.

3518
Cohen, Daniel
"Should we be Serious About UFO's? Unidentified Flying Object." Science Digest 57, June 4, 1965, pp. 41-44.

3519
"Cheerio There, Earthlings!" America 113, Aug. 21, 1965, p. 177.

3520
"Existence of Flying Saucers Discounted by U.S. Air Force Project Blue Book...Objects and Sightings put in 10 Categories." New York Times, Aug. 29, 1965, p. 79, col. 1.

3521
Buckner, H. T.
"The Flying Saucerians: A Lingering Cult." London, New Society 9, Sept. 1965, pp. 14-16.

3522
Cohen, Daniel
"The Return of Flying Saucers." The Nation 201, Sept. 13, 1965, pp. 131-134.

3523
Norman, Paul
"Electro-magnetic Effects of UFO's." London, Flying Saucer Review 11, Sept.-Oct. 1965, pp. 26-28.

3524
Ribera, Antonio
"Professor Hermann Oberth Revisits Barcelona." London, Flying Saucer Review 11, Sept.-Oct. 1965, p. 32.

3525
Wellman, Wade
"Sense and Speculation." London, Flying Saucer Review 11, Sept.-Oct. 1965, pp. 30-31.

3526
"Saucer Craze Continues Despite Facts." Air Force Times 26, Oct. 20, 1965, p. 3.

3527
"New Clues to UFO Electrical Interference." The UFO Investigator 3, Nov.-Dec. 1965, pp. 3-4.

3528
"Strange Effects From EM Waves." The UFO Investigator 3, Nov.-Dec. 1965, p. 5.

3529
Vallée, Jacques
"UFO Research in the U.S.A." London, Flying Saucer Review 11, Nov.-Dec. 1965, pp. 32-36.

NOTES:

## UNIDENTIFIED FLYING OBJECTS INCLUDING FLYING SAUCERS

3530
Adam, Frank
"UFOs Unter der Lupe. 300,000 Dollar für Ihre Wissenschaftliche Erforschung." Zürich, Weltwoche (34:1722), 1966, p. 31.

3531
Klass, Philip J.
"Plasma Theory may Explain Many UFO's." Aviation Week & Space Technology (85:8), 1966, pp. 48-50, 55, 56, 60, 61.

3532
Kantor, M.
"Why I Believe in Flying Saucers." Popular Science 188, Jan. 1966, pp. 72-74.

3533
Vallée, Jacques
"UFO Research in the U.S.A." London, Flying Saucer Review 12, Jan.-Feb. 1966, pp. 6-11.

3534
"Attention aux Soucoupes!" Paris, Phénomènes Spatiaux, March 1966, pp. 29-30.

3535
"BUFORA Research Officer's Annual Report - 27th Nov. 1965." London, BUFORA Journal and Bulletin 1, Spring 1966, pp. 9-12.

3536
"Professor Teller Calls UFO's Miracles on TV Interview." New York Times, April 4, 1966, p. 33, col. 8.

3537
"Gullible Experiment: Southern California." Time 87, April 8, 1966, p. 70.

3538
"A Hard Look at 'Flying Saucers'." U.S. News & World Report 60, April 11, 1966, pp. 14-15.

3539
Mingus, Ron
"Civilian Specialist to aid in Probe of 'Baffling' UFO's. (Project Blue Book)." Air Force Times 26, April 20, 1966, p. 10.

3540
"Flying Saucers?" Senior Scholastic 88, April 22, 1966, p. 13.

3541
Buckner, H. T.
"Flying Saucers are for People." Trans-Action 3, May-June 1966, pp. 10-13.

3542
"IATA Director Hammarskjold says UFO's may Actually be 'Neighbors in Space', at Aviation and Space Writers Association." New York Times, May 24, 1966, p. 94, col. 8.

3543
Arnell, J. C.
"UFO's: Figment, Fact or Fiction?" Ottawa, Sentinel-Canadian Forces Sentinel 2, June 1966, pp. 4-5.

3544
Fouéré, René
"Observations d'un Astronome Argentin." Paris, Phénomènes Spatiaux, June 1966, pp. 3-11.

3545
Luce, C. B.
"Without Portfolio: Flying Saucers and Mistaken Identities." McCalls 93, July 1966, p. 32.

3546
Fawcett, George D.
"UFO Repetitions." Flying Saucers From Other Worlds, Aug. 1966, p. 28.

NOTES:

**3547**
Rankow, Ralph
"The Disc With the Domed top." *Fate* 19, Aug. 1966, pp. 54-61.

**3548**
Stevens, Stuart
"Sighting Peaks and Planetary Oppositions." *Flying Saucers From Other Worlds*, Aug. 1966, pp. 20-21.

**3549**
Fuller, John G., et al
"Research in America — What are the Unidentified Aerial Objects?" *Saturday Review* 49, Aug. 6, 1966, pp. 41-52.

**3550**
Cleary-Baker, John
"The Significance of the UFO era." London, *BUFORA Journal and Bulletin* 1, Autumn 1966, pp. 1-10.

**3551**
Hatvany, Edgar A.
"Dr. Hynek and the UFO's." London, *BUFORA Journal and Bulletin* 1, Autumn 1966, pp. 5-8.

**3552**
Strentz, Herb
"Seeing Saucers." *Columbia Journalism Review*, Fall 1966, pp. 23-25.

**3553**
"UFO's or Kugelblitz?" *Popular Electronics* 25, Sept. 1966, p. 84.

**3554**
Lear, John
"The Disputed CIA Document on UFO's." *Saturday Review*, Sept. 3, 1966, pp. 45-50.

**3555**
"UFO's not From Mars." *Science News Letter* 90, Sept. 3, 1966, p. 165.

**3556**
Caputo, Livio
"Rapporto sui Dischi Volanti-2, Stanno per Invaderci?" Milan, *Epoca* LXIV, Sept. 4, 1966, pp. 32-37.

**3557**
"Flying Saucers: Fact or Fancy?" *Senior Scholastic, Teacher's Edition* 89, Sept. 16, 1966, pp. 4-7.

**3558**
"Northwestern University Astronomy Dept. Chairman, Dr. Hynek, Urges Scientific Study; Charges Scientists Have Avoided Study for Fear of 'Injuring Professional Standing'; Scores Assumption that all Reports Were Work of Hysterics, Cranks or Unreliable People." *New York Times*, Sept. 18, 1966, p. 62, col. 5.

**3559**
Vallée, Jacques
"A ten Point Research Proposal." London, *Flying Saucer Review* 12, Sept.-Oct. 1966, pp. 12-14.

**3560**
"Condon to Head UFO Study." *Science* 154, Oct. 1966, p. 244.

**3561**
Prytz, John
"The Air Force Opinion on UFO's." *Flying Saucers From Other Worlds*, Oct. 1966, pp. 26-28.

**3562**
Lear, John
"Scientific Explanation for the UFO's? P. J. Klass's Plasma Ball Theory." *Saturday Review* 49, Oct. 1, 1966, pp. 67-69.

**3563**
Klass, Philip J.
"Many UFO's are Identified as Plasmas." *Aviation Week & Space Technology* (85:14), 1966, pp. 54, 55, 59, 61, 65, 67, 69, 71, 73.

NOTES:

# UNIDENTIFIED FLYING OBJECTS INCLUDING FLYING SAUCERS

**3564**
Willems, Louis
"De Lieve Invasie." Amsterdam, <u>A.B.C.</u>, Oct. 8, 1966, pp. 14-16.

**3565**
"UFO's for Real? J. A. Hynek Calls for Serious Investigation." <u>Newsweek</u> 68, Oct. 10, 1966, p. 70.

**3566**
Greenberg, D. S.
"Condon to Head UFO Study." <u>Science</u> 154, Oct. 14, 1966, p. 244.

**3567**
"Unidentified Flying Objects (UFO's)." <u>TIG Brief</u> 17, Aug. 27, 1965, p. 17; <u>TIG Brief</u> 18, Oct. 14, 1966, p. 14.

**3568**
"Colorado to Study UFO's." <u>Aviation Week & Space Technology</u> 85, Oct. 17, 1966, p. 32.

**3569**
Hynek, J. A.
"UFO's Merit Scientific Study." <u>Science</u> 154, Oct. 21, 1966, p. 329.

**3570**
"Prof. McDonald Urges NASA, National Science Foundation or Like Organization Without Vested Interest Open Study of Reported Sightings; Says 18 Years of Administrative Foulup by U.S. Air Force and Deliberate Government Debunking Have Frightened Away Scientists, Confused the Public." <u>New York Times</u>, Oct. 21, 1966, p. 9, col. 1.

**3571**
"UFO's to be Probed: Study at the University of Colorado." <u>Science News Letter</u> 90, Oct. 22, 1966, p. 331.

**3572**
Simons, Howard
"American Newsletter." London, <u>New Scientist</u> 32, Oct. 27, 1966, p. 174.

**3573**
"Can Dr. Condon See it Through?" <u>Nation</u> 203, Oct. 31, 1966, p. 436.

**3574**
"1966 Tully." Moorabin, Australia, <u>Australian Flying Saucer Review</u> 9, Nov. 1966, pp. 16-18.

**3575**
"Important Discoveries." London, <u>Flying Saucer Review</u> 12, Nov.-Dec. 1966, p. 17.

**3576**
Winder, R. H. B.
"Design for a Flying Saucer." London, <u>Flying Saucer Review</u> 12, Nov.-Dec. 1966, pp. 21-36; 13, Jan.-Feb. 1967, pp. 13-18; 13, March-April 1967, pp. 20-23; 13, May-June 1967, pp. 9-12.

**3577**
Cohen, Daniel
"UFO's, What a New Investigation May Reveal." <u>Science Digest</u> 60, Dec. 1966, pp. 54-56.

**3578**
Dawson, Judith, & Carole Klemm
"Spies From Space." <u>Aerospace International</u> 2, Dec. 1966, pp. 44-45.

**3579**
Fouéré, René
"Sommes-nous à un Tournant?" Paris, <u>Phénomènes Spatiaux</u>, Dec. 1966, pp. 6-12.

**3580**
Hynek, J. A.
"UFO's Merit Scientific Study: Letter." <u>Science</u> 154, Dec. 2, 1966, p. 1118-

NOTES:

**3581**
Lear, John
"Research in America; Dr. Condon's Study Outlined." Saturday Review 49, Dec. 3, 1966, pp. 87-89.

**3582**
Hynek, J. Allen
"Da ist es! Es Bewegt Sich! Uber Fliegende Untertassen." Hamburg, Der Spiegel (21:17), 1967, pp. 162-171.

**3583**
"Pravda i Vynsysel ob UFO." Moscow, Tekhnika Molodiozki 8, 1967, pp. 23-27.

**3584**
Krooge, Harald
"UFO-Illusionisten. Auf der Suche Nach 'Unbekannten Flugkörpern'." Hamburg, Die Zeit (22:45), 1967, p. 11.

**3585**
Sagan, Carl
"Unidentified Flying Object." (In Encyclopedia Americana 27, New York, Americana Corp., 1967, pp. 368-369).

**3586**
Mallan, Lloyd
"What we are Doing About UFO's." London, Science & Mechanics 38, Jan. 1967, pp. 38-43, 62-67, 76.

**3587**
Salisbury, Frank B.
"The Scientist and the UFO." Bioscience, Jan. 1967, pp. 15-24.

**3588**
"UFO Inquiries set a Record." Air Force Times 27, Jan. 25, 1967, p. 22.

**3589**
Fuller, Curtis
"Air Force Grants $313,000 to Study UFO's." Fate, Feb. 1967, pp. 32-39.

**3590**
Fuller, John G.
"A Communication Concerning the UFO's." Saturday Review 50, Feb. 4, 1967, pp. 70-72.

**3591**
Pascalis, Bernardino
"Werkelijkheid of Fantasie?" Haarlem, Netherlands, Panorama, Feb. 21-27, 1967, pp. 31-38.

**3592**
Fouéré, René
"Leurres et Realités." Paris, Phénomènes Spatiaux, March 1967, pp. 13-17.

**3593**
Gray, Grattan
"The Town That Believes in Flying Saucers." Toronto, Maclean's 80, March 1967, p. 4.

**3594**
Hynek, J. A.
"Flying Saucers: are They Real?" Reader's Digest 90, March 1967, pp. 61-65.

**3595**
Lauritzen, Hans
"Flying Saucers, Superconducting Whirls of Plasma." Flying Saucers 51, March 1967, pp. 10-11.

**3596**
Rogers, Warren
"Flying Saucers." Look 31, March 1967, pp. 76-80.

**3597**
"Situation Report, What is the Unidentified Flying Object Situation These Days?" American Engineer, March 1967, p. 55.

NOTES:

UNIDENTIFIED FLYING OBJECTS INCLUDING FLYING SAUCERS

3598
Stanton, L. J.
  "The Final Word on Flying Saucers?" This Week, March 1967, pp. 6-7.

3599
Stine, G. H.
  "The Prowling Mind of Henri Coanda." Flying 80, March 1967, pp. 64-68.

3600
Wylie, Philip
  "UFO's: the Sense and Nonsense." Popular Science 190, March 1967, pp. 76-79.

3601
"New Light on Flying Saucers." U.S. News and World Report 62, March 20, 1967, p. 16.

3602
"More Intensive Scientific Study of Reported Sightings Urged by D. L. Morgan, Jr., of NYU and G. W. Earley of NICAP, Design Engineering Conference." New York Times, March 21, 1967, p. 74, col. 1.

3603
Keel, John A.
  "The 'Silencers' at Work." London, Flying Saucer Review 13, March-April 1967, p. 10.

3604
Miller, Stewart
  "On Scientific Dogma." London, Flying Saucer Review 13, March-April 1967, pp. 26-27.

3605
"UFO's Again." Ordnance 51, March-April 1967, p. 450.

3606
Lasco, Jack
  "Has the Air Force Captured a Flying Saucer?" Saga, April 1967, pp. 18-19, 67-68, 70-74.

3607
Vallée, Jacques, & Janine Vallée
  "Astronomers' Verdict: Flying Saucers are Real." Fate 20, April 1967, pp. 62-72.

3608
Hecker, Randall C.
  "Did UFO Sabotage Mariner IV?" Fate 20, May 1967, pp. 32-37.

3609
Mallan, Lloyd
  "These Saucers Nearly Fooled the Air Force." This Week, May 1967, pp. 4-5.

3610
"Latest UFO Report (on Project Blue Book) Reveals Nothing out of This World." Air Force Times 27, May 17, 1967, p. 28.

3611
Hughes, F. P.
  "Trained Eye on UFO's: Letter." Science 156, June 9, 1967, pp. 1311-1312.

3612
Young, Mort
  "The Flying Saucer Invasion That Panicked the Pentagon." Saga 34, July 1967, pp. 22-25, 70, 72, 74-76.

3613
Ogles, George W.
  "What Does the Air Force Really Know About Flying Saucers?" Airman, Official Magazine of the U.S. Air Force, (part 1: 11), July 1967, pp. 4-9; (part 2: 11), Aug. 1967, pp. 26-31.

3614
Fawcett, George D.
  "1966--the Year of Saucers." Flying Saucers From Other Worlds, Aug. 1967, pp. 6-11.

3615
"UFO Photographs, Anymore?" Science Digest 62, Sept. 1967, p. 73.

---

NOTES:

**3616**
Scott, David H.
"U and the UFO." *Flying* 81, Aug. 1967, pp. 81-82.

**3617**
"Fresh Look at Flying Saucers: Time Essay." *Time* 90, Aug. 4, 1967, pp. 32-33.

**3618**
Eberhart, Jonathan
"Flying Spacewatchers." *Science News* 92, Aug. 19, 1967, p. 179.

**3619**
"Dr. W. Markowitz Holds Control of UFO's by Extraterrestrial Beings Contrary to Laws of Physics, Article in *Science*...Agrees UFO's Exist." *New York Times*, Sept. 15, 1967, p. 49, col. 8.

**3620**
Markowitz, William
"Physics and Metaphysics of Unidentified Flying Objects." *Science* 157, Sept. 15, 1967, pp. 1274-1279; Discussion, 158, Dec. 8, 1967, pp. 1265-1266.

**3621**
"Colorado Horse Death Ruled no UFO Case." *The UFO Investigator* 4, Oct. 1967, p. 4.

**3622**
Shuldiner, Herbert
"The Great UFO Probe." *Popular Science* 191, Oct. 1967, pp. 120-123.

**3623**
Lear, John
"UFO's and the Law of Physics: Concerning Views of J. Allen Hynek and William Markowitz." *Saturday Review* 50, Oct. 7, 1967, p. 59.

**3624**
"UFO's and the Laws of Physics." *Saturday Review*, Oct. 7, 1967, p. 59.

**3625**
Hellman, Hal
"New Look at the UFO Enigma; Ball Lightning, Excerpt from *Light and Electricity in the Atmosphere*." *Science Digest* 62, Nov. 1967, pp. 9-15.

**3626**
Ruddy, John
"Look—There's a Flying Saucer." Toronto, *Mclean's* 80, Nov. 1967, pp. 34-36.

**3627**
Clark, Jerome
"Why UFO's are Hostile." London, *Flying Saucer Review* 13, Nov.-Dec. 1967, pp. 18-20.

**3628**
Fontes, Olavo T.
"A Suggested Scientific Investigation of the UFO Phenomenon." London, *Flying Saucer Review* 13, Nov.-Dec. 1967, pp. 22-24.

**3629**
Winder, R. H. B.
"Vehicle Stoppage at Hook." London, *Flying Saucer Review* 13, Nov.-Dec. 1967, pp. 6-7.

**3630**
Darby, Christian
"World's First UFO Murder Case." *Argosy* 365, Dec. 1967, pp. 23-25, 60-63.

**3631**
"Un Document Majeure: la Declaration du Dr. McDonald." Paris, *Phénomènes Spatiaux*, Dec. 1967, pp. 4-8.

**3632**
Hynek, J. Allen
"The UFO gap." *Playboy* 14, Dec. 1967, pp. 144-146, 267, 269-271.

**3633**
Munday, John C.
"On the UFO's." *Bulletin of the Atomic Scientists* 23, Dec. 1967, pp. 40-41.

NOTES:

# UNIDENTIFIED FLYING OBJECTS INCLUDING FLYING SAUCERS

**3634**
Sherman, Carl
"Why is the UN Censoring the Truth About UFO's?" *Saga* 35, Dec. 1967, pp. 32-35, 58-60, 62.

**3635**
"W. Sullivan Discusses UFO Studies in U. S." *New York Times*, Dec. 10, 1967, p. 3.

**3636**
"Sickles in the Sky: Communist UFO Observations." *Time* 90, Dec. 22, 1967, p. 28.

**3637**
Branscomb, Lewis M., et al
"UFO's: Irrational Public Debates." *Science* (new series), (161:3848), 1968, p. 1297.

**3638**
Buckner, H. T.
"The Flying Saucerians: an Open Door Cult" (In Truzzi, Marcello, ed. *Sociology and Everyday Life*. Englewood Cliffs, N.J., Prentice-Hall, 1968, pp. 223-230).

**3639**
Condon, Edward U.
"UFO Trouble in Science." *Science* (new series), (161:3844), 1968, p. 844.

**3640**
Marey, G.
"Soucoupes Volantes." Paris, *Forces Aériennes Françaises. Revue Mensuelle de l'Armée de l'air* (22:245), 1968, pp. 453-471.

**3641**
"Rymdbesok--Eller vad?" Stockholm, *Sökaren* 5, #1, 1968, pp. 8-10.

**3642**
Schütze, Alfred
"Nochmals: Die Fliegende Untertassen." Stuttgart, *Die Christengemeinschaft* 40, 1968, pp. 23-25.

**3643**
Schwartz, Berthold
"UFO's, Delusion or Dilemma." *Medical Times* 96, 1968, pp. 967-981.

**3644**
Stanford, Rex G.
"Brev Fran Rex Stanford." Stockholm, *Sökaren* 5, #3, 1968, p. 16.

**3645**
Binder, O. O.
"How Flying Saucers can Injure you." *Mechanics Illustrated* 64, Jan. 1968, pp. 64-66.

**3646**
Bowen, Charles
"The Spectre of Winterfold." London, *Flying Saucer Review* 14, Jan.-Feb. 1968, pp. 15-17.

**3647**
Binder, Otto O.
"'Oddball' Saucers That fit no Pattern." *Fate* 21, Feb. 1968, pp. 54-62.

**3648**
Settimo, Gianni
"Gli U.F.O. Preparano un 'Rapporto Kinsey' Interplanetario?" Turin, *Clypeus* 5, Feb. 1968, pp. 33-34.

**3649**
Lyustiberg, Villen
"Letaiushchie Tarelki? Mif!" Kiev, *Pravda Ukrainy* 40, Feb. 17, 1968, p. 2.

**3650**
"Snova 'Letaiushchie Tarelki'?" Moscow, *Pravda*, Feb. 29, 1968, p. 8.

**3651**
Hynek, J. A.
"UFO's and the Numbers Game." *Natural History* 77, March 1968, pp. 24-25.

NOTES:

3652
Lorenzen, Coral
    "The Great UFO Controversy: the Appaloosa From Alamosa." Fate 21, March 1968, pp. 34, 36-44.

3653
Merker, Donald
    "The Great UFO Controversy: the Appaloosa From Alamosa." Fate 21, March 1968, pp. 35, 45-52.

3654
Lyustiberg, Villen
    "Flying Saucers? a Myth!" Current Digest of the Soviet Press 20, March 27, 1968, pp. 12-13.

3655
"The Mysterious Chunk of Hardware at Ottawa." Ottawa, Topside, Spring 1968, pp. 1-4.

3656
McDonald, James E.
    "A Need for an International Study of UFO's." London, Flying Saucer Review 14, March-April 1968, p. 11.

3657
Clark, Jerome
    "The Roots of Skepticism." Flying Saucers From Other Worlds, April 1968, pp. 19-21.

3658
Fawcett, George D.
    "Flying Saucers, Explosive Situation for 1968." Flying Saucers From Other Worlds, April 1968, pp. 22-25.

3659
Keel, John A.
    "Secret UFO Bases Across the U.S." Saga 36, April 1968, pp. 30-33, 86, 89-90, 92-94, 96.

3660
Fuller, John G.
    "Flying Saucer Fiasco." Look 32, May 1968, pp. 58.

3661
Grescoe, Paul
    "This man Knows UFO's." Toronto, Canadian Magazine, May 1968, pp. 9, 11.

3662
Fuller, John G.
    "Flying Saucer Fiasco." Look 32, May 14, 1968, p. 58.

3663
"White House is Hiding the Truth About Flying Saucers." Midnight 14, May 20, 1968, pp. 14-15.

3664
Fawcett, George D.
    "Current UFO Status." Flying Saucers, June 1968, pp. 18-19.

3665
Steiger, Brad
    "What Price Silence?" Flying Saucers From Other Worlds, June 1968, p. 31.

3666
"UFO Study Credibility Cloud?" Industrial Research 10, June 1968, pp. 26-28.

3667
Zigel, F.
    "The UFO Problem--a Challenge to Science." Flying Saucers From Other Worlds, June 1968, pp. 25-26.

3668
"Saucers Down to Earth." London, The Economist 227, June 1, 1968, p. 30.

NOTES:

**3669**
Page, Thornton L.
"Photographic Sky Coverage for the Detection of UFO's." Science 160, June 14, 1968, pp. 1258-1260.

**3670**
Jones, R. V.
"The Natural Philosophy of Flying Saucers." Physics Bulletin, July 1968, pp. 225-230.

**3671**
Boffey, Philip M.
"UFO Project: Trouble on the Ground." Science 161, July 26, 1968, pp. 339-342; 162, Sept. 27, 1968, pp. 410-411.

**3672**
"6 Scientists Testify Before House of Representatives Committee That Objects Merit Serious Study, and Urge Federal Support to Collect Data." New York Times, July 30, 1968, p. 10, col. 1.

**3673**
Keel, John A.
"UFO's--the Statistical Problem." Flying Saucers From Others Worlds, Aug. 1968, pp. 18-19.

**3674**
Wilson, Harvey
"Found: Flying Saucer Base in Outer Space." The National Police Gazette 173, Aug. 1968, pp. 14, 24.

**3675**
Larson, Kenneth
"UFO's and Some Surprising Date Patterns." Flying Saucers 60, Oct. 1968, p. 11.

**3676**
Lingham, Tom
"Man Killed by Death Ray From Flying Saucer." Flying Saucers 60, Oct. 1968, pp. 12-13.

**3677**
Aggen, Erich
"UFO's, Enigma of the Skies." Flying Saucers 61, Dec. 1968, pp. 9-10.

**3678**
Löbsack, Theo
"Fliegende Untertassen Entpuppen Sichals Kugelblitze." Hamburg, Die Zeit (24:11), 1969, p. 68.

**3679**
Saunders, Richard, & Curtis Fuller, comps.
"Scientists Speak out on UFO's." Fate 226, Jan. 1969, pp. 85-96.

**3680**
Boffey, Philip M.
"UFO Study: Condon Group Finds No Evidence of Visits from Outer Space." Science 163, Jan. 17, 1969, pp. 260-262.

**3681**
"Saucers' End: Condon Report." Time 93, Jan. 17, 1969, pp. 44-45.

**3682**
"Colorado Team Headed by Dr. E. U. Condon Finds no Evidence That Objects or Sightings are Intelligently Guided Spacecraft From Beyond Earth..." New York Times, Jan. 18, 1969, p. 1, col. 2.

**3683**
"UFO's Grounded." London, The Economist 230, Jan. 18, 1969, p. 39.

**3684**
"Flying Saucers: not Real, but-." U.S. News & World Report 66, Jan. 20, 1969, p. 6.

**3685**
"Shooting Down the UFO's: Condon Report." Newsweek 73, Jan. 20, 1969, p. 54.

---

NOTES:

**3686**
"Lost Cause: Condon Report."
*The Nation* 208, Jan. 27, 1969, p. 100.

**3687**
"UFO Report Rejects Nonterrestrial Origin."
*Aviation Week*, 90, Jan. 27, 1969, p. 85.

**3688**
Lloyd, Dan
"Let's Take off our Blinkers." London,
*Flying Saucer Review* (15: 1), Jan.-Feb.
1969, pp. 9-11.

**3689**
"Edward Condon: a Physicist Never Afraid of a Fight." *Physics Today* 22, March 1969, pp. 66-67.

**3690**
Hooven, Frederick J.
"UFO's and the Evidence: Condon Report."
*Saturday Review* 52, March 29, 1969, pp. 16-17.

**3691**
Armagnac, Allen P.
"The Condon Report on UFO's, Should you Believe it?" *Popular Science* 194, April 1969, pp. 72-76.

**3692**
Condon, Edward U.
"Scientific Study of Unidentified Flying Objects. Review by J. Allen Hynek." *Bulletin of the Atomic Scientists* 25, April 1969, pp. 39-42.

**3693**
Keel, John A.
"The Myth of UFO Censorship." *Flying Saucers* 63, April 1969, pp. 5-10.

**3694**
Larson, Kenneth
"Top Scientists Warn Congressional House Committee That UFO Solution can no Longer be Delayed." *Flying Saucers* 63, April 1969, pp. 29-31.

**3695**
Krutch, Joseph W.
"If you don't mind my Saying so: Inexplicable Saucer Reports." *American Scholar* 33, Summer 1969, p. 370.

**3696**
Bell, Don
"The Mystery Behind the Investigation of the UFO Mystery." *Flying Saucers* 64, June 1969, pp. 10-12.

**3697**
Fawcett, George D.
"The Hills Revisited." *Flying Saucers* 64, June 1969, pp. 22-23.

**3698**
Ross, John
"The Condon Report: pro and con." *Fate* 231, June 1969, pp. 40-50.

**3699**
Vallée, Jacques
"A Catalogue of 923 Landing Reports."
London, *Flying Saucer Review* (15: 4), July-Aug. 1969, pp. 13-14.

**3700**
"University Professors Support Increased UFO Research." *Flying Saucers* 65, Aug. 1969, pp. 15-17.

**3701**
Rimes, Nigel
"The Pirassununga Landing." London,
*Flying Saucer Review*, special issue #3, Sept. 1969, pp. 39-44, 111.

**3702**
Aggen, Erich
"Exposed: the Myth of UFO Hostility." *Flying Saucers* 66, Oct. 1969, pp. 4-5.

**3703**
Prytz, John
"UFO Crashes." *Flying Saucers* 66, Oct. 1969, pp. 24-25.

NOTES:

# UNIDENTIFIED FLYING OBJECTS INCLUDING FLYING SAUCERS

3704
"UFO Skepticism: or Objectivity?" Flying Saucers 66, Oct. 1969, pp. 20-22.

3705
Woodruff, Michael
"University Research Team Confirms new Evidence of UFO's." National Enquirer, Nov. 23, 1969, back page.

3706
Michel, Aimé
"Paleolithic UFO Shapes." London, Flying Saucer Review (15: 6), Nov.-Dec. 1969, pp. 3-11.

3707
Aggen, Erich
"Further Aspects of the Hostility Theory." Flying Saucers 67, Dec. 1969, pp. 14-15.

3708
Condon, Edward U.
"UFO's I Have Loved and Lost: Adaptation of an Address, April 1969." Bulletin of the Atomic Scientists 25, Dec. 1969, pp. 6-8.

3709
Prytz, John
"UFO's: Theories in Time Travel, Dimensions and Anti-universes." Flying Saucers 67, Dec. 1969, pp. 4-7.

3710
"U. S. Air Force Ends its Investigation of Unidentified Flying Objects." New York Times, Dec. 18, 1969, p. 1, col. 4.

3711
"Closing the Blue Book: Air Force to Call off UFO Investigations due to Condon Report." Time 94, Dec. 26, 1969, p. 28.

3712
"Astronomers, Physicists and Social Scientists Participating in Symposium on UFO's ask U. S. Air Force to Preserve Secret Records.. Compiled by Project Blue Book." New York Times, Dec. 28, 1969, p. 18, col. 1.

3713
Dutton, T. W. S.
"The ion Engine and the UFO's." Sydney, Australasian Engineer (62:10), 1970, pp. 51-52.

3714
Farish, Lucius
"Some old UFO's." Fate 238, Jan. 1970, pp. 89-93; 239, Feb. 1970, pp. 101-107.

3715
Sharp, Peter F.
"Interstellar Refugees." London, Flying Saucer Review (16: 1), Jan.-Feb. 1970, pp. 21-22.

3716
Aguilar, Xavier F.
"Flying Saucers, the Money Making Myth." Flying Saucers 68, March 1970, p. 22.

3717
Larson, Kenneth
"Saucers and the Missing Topstone." Flying Saucers 68, March 1970, pp. 13-16.

3718
Hynek, J. Allen
"Commentary on the AAAS Symposium." London, Flying Saucer Review (16: 2), March-April 1970, pp. 3-5.

3719
Earley, George W.
"Scientists Urge new UFO Study." Fate 241, April 1970, pp. 38-48.

NOTES:

**3720**
Wolf, R.
"Reporters at Small." New York, <u>The Realist</u> (86:8), April 1970, p. 1.

**3721**
Heiserman, David L.
"Now, a do-it-yourself UFO." <u>Popular Science</u> 196, May 1970, p. 109.

**3722**
McGraw, Walter
"Science Advances on UFO's." <u>Fate</u> 243, June 1970, pp. 78-85.

**3723**
Maddock, D. P.
"New Classification Scheme for UFO Phenomena." <u>Flying Saucers</u> 69, June 1970, pp. 20-22.

**3724**
Prytz, John
"UFO's: Time and Distance." <u>Flying Saucers</u> 69, June 1970, pp. 10-11.

**3725**
Rothberg, Gerald
"UFO's: Réalité ou Fiction?" Paris, <u>Phénomènes Spatiaux</u> 24, June 1970, pp. 3-6.

**3726**
Keel, John A.
"Fables of the UFO age." <u>Fate</u> 245, Aug. 1970, pp. 90-96.

**3727**
Aggen, Erich
"The Absence of UFO Artifacts." <u>Flying Saucers</u> 70, Sept. 1970, pp. 20-21.

**3728**
Sarrantonio, Albert
"A Generation of UFO's." <u>Flying Saucers</u> 70, Sept. 1970, pp. 9-11.

**3729**
Fornwalt, Russell J.
"Space Travel-10,000 Years From now." <u>Search</u> 94, Nov. 1970, pp. 42-48.

**3730**
Kane, Jim
"UFO Kills Tree in Cincinnati." <u>Fate</u> 248, Nov. 1970, pp. 72-73.

**3731**
Brennan, Norman
"Some Critical Notes on the Condon Report." <u>Flying Saucers</u> 71, Dec. 1970, pp. 2-7.

**3732**
Schuessler, John F.
"New Theory of UFO Flight." <u>Fate</u> 251, Feb. 1971, pp. 96-102.

**3733**
"Who Sees Flying Saucers?" <u>Science Digest</u> 69, Feb. 1971, pp. 34-35.

**3734**
Schrafft, E. R.
"Project Doomsday: the UFO." <u>Flying Saucers</u> 72, March 1971, pp. 4-11.

**3735**
"Dr. J. E. McDonald Says UFO's Come From Outer Space and may Have Caused Electric Power Failures in New York City, House of Representatives Testimony." <u>New York Times</u>, March 3, 1971, p. 87, col. 3.

**3736**
Schönherr, Luis
"The Question of Reality." London, <u>Flying Saucer Review</u> (17:2), March-April 1971, pp. 22-25.

**3737**
Saunders, Alex
"The Untouchable Flying Saucer." <u>Search</u> 97, May 1971, pp. 58-63.

NOTES:

UNIDENTIFIED FLYING OBJECTS INCLUDING FLYING SAUCERS

**3738**
Clarke, Arthur C.
"Whatever Happened to Flying Saucers?"
Saturday Evening Post 243, Summer 1971, p. 10.

**3739**
Crain, T. S.
"Flying Saucer Casualties." Flying Saucers 73, June 1971, pp. 7-9.

**3740**
Keel, John A.
"A Question of Responsibility." Flying Saucers 73, June 1971, pp. 15-16.

**3741**
"Whatever Happened to UFO's?" Science News 99, June 26, 1971, pp. 435-436.

**3742**
"Saucer Diehards." Time 97, June 28, 1971, p. 49.

**3743**
Earley, George W.
"Poll Report: Scientists Remain Interested in UFO's." Fate 257, Aug. 1971, pp. 93-95.

**3744**
Cestling, Philip
"The Death Throes of Ufology." Caveat Emptor 1, Fall 1971, pp. 13-14, 26.

**3745**
Barker, Gray
"The Enigma of M K Jessup." Flying Saucers 74, Sept. 1971, pp. 8-12.

**3746**
Laprade, Armand
"What's up?" Flying Saucers 74, Sept. 1971, pp. 14-15.

**3747**
Prytz, John
"Believers, Skeptics and Non-Believers." Flying Saucers 74, Sept. 1971, pp. 25-26.

**3748**
Burt, Eugene H.
"Magnetic Explanation of UFO's." Flying Saucers 75, Dec. 1971, pp. 17-20.

**3749**
Prytz, John
"UFO's, Prospects for the Seventies." Flying Saucers 75, Dec. 1971, pp. 5-9.

**3750**
Toft, Ron
"Space Expert Denounces Flying Saucers." Flying Saucers 75, Dec. 1971, pp. 28-29.

**3751**
Keel, John A.
"Ufology in Retrospect." Caveat Emptor 2, Winter 1971-72, pp. 6-13, 27.

**3752**
Michel, Aimé
"Project Dick." London, Flying Saucer Review (18:1), Jan.-Feb. 1972, pp. 13-19.

**3753**
Dick, William
"Existence of UFO's can no Longer be Doubted." National Enquirer, Feb. 6, 1972, back page.

**3754**
Fox, James E.
"Space Drives and UFO's." Flying Saucers 76, March 1972, pp. 23-25.

**3755**
"The Problem of Witness Reliability." Flying Saucers 76, March 1972, pp. 36-37.

NOTES:

**3756**
Dick, William
   "Enquirer Offers $50,000 Reward for UFO Proof." National Enquirer, March 12, 1972, p.30.

**3757**
Baragiola, Antonio
   "A Remarkable Case from Mendoza, Argentina." London, Flying Saucer Review (18:2), March-April 1972, 7-11.

**3758**
Sutherly, Curtis K.
   "UFO's and the Post-Blue Book Air Force." Caveat Emptor 4, Summer 1972, pp. 16-17, 24.

**3759**
Ballester-Olmos, Vicente J., & Miguel Guasp
   "Quantification of the Law of the Times." Ben Lomond, Calif., Data-Net (6:6), June 1972.

**3760**
Zayin, Karl
   "The Validity of the Alien Spacecraft Hypothesis." Flying Saucers 77, June 1972, p. 19.

**3761**
Schwartz, Dr. Berthold E.
   "'Beauty of the Night'." London, Flying Saucer Review (18:4), July-Aug. 1972, pp. 5-9+.

**3762**
Hynek, J. Allen
   "UFO's and the Air Force." Fate 269, Aug. 1972, pp. 52-62.

**3763**
Fawcett, George D.
   "Flying Saucers Don't Exist, UFO's do, but..." Flying Saucers 78, Sept. 1972, pp. 47-49.

**3764**
Fawcett, George D.
   "New UFO Warning." Flying Saucers 78, Sept. 1972, pp. 26-37.

**3765**
Hartle, Orville R.
   "UFO Reports." Flying Saucers 78, Sept. 1972, pp. 50-52.

**3766**
Larson, Kenneth
   "UFO's and America's Geographic Center." Flying Saucers 78, Sept. 1972, pp. 2-10.

**3767**
Prytz, John
   "Flying Saucers Throughout Geologic Record." Flying Saucers 78, Sept. 1972, pp. 12-17.

**3768**
"Caldwell Report." Jonquière, Quebec, Canada, Cosmos Express (1:7), Oct. 1972, pp. 3-5.

**3769**
Bord, Janet
   "After Snippy-What Next?" London, Flying Saucer Review (18:6), Nov.-Dec. 1972, pp. 19-22.

**3770**
Creighton, Gordon
   "Gobbledygook." London, Flying Saucer Review (18:6), Nov.-Dec. 1972, pp. 25-27.

**3771**
Le Poer-Trench, Brinsley
   "You and me." Flying Saucers 79, Dec. 1972, pp. 45-48.

**3772**
Lindsay, Gordon
   "Some Questions About Flying Saucers." Flying Saucers 79, Dec. 1972, pp. 14-23.

NOTES:

## UNIDENTIFIED FLYING OBJECTS INCLUDING FLYING SAUCERS

3773
Silvers, Samula M. R.
 "Flying Saucers, 25 Years of Wasted Time."
 Flying Saucers 79, Dec. 1972, pp. 51-53.

3774
"Should we Advertise?" Flying Saucers 79, Dec.
 1972, pp. 24-27.

3775
"The Substance and the Shadow." Flying Saucers
 79, Dec. 1972, pp. 54-57.

3776
"UFO Problem: Call for Security Against Space
 War." Flying Saucers 79, Dec. 1972, pp. 32-34.

3777
Shaver, Richard S.
 "About UFO's." Caveat Emptor 6, Winter 1972-73, p. 9.

3778
Friedman, Stanton, & B. A. Slate
 "UFO's and the Electromagnetic Dragon." Fate 274, Jan. 1973, pp. 76-78.

3779
"UFO's: Scientific Debate, edited by Sagan
 and Page, Reviewed." Washington Post,
 Jan. 28, 1973, sec. BKS, p. 6, col. 1.

3780
Creighton, Gordon
 "But I Read It In a Book!" London, Flying Saucer Review (19:1), Jan.-Feb. 1973, pp. 24-27.

3781
Holiday, F.W.
 "Dragons and UFO's at Loch Ness." London,
 Flying Saucer Review (19:1), Jan.-Feb. 1973,
 pp. 15-17.

3782
Schwartz, Berthold E.
 "Woodstock UFO Festival, 1966-1." London,
 Flying Saucer Review (19:1), Jan.-Feb. 1973,
 pp. 3-6.

3783
Buchan, Vivian
 "UFO Damages Iowa Soybeans." Fate 275, Feb.
 1973, pp. 88-92.

3784
Burns, Donald A.
 "Maui Loa Talks to a Flying Saucer." Search
 108, March 1973, pp. 74-79.

3785
"L T E ... Jan. 28 Review of UFO's...Discussed." Washington Post, March 4, 1973,
 sec. BKS, p. 15, col. 1.

3786
Schwartz, Berthold E.
 "Woodstock UFO Festival, 1966-2." London,
 Flying Saucer Review (19:2), March-April
 1973, pp. 18-23.

3787
Hynek, J. A.
 "UFO Experience (Review by Jaques Vallee)."
 Bulletin of the Atomic Scientists 29,
 April 1973, pp. 49-52.

3788
"Fleetwood, Pa. Police Chief, E. R. Fox,
 Says on April 31 That Pile of Spent Fireworks Discovered During Patrol in Crystal
 Cave area of Berles County, may be Source
 of Nightly Sightings of UFO's...New York
 Times, April 1, 1973, p. 71, col. 8.

3789
Sigma, Rho
 "French to Study UFO's." Fate 278, May 1973,
 p. 51.

3790
"U. S. Air Force Consultant J. Allen Hynek
 Comments on UFO's." Los Angeles Times,
 May 22, 1973, sec. 1, p. 2, col. 6.

NOTES:

**3791**
Clark, Jerome, & Loren Coleman
"Mystery Airships of the 1800's." Fate 278, May 1973, pp. 84-94; 279, June 1973, pp. 96-104; 280, July 1973, pp. 61-67.

**3792**
Dufour, J.C.
"A French Repeater Case: Events At Les Nourrandons." London, Flying Saucer Review (19:3), May-June 1973, pp. 3-5.

**3793**
Edwards, P.M.H.
"A Few Coincidences, and Two Postscripts." London, Flying Saucer Review (19:3), May-June 1973, pp. 23-26.

**3794**
Sarrantonio, Albert
"UFO Photos, a Comparison." Flying Saucers 81, Summer 1973, pp. 38-42.

**3795**
Shaver, Richard S.
"Are UFO's Projections?" Flying Saucers 81, Summer 1973, pp. 20-21.

**3796**
Thurgood, Peter A.
"How to Observe UFO's." Flying Saucers 81, Summer 1973, pp. 43-45.

**3797**
Hynek, J. A., & B. Ford
"Science Takes Another Look at UFO's." Science Digest 73, June 1973, pp. 9-13.

**3798**
"A Propos du Symposium de Quimper: Deux Déclarations d'Astronomes." Paris, Phénomènes Spatiaux 36, June 1973, pp. 28-32

**3799**
"Blob Meets UFO." Newsweek 81, June 11, 1973, p. 32.

**3800**
Bowen, Charles
"The Annapurna-Pokhara UFO 'Ballet'." London, Flying Saucer Review (19:4), July-Aug. 1973, pp. 3-6.

**3801**
Creighton, Gordon
"These Cunning British: The Truth At Last!" London, Flying Saucer Review (19:4), July-Aug. 1973, pp. 23-26.

**3802**
Romaniuk, Pedro
"Rejuvenation Follows Close Encounter with UFO : An extraordinary case from Argentina." London, Flying Saucer Review (19:4), July-Aug. 1973, pp. 10-14.

**3803**
Kor, Peter
"The Radicalization of Dr. Hynek." Flying Saucers 83, Fall 1973, pp. 47-52.

**3804**
"UFO News 1973." Flying Saucers 83, Fall 1973, pp. 59-62.

**3805**
Slater, Tim
"UFO's Invade USA." Madison, Wis., Take-Over (3:18), Oct. 2-Nov. 7, 1973, p. 4.

**3806**
"Are Flying Saucers Real? Latest on an old Mystery." U. S. News & World Report 75, Nov. 5, 1973, pp. 75-76.

NOTES:

## UNIDENTIFIED FLYING OBJECTS INCLUDING FLYING SAUCERS

**3807**
"UFO's Real, Scientist Says." Milwaukee Journal, Accent section, Nov. 7, 1973, p. 30.

**3808**
"UFO's Visiting From Space, Goldwater Says." Milwaukee Journal, Accent Section, Nov. 9, 1973, p. 17.

**3809**
Williams, Edward D.
"UFO Expert Says he Believes." Milwaukee Journal, Nov. 14, 1973, p. 7.

**3810**
"Gallup Poll Finds 51% of 1550 Persons Interviewed Believe that UFO's, Sometimes Called 'Flying Saucers,' are Real..." New York Times, Nov. 29, 1973, p.45, col. 1.

**3811**
Fawcett, George D.
"UFO's Mark 26th Anniversary." Fate 285, Dec. 1973, p. 78.

**3812**
Michel, Aimé
"Ufot ja Historia." Turku, Finland, Argos 1974, pp. 13-16.

**3813**
"Ufologian Tila ja Puharich." Turku, Finland, Argos 1974, pp. 29-32.

**3814**
Zullo, Allan
"U. S. Senator Barry Goldwater Convinced UFO's are Real..." National Enquirer, Jan. 6, 1974, p.18.

**3815**
"Fenómeno OVNI". Buenos Aires, Revista Atom (I: 1), Feb. 1974-(see this section in each issue).

**3816**
"New Flying Saucer Story." Mechanics Illustrated 70, Feb. 1974, pp. 34-35.

**3817**
Crider, Bill
"UFO no one can Figure Alters 2 Lives." Milwaukee Journal, Green Sheet sec., March 9, 1974, pp. 1, 3.

**3818**
Adell, Alberto, & Casas Huguet
"Un Excepcional Caso OVNI: Observación en el Pántano Gabriel y Galán." Buenos Aires, Cuarta Dimensión 9, May 1974, pp. 8-15; 10, June 1974, pp. 14-21.

**3819**
Banchs, Roberto E.
"Conspiración Contra los OVNI?" Buenos Aires, CEFAI Revista, (3: 2), May 1974, pp. 3-6.

**3820**
Klein, David
"Jeane Dixon: The Secret of UFO's Will be Revealed in the Near Future." National Enquirer, May 19, 1974, p.32.

**3821**
"UFO Report." Saga, Summer 1974, entire issue.

**3822**
Burk, David
"I Have Proved UFO's Exist, Claims Top Space Scientist." National Enquirer, June 16, 1974, p.4.

**3823**
"UFO's: Military Watergate of Hell." Milwaukee, Bugle American (5:24, #165), July 17-31, 1974, pp. 10-11, 14-21.

**3824**
"U.S. to Accept Existence of Space Visitors?" Milwaukee Journal, July 21, 1974, Accent Section, p. 5.

NOTES:

**3825**
Ward, Bernie
   "Five Prominent Psychiatrists Predict how People on Earth Will React to Contact with Beings From Outer Space."
   National Enquirer, July 21, 1974, p.14.

**3826**
"Three Amazing Cases of UFO Healing."
   National Enquirer, Aug. 11, 1974, p.5.

**3827**
Mead, M.
   "UFO's, Visitors From Outer Space?"
   Redbook 143, Sept. 1974, pp. 57-58

**3828**
Hynek, J. Allen
   "Why we Should Keep an Open Mind on UFO's." The Science Teacher, December 1974, pp. 31-32.

**3829**
"J. J. O'Connor Reviews NBC News Program, UFO's: do you Believe?" New York Times, Dec. 13, 1974, p.91, col. 3.

**3830**
Hynek, J. Allen
   "The UFO Mystery." FBI Law Enforcement Bulletin (44:2), Feb. 1975, pp. 16-20.

**3831**
Pease, Harry S.
   "Scientist Says Saucers are Real."
   Milwuakee Journal, April 10, 1975, Accent sec., p.7.

**3832**
Tomorrow Show. (The Outer Space Connection, With Allen Landsberg.) NBC-TV Network August 12, 1975, 12 midnight - 1 A.M., Central Daylight Time.

NOTES:

# DIRECTORY SECTION

## ORGANIZATIONS

### ARGENTINA

3833
Centro de Estudios de Fenómenos Aéreos
  Inusuales
Casilla de Correo #9, Suc. 26
Buenos Aires, Argentina

3834
Centro de Investigación Sobre Fenómenos de
  Inteligencia Extraterrestres
Casilla de Correo 78, Sucursal 31
Buenos Aires, Argentina

3835
Centro de Investigaciones y Desarrollo
  Especial Argentino
c/o Raúl Roberto Podestá
Lorenzo Suarez de Figuero #167
B. M. de Sobremonte
Córdoba, Argentina

3836
Círculo Argentino de Investigaciones
  Ufológicas
Casilla de Correo 218
Córdoba, Argentina

3837
Círculo de Investigadores del Fenómeno OVNI
  M. Lebensohn #73
Mar de Ajo, Provincia de Buenos Aires
Argentina

3838
Comisión de Investigación de Objetos
  Voladores no Identificados (CIOVNI)
c/o J. Guillermo Monaco, Pres.
Ramírez de Velazco #377
San Salvador de Jujuy, Jujuy
Argentina

3839
Comisión Rastreadora de Bases Extrater-
  restres
Chiclana 226, Local 22, Galerías Galehot
Casilla de Correo 209
Bahía Blanca, Provincia de Buenos Aires
Argentina

3840
Grupo Investigación Vida Extraterrestra
Rafaela 4106, Depto. 1
Ciudadela
Buenos Aires, Argentina

3841
Sociedad Investigadora de Fenómenos
  Espaciales
Jacinto Ríos 1419
Córdoba, Argentina

NOTES:

DIRECTORY SECTION - ORGANIZATIONS

## AUSTRALIA

**3842**
Perth UFO Research Org.
President, W. Barton
Flat #5
52, Cunningham Terrace
Daglish
Western Australia

**3843**
SUFOS
26 Minnamurra Place (S)
Pymble NSW 2073
Australia

**3844**
Victorian UFO Research Society
P.O. Box 43
Moorabbin 3189
Victoria, Australia

## BELGIUM

**3845**
Belgian Federation of Ufology
255 Av. G. Henri
B 1200, Brussels, Belgium

**3846**
BUFOI/IGAP
13 Berkenlaan
B 2610, Brussels, Belgium

**3847**
CEFC
454 Av. G. Henri
B 1200 Brussels, Belgium

**3848**
G. Contact
26 Drève du Chateau
B 1630 Linkebeek, Belgium

**3849**
Société Belge D'étude des Phénomènes
M. Lucien Clerebaut, Secretary
Boulevard Aristide Briand 26
1070-Brussels, Belgium

## CANADA

**3850**
Saskatchewan Unidentified Phenomena Research
D.F. Clausen
1312 Garnet Street
Regina, Saskatchewan
Canada

## DENMARK

**3851**
Dansk UFO Center
Refsvej 62/2
Thisted, Denmark

NOTES:

DIRECTORY SECTION - ORGANIZATIONS

### ENGLAND

**3852**
Aerial Phenomena Enquiry Network
c/o Roy Thomas
23 Delamere Avenue
Clifton Green
Swinton, Manchester
Lancashire, England
M27 2Qn

**3853**
Cleethorpes & District Aerial Phenomena
(UFO) Society
c/o S. York
39, Wendover Rise
Cleethorpes, England

**3854**
Contact International
Mr. Brinsley Le Poer Trench
Flat 8
57 Drayton Gardens
London, S.W. 10, England

**3855**
Cosmology Newslink International
(C-N-K) (quarterly)
Magazine of the Cosmos
37, The Close
Great Dunmow, Essex
England

**3856**
WIRRAL UFO Society
3 Holmfield Drive
Great Sutton
Wirral, Cheshire L66 E88
England

### FRANCE

**3857**
Groupement D'étude des Phénomènes Aériens
Mr. René Fouéré, Secretary
69 Rue de la Tombe-Issoire
75014 Paris
France

**3858**
Ouranos Studies Committee
U.G.E.P.I.
F. 02110 Bohain
Becquigny, France

### ITALY

**3859**
Centro Ufologico Nazionale
Via Vignola 3
20136 Milan
Italy

**3860**
CNIFAA
Independent National Committee for the Study
of Anomalous Aerial Phenomena
Via Rizzoli 4
c/o Renzo Cabassi
40125 Bologna, Italy

NOTES:

DIRECTORY SECTION - ORGANIZATIONS

## SPAIN

**3861**
Centro de Estudios Interplanetarios (STENDEK)
Apartado 282
Barcelona, Spain

## SWEDEN

**3862**
Arbetsugruppen For UFO - Identifiering
Project U.R.D.
UFO Rapporterings O och Datasystem
Box 772
S-101 31 Stockholm, Sweden

## SWITZERLAND

**3863**
Swiss Federation of Ufology
U.G.E.P.I.
6 Rue Dassier
Ch. 1201, Geneva, Switzerland

## TURKEY

**3864**
Space Phenomena Research Group
P. O. Box 1157
Istanbul, Turkey

## UNITED STATES OF AMERICA

### ALABAMA

**3865**
North American Investigators of Strange Phenomena
Mr. Steven Elliott
Route #2
Box - 159
Vina, Alabama - 35593

### ARIZONA

**3866**
Aerial Phenomena Research Organization
(A.P.R.O.)
3910 E. Kleindale Road
Tucson, Arizona  85712

NOTES:

DIRECTORY SECTION - ORGANIZATIONS

**3867**
Ground Saucer Watch, Inc.
William H. Spaulding, Director, GSW-West
13238 North 7th Drive
Phoenix, Arizona 85029

DISTRICT OF COLUMBIA

**3874**
UFO Research Associates
P.O. Box 34252
Washington, D.C. 20034

CALIFORNIA

**3868**
Borderland Sciences Research Foundation
Dr. Riley H. Crabb (Director)
Box 548
Vista, California 92083

**3869**
Ministry of Universal Wisdom
George Van Tassel
(Proceedings Magazine)
P.O. Box 458
Yucca Valley, California. - 92284

ILLINOIS

**3875**
Ancient Astronaut Society
c/o Gene Phillips, President
600 Talcott Rd.
Park Ridge, Illinois 60068

**3876**
Center for UFO Studies
Dr. J. Allen Hynek (Director)
P.O. Box 11
Northfield, Illinois - 60093

**3870**
National Investigations Committee on Unidentified Flying Objects
7970 Woodman Avenue
Van Nuys, California 91402

**3871**
Senate of Wonders
701 Central Ave.
Sonoma, California 95476

**3877**
Midwest UFO Network
40 Christopher Court
Quincy, Illinois 62301

**3872**
UFO Research Institute
c/o Stanton Friedman
P.O. Box 941
Lawndale, California 90260

**3878**
Mutual UFO Network
40 Christopher Court
Quincy, Illinois 62301

**3873**
Worldwide Association for Anomalous Scientific
 Phenomena
P.O. Box 717
Huntington Beach, Calif. 92648

NOTES:

DIRECTORY SECTION - ORGANIZATIONS

**KENTUCKY**

**3879**
UFORG
Mr. Donald W. Krider, Jr.
609 Marengo Drive
Louisville, Kentucky - 40243

**MARYLAND**

**3880**
Isis UFO Project
P O Box 512
Silver Spring, Md. 20907

**3881**
National Investigation Committee on Aerial Phenomena (NICAP)
Suite 23
3535 University Blvd., West
Kensington, Maryland  20795

**MICHIGAN**

**3882**
Michigan UFO Center
of Project STRAT
c/o John Shepherd
R.R. 1, Box 313
Central Lake, Michigan 49622

**3883**
UFO Studies Center
c/o Lloyd F. Burton
2231 Briggs St.
Drayton Plains, Michigan 48020

**NEW JERSEY**

**3884**
Contact International
Brian Deloach, President
P.O. Box 575
Denville, New Jersey  07834

**3885**
Delval UFO
c/o Michael J. Campione
2202 New Albany Rd.
Cinnaminson, New Jersey  08077

**3886**
Foreign Objects Exchange Service (Foes)
August C. Roberts
25 Laurel Street
Paterson, New Jersey

**3887**
Saucer and Unexplained Celestial Events Research Society
P.O. Box 163
Fort Lee, New Jersey

**3888**
20th Century UFO Bureau
Dr. Carl McIntire, Director
756 Haddon Ave.
Collingwood, New Jersey  08108

NOTES:

DIRECTORY SECTION - ORGANIZATIONS

**NEW YORK**

**3889**
UFO Club
c/o Robert Chaffee
Pembroke Central School District
Routes 5 and 77
Corfu, New York  14036

**NORTH CAROLINA**

**3890**
Tarheel UFO Study Group
350 Pine Ridge Drive
Winston-Salem, North Carolina  27104

**3891**
UFONATION
George W. Fawcett
607 N. Main Street
Mount Airy, North Carolina  27030

**OHIO**

**3892**
Ohio UFO Investigators League, Inc.
Charles J. Wilhelm (Ex-Director)
5852 E. River Road
Fairfield, Ohio - 45014

**3893**
The Probers
2004 Meadowlawn
Fairfield, Ohio  45014

**3894**
UFO Research Bureau of Lorain
Norman Swinehart, Director
4435 Palm Avenue
Lorain, Ohio  44055

**OKLAHOMA**

**3895**
International UFO Bureau, Inc.
Hayden C. Hewes (Director)
P.O. Box - 1281
Oklahoma City, Oklahoma  73101

**OREGON**

**3896**
Understanding Inc.
Dr. Daniel W. Fry
(Understanding Magazine)
P.O. Box - 206
Merlin, Oregon  97532

NOTES:

DIRECTORY SECTION - ORGANIZATIONS

PENNSYLVANIA

**3897**
Pennsylvania Center for UFO Research
6 Oakhill Avenue
Greensburg, Pa.   15601

SOUTH CAROLINA

**3898**
A.P.I.O.
Mrs. Margaret Pine
Box - 663
Mauldin, South Carolina   29662

TENNESSEE

**3899**
Aerial Phenomena Education Committee
1234 Robin Hood
Memphis, Tennessee   38111

TEXAS

**3900**
MUFON
103 Old Towne Road
Sequin, Texas   78155

VIRGINIA

**3901**
INFO, The Fortean Organization
P.O. Box 367
Arlington, Virginia   22210

WASHINGTON

**3902**
UFO Newsclipping Service
3521 N.W. 104th St.
Seattle, Washington   98146

NOTES:

# PERIODICALS

## ARGENTINA

**3903**
Cuarta Dimensión, care of
Mas Allá de la Cuarta Dimensión
Gaona 1312
Buenos Aires, Argentina

**3904**
OVNIS, un Desafío a la Ciencia
Círculo Argentino de Investigaciones Ufológicas
Casilla de Correo 218
Córdoba, Argentina

## BELGIUM

**3905**
Inforespace
Boulevard Aristide Briand, 26
1070 Brussels, Belgium

## CANADA

**3906**
Canadian UFO Report
Box 758, R.R. 1
Duncan, B.C.
Canada

**3907**
Quest UFO Report
489 Krug Street
Kitchener, Ontario, Canada

**3908**
Saucers, Space and Science
c/o Gene Duplantier, Editor
17 Shetland Street
Willowdale, Ontario, Canada

## DENMARK

**3909**
UFO Contact.
 IGAP Information Service
 Vejle, Denmark

NOTES:

DIRECTORY SECTION - PERIODICALS

## ENGLAND

**3910**
British Unidentified Flying Object Research
 Association
 The BUFORA Journal. Ashford (Mrs. A.
Harcourt, 170 Faversham Rd., Kensington,
Ashford,
Kent, England

**3911**
Cosmology Newslink International (C-N-I)
(quarterly)
Magazine of the Cosmos
37, The Close
Great Dunmow, Essex
England

**3912**
Cos-Mos
95, Taunton Road
London S.E. 12, England

**3913**
Flying Saucer Review
Mr. Charles Bowen, Editor
c/o Compendium Books
281 Camden High Street
London N.W. 1, England

**3914**
Gemini, the Twin UFO Journal
95 Taunton Road
London S.E. 12 8PA

**3915**
Merseyside UFO Bulletin
53 Woodyear Road
Bromberg, WIRRAL, Cheshire
England L62 6AY

**3916**
Scan. Editor: Leslie Harris
 5 Grenfell Rd.,
 Moordown, Bournemouth, England

**3917**
Spacewise.
 Martec Pub. Group Ltd.
 61 Berners St.
 London, England

## FINLAND

**3918**
Argos (formerly Humanoidien Maailma)
Jyrkkalank. 4B141
20210 Turku 21
Finland

## FRANCE

**3919**
Groupement d'Etude de Phenomènes Aériens et
 Objets Spatiaux Insolites
 Phénomènes Spatiaux; Bulletin.
(69, rue de la Tombe-Issoire),
Paris, France

**3920**
Lumières Dans la Nuit
"Les Pins" - 43400
Le Chambon-sur-Lignon, France

**3921**
Ouranos
25 Rue Denfert-Rochereau
38000 Grenoble, France

NOTES:

DIRECTORY SECTION - PERIODICALS

## NETHERLANDS

3922
Life Sciences and Space Research.   Amsterdam,
  North-Holland Pub. Co.

3923
Space Life Sciences.   Reidel Pub. Co.,
  P.O. Box 17
  Dordrecht, Holland

3924
UFO Onderzoek
Korte Drift 13
Uithuizermeeden Gr. 822
Netherlands

## NEW ZEALAND

3925
UFO Bulletin
Auckland University UFO Research Group
Auckland University-Private Bag
Auckland, New Zealand

## SPAIN

3926
Stendek
Apartado 282
Barcelona, Spain

## UNITED STATES OF AMERICA

3927
Aerial Phenomena Perspectives
(An Owlexandrian Multimedia Publication)
P.O. Box 388
Main Post Office
Atlanta, Georgia   30301

3928
Anomaly
Box 351
Murray Hill Station
New York, N.Y. 10016

3929
ASA Confidential
Allied Saucers Assn.
P.O. Box 35
Eden, N.Y. 14057

3930
Caveat Emptor
G and G Enterprises
P.O. Box 688
Gatesville, PA. 19320

3931
UFOCAT
Center for UFO Studies
924 Chicago Ave.
Evanston, Ill. 60201

3932
Cosmic Newsletter
Clintonville, Wis.,

3933
The Data Net UFO Report
Mrs. Josephine Clark
7900 Harvard Drive
Ben Lomond, California - 95005

3934
Flying Saucer Digest
UAPA
P.O. Box 9811
Cleveland, Ohio 44142

NOTES:

DIRECTORY SECTION - PERIODICALS

**3935**
Flying Saucer News
346 W. 45th Street
New York, N.Y. 10036

**3936**
Flying Saucers
Palmer Pubs.
Route 2, Box 36
Amherst, Wis.

**3937**
Flying Saucers International
Amalgamated Clubs of America
P.O. Box 84
Northridge, Calif. 91324

**3938**
FSIC Bulletin
P.O. Drawer G
Akron, Ohio 44305

**3939**
Infinity Newsletter - Other Dimensions, Inc.
104 Washington Street
Decorah, Iowa 52101

**3940**
Info Journal - International Fortean Organization
801 N. Daniel St.
Arlington, Va., 22201

**3941**
The Kansas Newsletter
UFO Research Associates, Inc.
P.O. Box 1672
Topeka, Kansas 66601

**3942**
News Bulletin
Center for UFO Studies
924 Chicago Ave.
Evanston, Ill. 60201

**3943**
Occult Trade Journal
International Occult, Metaphysical News
2274 Como Avenue
St. Paul, Minnesota 55108

**3944**
Ohio UFO Reporter
Editor: Bonitan Roaman
Route 3, Yankee Rd.
Middletown, Ohio 45042

**3945**
Paraufologist
2875 Sequoyah Drive NW
Atlanta, Ga. 30327

**3946**
Phenomena Research Newsletter
Mr. Robert Gribble
P.O. Box - 1807
Seattle, Washington 98111

**3947**
Police and the UFO Experience.
P.O. Box 11, Northfield, Ill. 60201

**3948**
Proceedings
P.O. Box 458
Yucca Valley, California 92284

**3949**
Pursuit; the Journal of the Society for the Investigation of the Unexplained.
Columbia, New Jersey 07832

**3950**
Saucer News
P.O. Box 2228
Clarksburg, W.Va.

**3951**
Saucer News
Saucer and Unexplained Celestial Events Research Society
P.O. Box 163
Fort Lee, New Jersey

NOTES:

DIRECTORY SECTION - PERIODICALS

**3952**
Saucer Review
Mr. Randall Grant
104 Hickory Lane
Mauldin, South Carolina  29622

**3953**
Scientific Sauceritis Review
Perkins Lane
Hallowell, Maine  04347

**3954**
The Silver Bridge
Gray Barker
Saucerian Publications, Inc.
Box 2228
Clarksburg, West Virginia  26301

**3955**
Skylook
Dwight Connelly
26 Edgewood Drive
Quincy, Illinois  62301

**3956**
Solar Space Letter
Box 662
Joshua Tree, California  92252

**3957**
Spacemen
1426 E. Washington
Philadelphia, Pa.

**3958**
Spaceview
Paradice International Co.,
Beaumont, Texas

**3959**
Spectrum: Journal of the Occult
Ms. Barbara Lucas
6293 34th Avenue North
St. Petersburg, Florida - 33710

**3960**
UFO Central Annual Report
Center for UFO Studies,
924 Chicago Ave.
Evanston, Ill. 60201

**3961**
UFO Chronolog
c/o J.M. Erhardt, Editor
43 Richmond Dr.
Newport News, Va. 23602

**3962**
UFO Commentary
72 Jeffreys Dr.
Newport News, Va. 23601

**3963**
The UFO Investigator - NICAP
Suite 23
3535 University Blvd., West
Kensington, Maryland  20795

**3964**
UFO Magazine News Bulletin
Mr. Rick R. Hilberg
3403 W. 119th Street
Cleveland, Ohio - 44111

**3965**
UFO Research Newsletter
Mr. Gordon I.R. Lore
10013 Erlitz Drive
Albuquerque, New Mexico - 87114

**3966**
UFO Sightings Newsletter
Editor: Janice M. Croy
512 S. Logan
Denver, Colorado  80209

**3967**
The Unidentified
Editor: Gilbert J. Ziemba
Box 515 - #73749
Joliet, Illinois - 60434

**3968**
Unknown
Paul Doerr
Box 1444
Vallejo, California 94590

NOTES:

AUTHOR INDEX

Note: Numbers immediately following names of personal and corporate authors, as well as titles entered as authors, refer to individual, numbered bibliographic items, and not to pages. In cases where a given author has also been editor (or compiler, translator, etc.) of additional works, his name is re-entered in the index, followed by the appropriate designation. The names of joint authors (not exceeding a total of two individuals) have also been included. It should be noted that periodicals have been identified as such.

A

"A. Oparin Sees Primitive Life Forms Possible on Moon...", 2214
Abbot, C. G., 2016, 2061
"ABC-TV Presents Special Show, We are not Alone...", 1367
Abel, R. C., 1799, 3404
Abelson, Philip H., 1260, 2186
Aberg Cobo, Azel, 765
Adam, Frank, 3530
Adams, John B., 2239
Adamski, George, 199, 728, 877, 2419, 2451, 2846, 2912
"Adamski's Hieroglyphics", 1279
Adell, Albert, 2394, 3818
Adler, Bill, 3032, 3148
Adler, Bill, comp., 3033
"Adler Planetarium Director Johnson and Aide Recall Tracking Reddish Object Moving Over Chicago Aug. 26...", 321
"Admiral Dufek Sees 'Flying Saucer' Possible", 3434
Advances in Space Science and Technology, 1661

"Aerial Phenomena Investigations Committee Urges Secretary Sharp Clarify Report on Flame-Spouting Object...", 317
"Aero Club Asks U.S. Air Force to Reopen Probe, 2404
"Aeronutronic Studying Automated bio lab: Search for Life on Mars", 2654
Aerospace Corp. Lab Operations, El Segundo, Calif., 1470
"L'affaire de Valensole", 867
Afshar, H. K., 2785
"After Tully", 457
Aggen, Erich, 944, 1410, 3677, 3702, 3707, 3727
Agnew, Irene, 1414, 2737
Aguilar, Xavier F., 1416, 3716
Aharon, Y. N. ibn, 1336
Ahmad, I. I., 1039
Aho, Arthur C., 2615
"Air Force Demands Respect for 'UFO Report'", 3448
"Air Force has 'Jet Saucer'", 3374
"Air Training (U.S. Air Force Publication)," 223

Aitken, R. G., 2064, 2071
Akademski Astronomsko-Astronautitski Klub, Sarajevo, Yugoslavia, 167
"Alaska Air Command Probes Reports of Silvery Objects Trailing Flames...", 314
Albanesi, Renato, 846
Alceoli, 2776
Alemán Velasco, Miguel, 2980
Alessandri, Michelangelo, 1054
Allemann, Theodor, 150
Allen, T. B., 1055, 1694
Allen, W. Gordon, 299, 2965
Allingham, Cedric, 2925
Allison, Anthony, 1961
Alvarez Lopez, José, et al, 1077
Ambartsumyan, V., 112
"Ambassador H. J. Taylor Warns Senate Foreign Relations Committee not to Dismiss Saucer Reports", 3394
"American Astronautical Society Sponsor Symposium on Search for Extraterrestrial Life Anaheim, Calif.", 1951
"American Astronomers at Arecibo Ionospheric Observatory Detect Radio Signals From Space That Might be From Other Civilizations", 1606
"American Biological Sciences Institute Publication...", 2652
American Chemical Society, 1729
"Ames Exobiologists Provide Major Link in Understanding Origin of Life on Earth", 108

"Ames Researchers Find Mars Produces Life Forms", 2203
"An Account of the Michigan Incident Through the Experts and Witnesses", 492
"An Airline Captain Speaks Out", 346
Anchor, 149
"Ancient Records of UFO in Japan", 1284
Anders, Edward, 1857
Anderson, Dave, 434
Anderson, Kenneth V., 779
Anderson, Poul, 1040
Andrews, F. F., 1879
Anfinsen, C. B., 1649
"The Angels", 1176
Angelucci, Orfeo M., 731, 739, 2926
Anglada Font, Luis, 3077
"Animal Life on Jupiter", 2261
"Animal, Vegetable, or...", 1847
"Ann Arbor, Hillsdale et Autres Lieux", 426
"Another Speech by Wilbert B. Smith", 849, 3497
Antoncadi, E. M., 2078
"Anybody out There?", 1225
"Any Life on Venus is Declared Impossible...", 2083
"Apparaissaient des Soucoupes Volantes", 3320
Appel, T., 2486
Araujo, Hernani Ebecken de, 3192
Archer, J. L., 2591
Archers' Court Research Group..., 2400
"Are 'Contact Group' Sightings Metaphysical?", 2357, 2515

"Are Flying Saucers Real? Latest on an Old Mystery", 3806
"Are There Other Worlds?", 1166
"Are There Spores in Space?", 1843
Arejula, Francisco, 3219
"ARIC Begins Study of Saucer Reports", 3342
Armagnac, Allen P., 2191, 3691
Armand, John, 169
Arnell, J. C., 3543
Arnold, Kenneth A., 183, 184, 2902, 2910
"Arrhenius Again Cites Doubt That Living Beings Inhabit Mars", 1135
Arrhenius, Svante, 6, 69, 2018, 2431, 2434
"Un Article d'Antonio Ribera", 558
"Article on UFO's Notes for Most of Past 25 Years...", 639
Ascher, Marcia, 1555
Ascher, Robert, 1555
Ash, M. E., et al, 2244
Ashbrook, J., 1869
Ashtar. In Days to Come, 1004
Asimov, Isaac, 1030, 1262, 1340, 1576, 1580, 1940, 2038, 2371, 2450
Astapovich, I. S., 2001
"Astrobotany", 1831
"Astronautics, Development and Goals", 2842
"Astronomer, Dr. Gadomski of Poland, in Recent Paper, Says High Developed Civilizations on Planets of Suns in our Galaxy...", 1271
"Astronomers, Physicists and Social Scientists Participating in Symposium on UFO's ask U.S. Air Force to Preserve Secret Records...", 3712

"Astronomy", 116
Aswell, J. R., 3302
"At Least 40 Report Flying Object and 4 Sister Ships Landed in Swamp near Ann Arbor, Michigan...", 410
Atmananda, 3169
"The Atmosphere of the Planets", 1765
"Attention aux Soucoupes!", 3534
"Atterrissages de 'M.O.C.' en 1967", 529
"Aurora, Texas Site of 1897 UFO Crash Investigated", 668
Austin, R. R., 1287
"Australian Scene, 1963-64", 392
"Australian Scene 1965", 458
Avenal, A., 2763
Avignon, Andre, 854


B


Babcock, Edward J., 449, 3007
Babcock, Edward J., ed., 159
"Bacteria able to Survive Mars-like Conditions", 2147
"Bacteria Thrive in Jupiter-like Atmosphere", 2292
"Bahia Blanca; l'Extraordinaire Rencontre du Camionneur Dionisio Llanca", 973
Bailey, A. C., 1452
Bailey, James O., 995
Baker, B. L., 1981
Baker, Robert M. L., 3078
Balagezyan, Jivan, 1417
"Baldwin Article Discusses use of jet Stream Lift Principle in Wingless Craft", 3383

Balfour, Malcolm, 651, 1617
Ball, G., 1523
Ball, J. A., 1424
Ball, R. S., 1126
Ballester-Olmos, Vicente J., 594, 597, 601-603, 617, 620, 641, 644, 650, 958, 960, 977, 3759
Balta Elias, J., 1501
Banchs, Roberto E., 171, 711, 3819
Bandini, Franco, 3170
Banerji, A. C., 75
Barabashev, G., 2401
Baragiola, Antonio, 3757
Barker, Gray, 296, 301, 308, 311, 327, 753, 805, 807, 830, 3001, 3034, 3427, 3745
Barker, Gray, ed., 758, 2936
Barnes, Sam, 1307
Barr, R., 1272
Barricelli, N. A., 2142
Barth, Charles A., 1980
Barton, W. H., 2085
Barton, Michael X., 736, 737, 740, 2309, 2317, 2319, 2320, 2945, 2970
Bauchau, A., 1913
Baur, Franz, 110
Beach, David, 2836
Beard, D. P., 2060
Beard, Robert B., 143
Beauchamp, T., 507
Beckley, Timothy G., 449, 479
Beckley, Timothy G., ed., 159
Becquerel, P., 1147, 1751
Beer, Lionel, 2988
Beere, D. C., 788
Beischer, D. E., 1800
"Belated Explanation on Flying Saucers", 3285
"Belgische UFO-Waarnemingen", 1379
Belitsky, Boris, 1615

"Bell Aircraft Engineer Says Objects Come From Outer Space; Harvard Observatory Director Menzel Calls Them Mirages", 3408
Bell, Don, 3696
Bellcomm, Inc., Washington, D. C., 1704
Belle, Jack van, 2344
Beller, William, 1529, 1892, 2625
Bemelmans, Hans, 561, 3149
Bender, Albert K., 747
Bendix Aviation Corp. Aerospace Systems Div., 2587
Benedict, W. R., 693
Benham, Wilfrid E., 3079
Bennett, Dorothy A., 1158, 2076
Bergamini, David, 1860, 1876, 1878
Bergendahl, P. E., 473
Berger, Rainer, 1382, 1674
Berget, Alphonse, 72
Bergier, Jacques, 775, 1092, 1456, 3193
Bergier, Jacques, & INFO, 782
Bergrun, Norman R., 3194
Berkner, L. V., 2810
Berland, Theodore, 1870
Berlitz, Charles, 1104
Bernal, J. D., 1286, 1884, 1891
Bernard, Raymond, 1435, 1436, 3171
Bernstein, J., 1358
Berrill, N. J., 51, 1062, 1911, 2627
Berry, Bruce D., 279
Bersadski, Rudolf, 1431
Besset, Henry J., et al, 661
Bessor, John P., 239, 817, 908
Bethurum, Truman, 724, 2845

"Better Life Spies for Space", 2694
"Biblical Flying Saucers", 2503
Biemann, K., 1988
Bieri, R., 1328
Binder, Otto, 1247, 2136, 3035, 3466, 3645, 3647
"Biomedical Space Technology Lagging", 1949
Biospherics, Inc., 2595
Biraud, François, 1093, 1105, 1402
Birch, A., 349
Black, V., 3319
Blaher, Damian J., 1309, 2536
Blei, Ira, 2656, 2565
"Blips on the Scopes", 206
Bliven, B., 1787
"Blob Meets UFO", 3799
Bloecher, Ted, 162, 3036
Blum, H. F., 78
Blumrich, Josef F., 2481
Bobrovnikoff, M. T., 1383
Boehm, A. W., 1537
Boffey, Philip M., 3671, 3680
Bok, B. J., 1917
Bonabot, J., 650
Boncompagni, Solas, 1385
Bongers, Leonard H., 1308, 2169
Bonzhich, Serge P., 2293
"Book, Assault on the Unknown, Reviewed", 3467
The Book of Spaceships and Their Relationship With Earth..., 2333
Bord, Colin, 2550
Bord, Janet, 2393, 3769
Bordeleau, Henri, 3008, 3150, 3172
Botan, E. A., 2655
Botan, E. A., et al, 1934, 2658
Bourquin, Gilbert A., 2474, 3080
Bourret, Jean C., 3239

Bova, Benjamin, 1716
Bowen, Charles, 378, 389, 393, 398, 423, 445, 474, 508, 530, 647, 656, 797, 855, 861, 901, 911, 922, 955, 2381, 3646, 3800
Bowen, Charles, ed., 776
Bowen, Mollie, 3120
Bowman, Norman J., 1783, 2100, 3274, 3345
Boyd, L. G., 2986, 3493
Boyer, Samuel H., 550
Bracewell, R. N., 1242, 1526, 1552, 1897
Bradbury, Ray, 2610, 2612
Brahe, T., 1, 4
Brandt, Ivan, 898, 2532
Branscomb, Lewis M., et al, 3637
Brasington, Virginia F., 2456
Bray, Arthur R., 3037
Bray, Derek R. M., 3485
"The Brazilian Abduction", 2874
"A Brazilian Contact Claim", 841
Breig, Joseph A., 1210, 1211, 1253
Brennan, Norman, 144, 3731
Brewer, Fred, 2622
Brewster, David, 983, 2429
"Brief Saucer Review", 285
Briggs, Michael H., 83, 124, 1031, 1041, 1220, 1230, 1250, 1263, 1289, 1530, 1542, 1695, 1798, 1894, 2611, 2635
Briggs, Michael H., ed., 1056
Brissenden, R., 351
British Unidentified Flying Object Research Association, 3195
Brock, Rudolf, 2786
Brooks, Angus, 519
Brooks, J., 1982

"Brookings Institute Study for NASA Warns on Implications of Possible Discovery of Intelligent Life in Space", 1256
Brookings Institute, Washington, D. C., 2771
Brovetto, P., 3240
Brown, A. H., 2645
Brown, William F., 3010
Bruns, J. E., 2522
Bryant, Larry W., 537, 2889, 3500
Buchan, Vivian, 3783
Bucheim, R. W., et al, 2767
Buck, Richard M., 2297
"Buck Rogers Baedeker", 1500
Buckle, Eileen, 598, 964, 3038
Buckner, H. T., 2414, 3521, 3541, 3638
Buelta, Eduardo, 331, 732
"BUFORA Research Officer's Annual Report - 27th Nov. 1965", 3535
Buhl, D., 117
Buhler, W., 588
Bujanda, Jesús, 1008
Bull, F. M., 2989
Bun, Thomas P., 1801
Burbidge, E. M., 1957
Burbidge, G. R., 1957
Burk, David, 3822
Burns, Donald A., 3784
Burns, J. A., 1594
Burr, E. F., 2494
Burr, Frank, 847
Burridge, Gaston, 2802
Burt, Bill, 652-654
Burt, Eugene H., 3151, 3748
Busson, Bernard, 2937
Busson, Le Roy, 2927
Butcher, Lee, 551
Bylinsky, C., 2544
Bylinsky, Gene, 1983

C

"C. F. Jankins Takes 30 Feet of Radio Film of Mysterious Signals During Closest Approach to Earth With Device he Invented", 1491
"C. Flammarion Predicts That we Will Talk With Mars", 1487
"C. G. Abbott Discusses Habitability of Other Worlds", 1750
"C. G. Abbott Says Venus not Mars, is abode of Life", 2062
Cade, C. M., 1063, 1301, 1374, 1375, 1560, 1565, 1595, 1598, 1600
Cadman, A. G., 3504
Cahn, J.P., 803
Caidin, Martin, 2285
Caldbeck, David, 258
Calder, N., 1522
"Caldwell Report", 3768
California Institute of Technology. Jet Propulsion Laboratory, 2589, 2590, 2719, 2780, 2841
California. University at Berkeley. Dept. of Soils and Plant Nutrition, 2048
California. University at Berkeley. Space Sciences Lab., 1471
California. University at Los Angeles, 1721
California. University at San Diego, 1675
Calvillo Madrigal, Salvador, 2916
Calvin, Melvin, 33, 92, 93, 1558, 1562, 1650, 1836, 2805

"The Camera Sees Flying Saucers", 195
Cameron, A. G. W., 101, 1568, 1676, 1898
Cameron, A. G. W., ed., 49, 1457
Camlin, Edward B., 708
Campbell, John W., 3322
Campione, Michael J., 3081, 3152
"Can Astronomy Ever Say Positively That Other Planets are Inhabited?", 1138
"Can Dr. Condon See it Through?", 3573
"Can Life Exist on any Planet but the Earth?", 1154
"Canada Building Lab to try to Prove or Disprove Existence", 3346
Canada. Deputy Minister of Transport for Air Services, 2908
"Canadian 'Saucer' may be Delta Fighter", 3336
Capen, Charles F., 2254
Cappa, Fidel A., 338
Caputo, Livio, 438, 883, 3556
Cardinale, Quixe, 777
Carey, G. C. S., 2517
Carlisle, M., 1164
Carpenter, Mark, 432
Carr, A., 2539
Carr, Aidan M., 856
Carriker, A. W., 2155
Carrión, Felipe Machado, 3082
Carrougès, Michel, 748, 1042
Carson, Anthony, 500
Carson, C., 800
Carter, Joan F., 2334, 3009
Casault, Jean, 3196
"Case of Dallas, Texas 'Blob' Tied to UFO Sighting", 673

Cassens, Kenneth H., 2505
Casserly, J. J., 2518
Castells Adriaensens, Francisco do, 2803
Cathcart, John M., 2753
Cathie, Bruce L., 1082, 3173
Catoe, Lynne, 141
"Cause of Flying Saucers", 3308
"Cautious Mars Landing", 2815
Cestling, Philip, 3744
Chaikin, G. L., 1782
Chaloupek, Henri, 480, 524
Chamberlin, Ralph V., 115, 1757
Chambers, Howard V., 3039
"Chances of Contacting Extraterrestrial Civilizations Seem Poor", 2748
Chapman, Clark R., 2241
Chapman, Robert, 3121, 3153
Chappell, E. W., et al, 1955
Charles, Jules, 46
Charles, R. H., editor, 2430
Charroux, Robert, 1111, 2345
Chartrand, Robert L., 3010
Chase, Frank M., 2777, 3083
"Cheerio There, Earthlings!", 3519
Chernenko, Mikhail, 1241
Chernigovski, V. N., ed., 2044
Cheville, R. A., 2453
Chibbet, H. S. W., 2382
"Chlorophyll and the red Spot", 1960
Ciardi, John, 2731
"Civil Aeronautics Authority Reports Saucers are Reflections Caused by Temperature Inversions...", 3329

Clark, J. J., 595
Clark, Jerome, 433, 450, 464, 499, 586, 712, 858, 862, 869, 956, 3627, 3657, 3791
Clark, Josephine, 606, 612, 613
Clarke, Arthur C., 1193, 1194, 2554, 2839, 2840, 3337, 3432, 3738
Cleary-Baker, John, 865, 866, 2367, 3516, 3550
Cleator, P. E., 1156
Clerke, A. M., 1129
"Closing the Blue Book: Air Force to Call off UFO Investigations due to Condon Report", 3711
"Coachella Valley Residents Report UFO's in Riverside, Cal", 671
Coarer-Kalondan, Edmond, 1110
Coblentz, W. W., 1754, 2063
Cocconi, G., 1515
Codr, Milan, 3174
Coelho Netto, Paulo, 3011, 3040, 3084
Cohen, D., 891, 1950, 3518, 3522, 3577
Cohen, Gaston, 1922
Cole, Dandridge M., 2808
Cole, George H. A., 2045
Coleman, Loren, 956, 3791
Collyns, Robin, 1116
"Colorado Horse Death Ruled no UFO Case", 3621
"Colorado Team Headed by Dr. E. U. Condon Finds no Evidence That Objects or Sightings are Intelligently Guided Spacecraft From Beyond Earth...", 3682
"Colorado to Study UFO's", 3568
Colorado. University, 3085
Colthup, N. B., 2140

"Columbia University Scientists Drs. G. Wollin and D. B. Ericson Hold... That Life Might Arise on Waterless Planet...", 1998
Comella, Tom, 295, 2873, 3379
"Comment by Prof. Einstein on Signals", 1481
"Comment on Attempt to Signal to Them", 1496
"Comment on Dr. W. S. Adams's Theory of no Life on Mars", 2082
"Comment on F. Hoyle Theory That Life Exists Outside our Solar System", 1165
"Comment on Possibilities of Life on Mars", 2074
"Committee on Space Research (COSPAR): Organization Meeting, London...", 2409
"Communication to Mars (and Interplanetary Communication)", 1509
Concilio, J. B. de, 2492
Condon, Edward U., 3639, 3692, 3708
"Condon to Head UFO Study", 3560
Confusion in the sky", 3364
Congress of Scientific UFOlogists, New York 1967), 2402
Conniff, James G., 2528
"Conserving Moondust", 2122
Constance, Arthur, 250, 2946
Conti, Sergio, 648
"Convegno Nazionale di Studi fra i Gruppi di Ricera Ufologica", 2427

Conway, J., 2538
Cooper, J., 1168
Copernicus, N., 15
Copland, Alexander, 982
Corliss, William R., 2659
Cornford, F. M., 2498
Cort, D., 3442
Cortright, Edgar M., 2690
Coupe, Charles, 2496
Couten, François, 374
"Couzinet's Saucer", 3371
Cove, Gordon, 1005, 2442
Cowgill, G. L., 576
Cowie, D. B., 2121
Cox, Adrian, 2363, 3501
Cox, Donald W., ed., 3041
Crain, T. S., 3739
Cramp, Leonard G., 94, 515, 2775, 2821, 3012
Crandall, L., 733
Crash go the Chariots, by C. Wilson. Review, 2552
Creighton, Gordon W., 340, 347, 369, 375, 424, 520, 530, 531, 599, 614, 632, 633, 649, 848, 890, 899, 903, 926, 1280, 1337, 1574, 2374, 2384, 2878-2880, 2885, 2886, 2891, 2893-2895, 3490, 3514, 3770, 3780, 3801
Cremaschi, Inisero, 3042, 3086
Crenshaw, James, 879, 1592
Cresson, J., 703
"Crew and Passengers on American Airlines Plane Report 3 Objects Acccompany Plane for 45 minutes...", 3431
Crexells, Joan, 976
Crick, F. H. C., 2008
Crider, Bill, 3817
Croissette, Dennis H., 2649
Croome, A., 2166
Cross, Charles A., 2050, 2158

Crum, Norman J., 3354
Crum, W. L., 3002
Cruttwell, Norman E. G., 324, 604
Cyr, Guy J., 1376

D

"D. E. Keyhoe Charges U.S. Air Force Conceals Facts", 3373
"D. E. Keyhoe cut off air During TV Discussion of Flying Saucers When he Digresses From Script", 3414
"D. E. Keyhoe Says Public Congressional Hearing on Information Supplied by Aerial Phenomenon Investigating Committee Would Have Proved Reality of Flying Saucers", 3415
Dagenis, Arleigh J., 309, 3447
Dana, Gwezenn, 1110
"Danger From Space?", 1274
"Danger From the Stars...", 3469
Daniel (pseud.), 3197
Daniel-Rops, Henri, 3468
Däniken, Erich von, 1094, 1101, 1102
Danyans, Eugenio, 3043
Darby, Christian, 3630
Darden, Mona, 493
Darrach, H. B., 802
Darwin, Charles G., 20, 22
Dauvillier, A., 25, 1705, 2094
David, H. M., 2108
David, Jay, ed., 3044
Davidson, Leon, 3013, 3175, 3449, 3452, 3474
Davies, John O., 1418

Davis, C., 2524
Davy, J., 1362
Dawson, Judith, 3578
Dawson, William F., 2379
Day, Lanston, 3279
De Araujo, Hernani Ebecken, 53
De Marcus, W. C., 1646, 2732
De Tastelero, Mira, 3395
De Vaucouleurs, G., 2020, 2159
"Dean Inge Thinks They are Peopled", 1152
Dean, J. C., 70
Dean, John W., 2458, 3122
Deerwester, Jerry M., 2787
Defense Documentation Center, 140
del Carmen Tamayo, Maria, 713
Delboeuf, J. R. L., 1127
Dello Strologo, Saulla, 3154
Demoulin, Maurice, 1770
Dempster, Derek D., 28, 1010, 1200
Denaerde, Stefan, 2043
Dendle, Brian J., 2888
Deneault, Harold H., 3098, 3129
"Deputy Premier Pervukin (USSR) Links Saucers to U.S. war Jitters", 3325
Derenberger, Woodrow W., 780
Derpgolts, Vladimir, 1724, 2005
Dethier, V. G., 2537
"Developments of International Effects to Avoid Contamination of Extraterrestrial Bodies", 1802
"Device (Biotelescanner) Unveiled to Relay Data on Life in Space", 2630
Deville, Gerard, 2106
Dewey, Mark, 2335
Diamond, E., 1234

Dick, William, 610, 630, 655, 663, 674, 682, 2378, 2386, 2744, 3753, 3756
Dickinson, Terence, 2843
Dickson, Stewart, 700
"Did Life Reach Earth on a Comet's Bright Tail?", 73
"Did UFO Bring Death?", 451
Diez Gomez, J. M., 2990
Digard-Comet, S., 2917
Dingle, H., 1793
"Dinner Time", 3410
Dinotos, Sábado, 2466, 3045
Dione, R., 2471, 2478
Dixon, William A., 201
Dmitriyev, A. L., 2672
"Do Flying Saucer Ever Land?", 452
"Do Other Humans Live?", 1204
Dockter, K. W., 2674
"Dr. Abelson Decries Search for Life on Mars...", 2187
"Dr. B. Commoner Says Possibilities of Life on Other Planets are too Small to Justify Scientific Exploration of Then...", 2181
"Dr. B. M. Oliver Reports he has Devised Antenna System That he Believes can Detect TV Emissions of Civilizations up to 200 Light Years Away...", 1588
"Dr. C. C. Wylie on Mass Hysteria; Urges sky Patrol to Guard Against Repetition", 3257
"Dr. C. Sagan Says Future Spacecraft That Land on Mars are Likely to Find Evidence of Some Forms of Life...", 2284

"Dr. Calvin on Probability That Plants, Animals and Other Forms of Life Exist on Millions of Earthlike Planets...", 1814

"Dr. Calvin's Discovery of pre-Biological Chemical Compounds in Meteorite Molecules...", 1229

"Dr. D. B. McLaughlin Says Mars is Barren, no Life Exists", 2109

"Dr. E. J. Opik Sees Craters (on Mars) Dating From Earliest Period...", 2210

"Dr. F. B. Salisbury Holds Flourishing Plant Life Probable...", 2148

"Dr. F. J. Dyson Says Civilizations Which may Exist on Planets in Orbit Around Pairs of dim Stars Called 'White Dwarfs'...", 1320

"Dr. F. J. Dyson Suggests Astronomers Seek Signs of Stars Surrounded by Gigantic Habitable Shells...", 1244

"Dr. H. C. Urey Sees Life on Other Planets", 2093

"Dr. H. K. Debus Predicts Astronauts Will one day Find Living Things in Space", 2218

"Dr. Huang Holds 2 Stars Within 16 Light Years of Earth may Have Planets With Intelligent Life", 1218

"Dr. J. E. McDonald Says UFO's Come From Outer Space...", 3735

"Dr. J. Huxley Holds he Cannot Dismiss Reports", 249

"Dr. J. W. Schopf's Search for Primitive Life on Earth...", 2224

"Dr. Jung and the Saucers", 2516

"Dr. Jung Says Flying Saucers are not Quirk of Imagination...", 3422

"Dr. L. B. Andrews on Evidence That There is no Life on Mars", 2073

"Dr. Lederberg Reports Device Being Developed at Stanford University Medical School to Determine if Life Exists...", 2603

"Dr. Menzel Maintains Saucers are Natural Phenomena", 3370

"Dr. N. Horowitz of Jet Propulsion Lab Says Increasing Evidence of Hostile Environment on Mars...", 2676

"Dr. Newell Defends Search for Life on Mars...", 2662

"Dr. O. Struve's Speech on Life Possibilities on Other Planets and Solar Systems...", 3311

"Dr. P. Lowell Holds... Organized Life Exists on Mars", 1134

"Dr. P. Lowell Presents Arguments to Prove Mars Canals Artificially Made", 2057

"Dr. P. Morrison Proposes Devising Easily-Broken Codes for Contacts With Intelligent Beings on Other Planets", 1551

"Drs. Plummer and Strong Believe Extensive Areas of Venus Have Temperature Comfortable to Man", 2206

"Dr. R. Jastrow Article Holds... Exploration of Mars in Next 30 Years...", 2717

"Dr. S. Jones Speculates on Life on Venus", 2081

"Dr. S. M. Siegel and C. Giumarro Report Some Bacteria and Fungi Thrive in Atmosphere with High Concentrations of Methane and Ammonia Life That on Jupiter...", 1939

"Dr. S. M. Siegel Reports Union Carbide Research Institute Show Many Life Forms Have Ability to Adapt to Conditions Similar to Those on Mars", 1936

"Dr. Sagan Sees Jupiter Better Possibility Than Venus for Life", 2139

"Dr. Schlovsky Says Most Logical Place to Look for Signs of Intelligent Life Outside Earth is in Spiral Nebula of Andromeda...", 1292

"Dr. Schlovsky (USSR) Holds 2 Moons are Probably Artifical Satellites put into Orbit Millions of Years ago by Then Existing Martians", 1215

"Dr. Shapley Doubts Other Worlds Seek Contact With Earth", 1246

"Dr. Shapley Sees at Least 100,000,000 Planets in Universe Suitable for Human Life", 1811

"Dr. Shapley Sure no Life Above Vegetal Exists on Mars", 2134

"Dr. Steinmetz Predicts Communication With Mars", 1486

"Dr. Struve's Letter Expands on Laurence Comment on Possibility of Life in Distant Systems", 1228

"Dr. Urey Estimates 1 Quadrillion Worlds in Known Universe Might Originate and Support Living Organisms...", 1785

"Dr. Von Braun Holds Existence of Life Elsewhere in Space a 'Logical Assumption'", 1239

"Dr. W. Coblentz Will Make Further Investigation of Existence of Life on Mars", 1755

"Dr. W. Markowitz Holds Control of UFO's by Extraterrestrial Beings Contrary to Laws of Physics...", 3619

"Un Document Majeure: la Declaration du Dr. McDonald", 3631

Doebel, Gunter, 1064

"Does Life Exist in Space?", 1863

"Does Life Exist on Mars?", 2190

Dohmen, J. G., 2865, 3198
Dole, Stephen H., 1464, 1777, 2037, 2038
Dolezol, Theodor, 1083
Dollfus, A., 1651, 2026, 2129
Dooling, D., 2738
Dorsey, Tom, 3382
Douglas, A. V., 3253
Douglas Aircraft Company, 2033
Douglas, Ulysee, 1084
Dowding, Hugh C., 3356
Dowling, J., 242
Downing, Barry H., 1426, 2467, 2469
Drake, Frank D., 1032, 1415, 1510, 1519, 1540, 2616, 2628
Drake, Walter R., 1021, 1050, 1078, 1299, 1317, 1323, 1354, 1389, 1393, 1397, 1403, 2125, 2132, 2459, 2470, 2541, 3459, 3487
Drake, Walter R., tr., 545
Drake, William R., 2482
Draper, H., 838
Drawert, F., 1855
Dreyer, H. R., 1514

Drury, Neville, 459, 886
"Du Côté de la mer", 634
Duchêne, J. L., 387, 850, 1977
Duckwall, W. E., 2079, 2084
Duclout, J. A., 2918
Dufay, J., 1643
Dufour, Jean C., 532, 3792
Dunn, William J., Jr., 427
Duplantier, Gene, 435, 538
Durham, Anthony, 489, 501
Durrant, Henry, 3155, 3220
DuShane, G., 1206, 1518
Dutta, Rex, 783, 3156, 3199
Dutton, T. W. S., 3713
Dutuit, J. M., 963
Dye, Clarkson, 1502
Dyson, F. J., 2609

E

"E. W. Barnes Believes There is Life on Others Besides Earth", 1761
Earley, George W., 1449, 3046, 3719, 3743
"Earth: zoo or Petri Dish?", 2291
Easley, Robert S., 766
Eastern UFO Symposium, Baltimore, 1971 Proceedings, 3176
Easton, W. B., 2525
Ebel, Hans F., 1972
Eberhart, J., 2195, 3618
"Ecosphere may Shape Life on Distant Planets", 96, 1874
"...ed ora Parliamo di Allucinazioni Collecttive", 2417
Eddington, A. S., 1639
Eden, Jerome, 3221
"Editorial Denounces E. Perrier Claim That Huge Animals and Plants Exist on Mars", 2054

"Editorial on Brookings Institute Study of Intelligent Life in Space", 1257
"Editorial on Dr. C. C. Wylie...", 3258
"Editorial on Life's Account of Saucers", 3296
"Editorial on Lowell Observatory Report on Possibilities of Life", 1766
"Editorial on Probability That the Earth is the Only Inhabited Planet in the Solar System", 1145
"Editorial on Probable Illusion of Wireless Communication...", 1490
"Editorial: Presence of UFO's Near Midwest Lead Fields Viewed", 667
Edson, J. B., 1521, 1525
"Edward Condon: a Physicist Never Afraid of a Fight", 3689
Edwards, Allan, 2526
Edwards, Frank, 254, 271, 276, 286, 297, 322, 335, 343, 363, 812, 823, 828, 933, 1182, 1192, 1261, 2123, 2387, 3047, 3222, 3393, 3396, 3397, 3400, 3406, 3411, 3417, 3419-3421, 3423, 3425, 3433, 3435, 3438, 3443, 3454, 3456, 3462, 3489, 3793
Ehrensvard, G. C. G., 54
Ehricke, Kraft A., 121, 2003
Eidman, V. I., 1607
"Eight Eyes on Strange New Worlds; Opinions of Science-fiction Writers", 1348
Eisely, Loren C., 806, 1178
Eisnau, M., 1251

"Elizabeth Klarer's Flying Saucer", 821
Ellerby, Christopher, 3482
Elliott, G., 1605
Elliott, L., 3323
Ellwood, Robert S., 2479, 2546
Elsasser, Hans, et al, 1095
Emenegger, Robert, 3241
Encrenaz, P., 1997
"End of the Myths About men on Mars", 1339
England, J., 1173
"Enoch and Other Cosmonauts; Soviet Theories", 1235
"Enquête a Valensole", 868
"Enquêtes et Observations Diverses", 402
Ensanian, Minas, 1958
"Entities Associated With Type I Sightings...", 852
Eshleman, Von R., 1978, 2255
Esnaola Auzmendi, Francisco M., 1236
Espigador, 3366
"Etranges Creatures", 888
"L'Etrange Histoire des Satellites de Mars", 1333
Eugster, Jack, 1085, 2825
European Space Research Organization, 1718
Evans, Gordon H., 465, 1319, 1372, 2209, 2835, 3496
Evans, Livio, 685
Evans, S., 1201
Eveno, Maryvonne, 624
Everett, Eldon K., 292
"Evergreens Will Greet Visitors to Mars", 2089
Everson, Vincent, 3418
Ewing, A., 1846, 1888, 1890, 1926, 2151, 2179
Ewing, A., 1926
"Existence of Flying Saucers Discounted by U.S. Air Force Project Blue Book...", 3520
"Experimental Exobiology", 1914

"Exploring the Planets; Mariner V Probes Venus Secrets", 2235
"Exploring Venus by Radar", 2618
"Extraterrestrial Life?", 1859, 1916

F

"Fanciful Preview to new Facts", 1794
"Le Fantastique Incident de Catanduva", 966
Faria, J. E., 742
Farish, Lucius, 390, 460, 878, 1343, 3714
"Farmer Trent's Flying Saucer", 194
Farris, Joseph, 3087
"Fatous Season: Ann Arbor and Hillsdale Sightings", 416
Faust, H., 1183
Fawcett, George D., 336, 535, 571, 582, 635, 924, 951, 2745, 3465, 3546, 3614, 3658, 3664, 3697, 3763, 3764, 3811
Fears, Francis R., 1781
"Feasible Flying Saucers", 3330
"Feature Article on Possibility of Life on (Planets)", 1759
Feldman, G. J., 3003
Fenn, W. O., 1803
"Fenómeno OVNI", 3815
Ferber, Adolph C., 1009
Ferguson, Jean, 3200
Ferguson, William, 1006, 2855
Fernandez, Juan M., 1223
Fernandez Luna, Walter, 3088
Ferrier, E., 2055
Ferriere, Joseph L., 481
Feryer, Rich (pseud.), 2919

Fesenkov, V. G., 97, 1051, 1656, 1984, 2171, 2236
Festinger, Leon, 2445
"Fifteen-Year Air Force Flying Saucer Verdict: They Exist Only in Domestic Spats", 3473
Finch, Bernard E., 913, 2376, 2540, 2813
"Finding Life on Mars may Depend on Season", 2188
"Finds Saucers Exist Solely in Imagination", 3360
Fink, Herschel P., 525
Firsoff, V. A., 58, 1043, 1072, 1096, 1207, 1905
Fishbach, Laurence H., et al, 2788
Fishbein, J., 1170
Fitch, C. J., 1493
Fitch, C. W., 3505
"532 UFO Sightings Checked During 1964; 16 Remain Unidentified", 383
Flammande, Paris, 3177
Flammarion, Camille, 5, 984, 1125, 1477, 2299
"Flammarion, Camille, on Canals of Mars", 2056
Flanders, C. M., 1221
"Fleetwood, Pa. Police Chief, E. R. Fox, Says on April 31 That Pile of Spent Fireworks Discovered During Patrol in Crystal Cave area of Berles County may be Source of Nightly Sightings of UFO's...", 3788
Fleissig, Josef, 3123
Flick, David, 3363
Flight, W., 1732
"Fly From Venus?", 2088
"Flyby Hints Life", 2705
"Flygande Tefat Over Lappland", 471
"Flying Disks Break Out Over the U.S.", 180
"Flying Saucer Buffs Discuss Ways to Improve Communications with Objects, Convention, Los Angeles", 2415

"Flying Saucer Controversy...", 227
"Flying Saucer Controversy; Meteorological Balloons, and Weather Conditions Which May Provide Explanations of the Phenomenon", 3347
The Flying Saucer Menace, 3048
"The Flying Saucer Mystery", 3266
"The Flying Saucer Myth", 3426
'Flying Saucer Review' Case Histories, 3157
Flying Saucer Review; World Roundup of UFO Sightings and Events, 2956
"Flying Saucer Roundup: Saucer Report no. 1", 280
"Flying Saucer Spots Before Their Eyes", 179
"Flying Saucers", 3277, 3313, 3362, 3540
"Flying Saucers Again", 3271, 3284
"Flying Saucers and Science", 2795, 3401
Flying Saucers and UFO's 1968, 3089
"Flying Saucers are old Stuff", 3297
"Flying Saucers: Ballooney, not Baloney", 3283
"Flying Saucers Definitely Proved", 186
"Flying Saucers (Editorial)", 3288
"Flying Saucers: Fact or Fancy?", 3557
"Flying Saucers: Illusions or Reality?...", 418
Flying Saucers Illustrated, 3049
"Flying Saucers: not Real, but-", 3684

Flying Saucers Pictorial, 163, 3050
"Flying Saucers Ruled Out by Latest UFO Fact Sheet", 3457
"Flying Saucers: the Some-things", 3255
Flying Saucers; Twenty-one Years of UFO's..., 3090
Flying Saucers, UFO Reports #1-4, 3051
"Flying Saucers yet...but", 3377
"Flying Something Touches Down in Brazil", 456
Foght, Paul, 2361, 3470
Fogl, T., 303
Fontenelle, B., Le Bovier de, 993, 2443
Fontes, Olavo T., 164, 305, 318, 328, 835, 904, 2602, 3628
Ford, Brian J., 789, 3797
Fornwalt, Russell J., 3729
Forschung in Fessein: Elektro-Gravitation UFO-Phänomen..., 3201
Fort, Charles, 2300-2303
"Fort-Lamy, 27 Mars 1955", 482
Fortenberry, William H., 214
Foster, G. V., 2007
Fouéré, Francine, 1446
Fouéré, René, 428, 483, 485, 494, 547, 697, 853, 880, 974, 1353, 1444, 1446, 3510, 3544, 3579, 3592
Fournier, G., 2077
Fowler, Raymond E., 3242
Fox, C. V. J., 2955
Fox, James E., 3754
Fox, Sidney W., 90
Fox, Sidney W., et al, 1662
Fragner, Wolfram, 2985
Francis, V., 1745, 1746
Fredrickson, Sven-Olaf, 579, 607, 631, 715, 938

"French War Ministry Investigates Radio Signals Which Experts Think may Have Been Sent From Mars", 1485
"Fresh Look at Flying Saucers: Time Essay", 3617
Freudenthal, Hans, 743
Freundlich, M. M., 2582
Friedman, Bruno, 1318
Friedman, Stanton T., 943, 3178, 3778
Friedmann, Eugenio, 1086
Frisch, B. H., 109, 1968, 2192
Fritts, C. A., 2766
From Jupiter, Planet of Joy, 2338
From Planet Pluto With Brotherly Love, 2339
"From Surveyor I Findings, Dr. J. J. Gilvarry Believes Moon Once had Seas and Life", 2208
"From Saucers Investigation Group", 3390
Fry, Daniel W., 160, 730, 754, 2329, 2856, 2938
Fuhs, Allen E., 1243
Fuller, B. A. G., 3305
Fuller, Curtis, 232, 263, 304, 329, 334, 809, 814, 3275, 3326, 3331, 3367, 3458, 3589, 3679
Fuller, John G., 422, 755, 756, 870, 873, 875, 881, 885, 905, 2858, 3590, 3660, 3662
Fuller, John G., ed., 3091
Fuller, John G., et al, 3549

G

Gaddis, Vincent, 1377, 3052
Gaffron, H., 85
Gale, W. A., 2265

Galindez, Oscar A., 446, 475, 589, 723, 945, 970, 975, 2890, 2896, 2897, 3092
Gallagher, B. J., 2768
Gallup, George H., 3202
Gallup, Jr., George, 1418
"Gallup Poll Finds 51% of 1550 Persons Interviewed Believe that UFO's, Sometimes Called 'Flying saucers,' are Real...", 3810
"Gallup Poll Survey of World Leaders in Various Fields Shows 53% Believe in Existence of Life on Other Planets", 1408
Galton, F., 1478
Gambling, W. A., 1578
Gamow, G., 1775
Gardner, M., 1582, 2911
Garreau, C., 2939, 3179
"Gasbags of Venus", 2227
Gaspa, Pietro, 2454
Gaston, Henry A., 2296
Gaston, Patrice, 3223
Gastonguay, P. R., 1429
Gatland, Kenneth W., 28, 1010, 1200
Gauroy, Pierre, 35, 1652
Gazenko, O. G., 2699
Gearhart, Livingston, 3512
Geiser, Chester S., 3259
Geiger, E. E., 3380
Gelatt, R., 2866, 3328
Genet, 3355
George Washington University, Washington, D. C., 129
Georgetown College Observatory, 2025
Getmantsev, G. G., et al, 2701
Gheorgita, Florin, 557, 3224
Gibbons, G., 2850, 2957
Gibbons, Russell W., 224
Gibbs-Smith, Charles H., 2781, 3358
Gibney, F., 3003

Gibson, E. H., 2851
Giffin, C. E., 1973
Gifford, Dennis, 3180
Gigi, Robert, 792, 3210
Gilbert, D. L., 86
Giles, Gordon A., 896
Gilman, Peter, 2542
Gilvarry, J. J., 1931
Ginna, R. E., 3299
Gindilis, L. M., et al, 1087
Giordano, Alberto, 3143
Giret, A., 2460
Girvan, Ian W., 2928
Girvan, Waveney, 358, 2940
Girvin, Calvin C., 1013, 2958
Glasby, J. S., 63
Glemser, Kurt, 546, 596, 778, 931, 959, 3158, 3203
Goble, H. C., 261, 1439, 3391
Goddard, J., 1315, 2679
"Goddard Space Flight Center's Dr. R. A. Hanel Says That Although Prospect of Life on Mars is not Certain...", 1995
Godefroy, Roland, 698
Godwin, John, 2346
Goff, Kenneth, 2444
Going Up. Practical Methods of Astral Projection, 2324
Golay, M. J. E., 1534, 1543
Gold, Thomas, 1840
"Gold Mine in the Sky", 1994
Golomb, S. W., 839, 1300, 1536, 1544
Golowin, Sergius, 1073
Gonzales Ganteaume, Horacio, 153, 2975
Good, I. J., 1341
Gordon, Samuel, 1825
Gossner, Simone D., 2167
Gosta Rehn, K., 3243

Gould, Bernard H., 1425
Gould, Rupert R., 2330
Goupil, Jean, 894, 2833, 2834
"Government to Cooperate With Prof. Todd in his Effort to Obtain air Silence for Attempted Communication", 1489
Gradecak, Vjekoslav, 2816
Graf, E. R., et al, 2682
Graham, David, 173
Grant, Walter V., 725, 726, 2441
Gray, G. W., 2067, 3593
"Great Ball of Fire: Philip Klass Theory", 444
"Great Lakes Identified Flying Objects Association Meets in Ill.", 2426
Green, Gabriel, 502, 3053
Green, Timothy, 3007
"Green Salad on the red Planet: What Next?", 2233
Green, Vaughn M., 813
"Green UFO Sighting in New Orleans Said to be Advertizing Gimmick", 681
Greenbank, Anthony, 1074
Greenberg, D. S., 3566
Greenfield, Irving A., 1075, 3054
Greenspan, Jack A., 2232
Greenstein, Jesse L., 2468
Greenwald, H., 2882
Greenwood, S. W., 2811
Gregg, Doris D., 332
Gregory, J. C., 1167
Grescoe, Paul, 3661
Griffin, E., 230
Gris, Henry, 1624, 2744, 3252
le Groupe INFO, 3193
Grove, Carl, 575, 580
Grumman Aircraft Engineering Corp., 2042
Guasp, Miguel, 718, 3225, 3759
Guieu, Jimmy, 2941, 3204, 3205

Guillemin, A. V., 1476
"Gullible Experiment: Southern California", 3537
Gurney, Clare, 3159
Gurney, Gene, 3159
Guttilla, Peter, 2545

H

"H. E. Newell Discusses NASA Plans to Probe Mars for Signs of Life", 2637
"H. K. Kaiser Holds Mars has Primitive Vegetation and Venus may Have Organic Life in Millions of Years", 2097
Haas, Ward J., 2213
Haber, Heinz, 1097
"Habitability and Uninhabitability of Earth's Sister Planets", 1131
"Habitability of Worlds", 2487
"Habitable Worlds", 1121
Hagen, Charles A., 1677
Hald, T. S., 1807
Haley, Andrew G., 749
Hall, Richard, 2870, 2977, 3398
Hall, Richard, ed., 2991
Halpern, B., 2707
Halsey, Wallace C., 2331
Hamilton, Jared, 243
Hampton, Wade T., 2353
Handelsman, M., 1454, 1550
Hankow, Ralph, 3547
Hanlon, Donald B., 411, 447, 476
Hansen, L. T., 187, 2457
"Happening at Hoogdal: an Unidentified Beeping Object", 505
"A Hard Look at 'Flying Saucers'", 3538
Harder, James A., 2778
Harford, J., 1282

Harkins, R. R., 3106, 3140
Harney, John, 906, 923
Harrison, Philip, 1956
Hartlaub, G. F., 999
Hartle, Orville R., 3765
Hartmann, William K., 1967
"Harvard University Scientist, C. Z. Sagan, Says Astronomer on Mars Would Probably be Unable to Prove Existence of Life...", 2634
Hatcher, Judi Anne, 1380
Hatvany, Edgar A., 3551
Have, P. J. ten, 1088
Haviland, R. P., 2739
Hawrylewicz, Ervin J., et al, 1932
Hayatsu, Ryoichi, 1930
Haythornthwaite, P. K., 361, 394
Heard, Gerald, 997
Heard, J. F., 82
Heath, David, 453
Heath, T., 8
Hecker, Randall C., 3608
Heinecken, M. J., 2449
Heinlein, R., 1868
Heinrich, M. R., 2279
Heintze, Carl, 2573
Heiserman, David L., 3721
Heline, Corinne, 2899
Helland, Albert E., 826, 1198
Hellman, Hal, 3625
Henderson, L. J., 30
Henkel, F. W., 1739, 2053
Henry, Dick, 572
Henry, Robert W., 127
Hermann, Joachim, 41, 1678, 2323, 2981
Herschel, W., 2310
Hervey, Michael, 2352, 3124
Hess, H. H., et al, 2039
Hess, S. L., 1813, 2027
Heuer, Kenneth, 1000, 1172
Hewes, Hayden C., 495, 790
Hibben, R. D., 1925, 2189
Hickman, Warren, 225
Hicks, Clifford B., 1531

Hide, Raymond, 2234
Hieronymus, W. S., 2278
Hiestand, Edgar W., 3450
Hilberg, Rick R., 766, 3103
Hill, Arthur, 2696
Himmel, N. S., 2263
Hinman, Grace, 506
"Hint of Martian Life Found", 1979
Hirano, Imao, 3125, 3226
Hitchcock, Dian R., 2692
Hitchcock, Dian R., et al, 2686
Hoag, Amey, 262
"Hoax and Hallucination: the Evidence", 2389
Hobana, Ion, 172, 3181, 3191, 3218
Hodgson, G. W., 1981
Hodgson, Richard G., 2242
Hoerner, S. von, 1458
Hogben, Lancelot, 804
Hohmann, R. E., 1033, 1455
Holden, A. R., 2799
Holden, C., 2728
"Holds Life on Other Planets Unlikely", 1858
Holiday, Frederick W., 948, 3227, 3781
Holledge, James, 157
"Hollywood Builds Flying Saucers", 3324
Holm-Hansen, Björn, 169
Holmes, David C., 1089, 2574
Hood, Joseph F., 3093
Hooven, Frederick J., 3690
Hord, C. W., et al, 2716
Horowitz, Norman H., 79, 98, 1259, 1856, 1864, 1880, 1948, 2175, 2575, 2730
Horswell, Christine, comp., 135
Hotchin, John, 1969
Housden, C. E., 989
Hovis, W. A., 1924
"How to Find Life on Mars", 2185

"How to fly a Saucer", 3321
"How to Tell a Martian", 1332
Howard, Dana, 2311, 2321
Howells, W., 1141, 1208, 1232, 1264
Hoyle, F., 1929, 2627
"Huaco: Los Foo-Fighters en la Argentina", 709
Huang, Su-Shu, 42, 47, 84, 1044, 1689, 1827, 1837, 1848, 1865
Hudson, Jan, 759
Hughes, F. P., 498, 3611
Hughes, T. A., 9, 2493
Hugill, Joanna, 533, 914
Huguet, Casas, 702, 3818
Hulley, John, 1816
"The Humanoids", 884
Hunt, Gerry, 701
Hunt, Richard, 412
Hunten, Donald M., 2266
Hutchings, F., ed., 31
Huygens, C., 2
Hynek, J. A., 526, 563, 573, 3094, 3206, 3333, 3509, 3569, 3580, 3582, 3594, 3632, 3651, 3718, 3762, 3787, 3797, 3828, 3830

I

"I. Asimov Article on Possible Life on Other Planets and Ways of Communicating With Such Planets", 1327
"IATA Director Hammarskjold says UFO's may Actually be 'Neighbors in Space'...", 3542
Ibbotson, B., 368
Ibson, Jack, 816
"Icarus, International Journal of the Solar System...", 1889
"If You're Seeing Things in the sky", 273

Illinois Institute of Technology. Research Institute. Astro Sciences Center, 2586
"Illusions of Nature", 181
"Immense Triangular Object Over Majorca", 477
"Important Discoveries", 2881, 3575
"Improbability of Life on the Planets", 1749
Imshenetskii, A. A., 102, 1291, 1293-1295, 1384, 1899, 1919, 2580
Imshenetskii, A. A., ed., 1098, 2596
Imshenetskii, A. A., et al, 1345, 2721
In Search of Ancient Astronauts, 1112
In Search of Ancient Mysteries, 2350
"Incident en Norvège", 583
Inge, W. R., 2437
Ingle, Bob, 695
International Space Science Symposium, 48, 1690
Invitation From the Planet Venus, 2340
"Is Anybody out There Sending?", 1614, 1623
"Is the U. S. Government Expecting Invaders From Space?", 1191, 3403
"Is There Life on Mars or Beyond?", 2000
"Is There Life on Mars or Earth?", 2202
"The Italian Scene, Part 2", 3488
"It's Official: UFO's not Flying Saucers; Most Reports Traced to Aircraft, Stars", 3451

## J

"J. B. Edson Article on Exploring Possible Existence of Earthlike Planets by Means of Radio Astronomy", 2605
"J. H. C. Macbeth Says W. Marconi is Sure That Mars Flashes Messages and That it Would Be Simple Matter to Arrange Code", 1483
"J. Hillaby Series on Fifth International Astronautical Federation Congress, Innsbruck...Sees Life Impossible...", 1180
"J. J. O'Connor Reviews NBC News Program...", 3829
"J. N. Wilford Reports USSR Plans to Land a Robot-Like Detection Lab on Mars...", 2724
"J. V. Miller Letter on Nov. 16 on Prof. Urey's Theory of Possibility of Life on Other Planets", 1177
"J. W. Macvey's Book, Alone in the Universe, Review", 3507
Jacchieri, Carlos, 2476
Jack, Homer A., 2504
Jackson, C. D., 1033, 1455
Jackson, F. L., 1663, 1933
Jacobs, David, 3228
Jain, Sushil K., comp., 135
James, Michael, 611
James, Trevor, 834, 837, 842, 1014, 2359, 2360, 2520
Jansen, Clare J., 874
Janssen, John H., 189
Jastrow, Robert, 2223, 2247, 2260, 2629
Jastrow, Robert, ed., 49

Jeans, J., 10, 1629, 1635, 1776
Jensen, Erling, 584
Jessup, Morris K., 2446, 2929, 2947
Jessup, Morris K., ed., 2942
"Jet Propulsion Lab Reports Preliminary Analysis of Mariner 6 Data has Shown Mars to be Unlike Earth...", 2711
"Jet Propulsion Lab Scientists...Report Lab Experiments Conducted With Soil...", 2272
Jethon, Zbigniew, 2274
Johannes, Lewis W., 953
Johnson, De W. B., 2903
Johnson, R. D., 364, 1706
Jones, Bob, 2827
Jones, R. V., 3670
Jones, S. H., 246, 1632, 1772, 1817, 3280
Jonsson, Ake, 534
Jully, John 3339
"Jung and the UFO's", 2390
Jung, Carl G., 2519, 2966, 2992, 2993
Junt, Graham R., 2249
Jurgens, R. F., 2253

## K

Kaempffert, W., 203, 2065, 3295
Kahn, F. D., 2280
Kampen, Hans van, 3229
Kane, Jim, 3730
Kantor, M., 3532
Kaplan, S. A., 1599, 1607
Kaplan, S. A., et al 1103
Kapp, R. O., 36
Karavieri, Ahti, 608
Kardashev, N. S., 1316, 1572
Kasantsev, Aleksandr, 510
Katterfeld, Gennady, 2005

Kaufmann, Richard, 1346
Kavanau, J. L., 76
Kazantsev, Aleksandr, 1370
Keel, John A., 161, 488, 570, 609, 781, 887, 892, 893, 895, 900, 907, 909, 910, 915, 917, 921, 1447, 2884, 3160, 3603, 3659, 3673, 3693, 3726, 3740, 3751
Keenan, C., 2110
Keilin, D., 1819
Kelley, Jack, 640
Kellogg, W. W., 1658, 1684, 2028, 2176
Kelly, Maryellen, 2380
Kelly, Peter J., 384
Keosian, J., 52
Kepler, J., 14, 16
Kettelkamp, L., 3207
Keyhoe, Donald E., 197, 216, 377, 472, 1113, 1349, 1626, 2111, 2794, 2826, 2872, 2904, 2930, 2971, 3230, 3340, 3343, 3352, 3388, 3455
Kiely, John R., ed., 2571
Kilson, Walter K., 2149
Kind, S. S., 1174, 1202, 1216, 1812
King, David W., 1394
King, George, 2313, 2325, 2327, 2328, 2332, 2994
King, R. S., 245
Kjems, Iver O., 3135
Klass, Philip J., 3095, 3244, 3531, 3563
Kleczek, Josip, 1297, 1679, 2160
Klein, David, 610, 3820
Klein, Harold P., 1706, 1707, 1959, 1971, 2702, 2720
Klein, Harold P., et al, 2734
Kleinz, John P., 2523
Klemin, Alexander, 3265
Klotz, J. W., 2452
Knaggs, Oliver, 1065, 3014
Kneitel, Thomas S., 1520
Knight, C., 237
Knight, Oscar F., 750

Knowlson, James R., 918
Koch, Howard, 1258
Kocherhans, Ernst, 1680
Kocivar, B., 3365
Koestler, A., 1527
Kohler, John, 2406
Kok, B., 2567, 2683, 2729
Kolosimo, Peter, 791, 1017, 1106, 3161
Kon, Alejandro, 3055
Kopyev, V. Y., 2819
Kor, Peter, 2372, 3444, 3803
Korcsmaros, Jesse, 2796, 2807
"Korean Saucers", 204
Korell, Federico, 2462
Kotsakes, Demetrios D., 2920
Kotulak, Ronald, 1627
Koval, I. K., 2281
Kowalezewski, Stanislaw, 312
Kozel, Carlos, 3096
Kozlov, B. L., 2592
Kraft, Carl F., 1634, 2797, 2806
Krasovskii, V. I., 1268, 1324, 1329
"Kraus Reports new Radio Signals From Venus", 1505
Krebs, Columba, 1405
Kreifeldt, J. G., 1406
Kremp, Gerhard O. W., 1962
Krmelj, Milor, 716
Kronos, 3056
Krooge, Harald, 3584
Krupenio, N. N., 1715
Krutch, Joseph W., 3695
Kublin, D., 1601
Kuiper, G. P., 2029, 2790
Kumar, S. K., 119
Kuningas, Tapani, 3162, 3182
Kunkel, Wallace, 810
Kuttner, Robert E., 1398

# L

"L. Campbell Believes Life on Planets of This and Other Solar Systems may Exist", 1756
"L. J. Lorenzen Calls Aurora, Texas Story 'Hoax!'", 680
"L T E...Jan. 28 Review of UFO's...", 3785
Ladin, V., 2204
Lafleur, L. J., 1777
Lafonta, Paul 2673, 2822
Lagarde, Fernand, 581
Lagarde, Fernand, et al, 3231
Laine, Juliette, 2354
Lamb, I. M., 1805
Lamb, Peter, 2688
Lambert, Richard S., 1175, 3298
Landsburg, Allan, 2483
Landsburg, Sally, 2483
Lang, Daniel, 294, 2769, 3318, 3357
Langer, R. M., 1160
Lapp, Ralph E., 1026, 1538
Laprade, Armand, 2718, 3746
Larkin, L. S., 2706
Larson, Kenneth, 621, 637, 1079, 1366, 2358, 3163, 3676, 3694, 3717, 3766
Lasco, Jack, 3606
Last, Cecil E., 148
"Latest on the Flying Saucer", 3384
"Latest Theory Regarding Life on the Planet Mars", 2058
"Latest UFO Report (on Project Blue Book) Reveals Nothing out of This World", 3610
Lau, G., 2761
Lauritzen, Hans, 2828, 2829, 2831, 2832, 3595
Lavrentyev, G. A., 1722
Lawrence, L. G., 1613
Lawton, A. T., 1611, 1618
Lax, Soini, 3208

Layne, Meade, 751, 2905, 2995
Layne, Stan, 2314
Lear, J., 1231, 2741, 3554, 3562, 3581, 3623
Leavitt, J., 2488
Lederberg, J., 1655, 1835, 1845, 1861, 1873, 1918, 2121, 2154, 2604, 2668, 2708
Ledger, Joseph R., 1288
Lee, B. S., 3301
Lee, Gloria, 2336
Lehr, Georges, 3127
Leiber, Fritz, 395, 1342
Leighton, Robert B., et al, 2713
Lemaitre, Jules, 833
Leonard, F. C., 1153, 1760, 2070
Leonard, R., 2472
Le Poer-Trench, Brinsley, 774, 857, 1022, 2349, 3126, 3506, 3771
Leslie, Desmond, 728, 863, 2912, 3164
Levin, Gilbert V., 1935, 2155, 2624
Levin, Gilbert V., et al, 2559, 2560
Levin, V. L., 2246
Levitt, I. M., 3314
Lewin, Harold, 2396
Lewin, Roger, 1432
Lewis, C. S., 2514, 2535
Lewis, I. M., 2072, 2090
Lewis, R. S., 1585
Ley, Willy, 288, 1197, 1199, 1900, 3289
Ley, Willy, et al, 2307
Li Ch'u, 1023
Libby, Willard F., 1963, 2250, 2269
Liddel, Urner, 221, 354
Liebl, V., 1970
Lieblova, J., 1970
"Life Beyond Earth", 1773
"Life Beyond Earth Sought: Project Ozma", 2640
"Life Detector: Multivator", 2638

"Life Detectors: Wolftrap and Multivator", 2641
"Life Elsewhere", 1400
"Life Forms may Exist in hot Spots on Mars", 2156
"Life in Meteorites?", 1915
"Life in the Clouds: Conditions on Venus", 2225
"Life in Time and Space", 1877
"Life may Exist on Jupiter", 2141
"Life on a Billion Planets?", 1806
"Life on Jupiter", 2222
"Life on Largest Planet?", 1875
"Life on Mars?", 2217, 2118, 2124, 2130
"Life on Mars Hardy", 2114
"Life on Mars Talk Dealt a Setback", 2172
"Life on Other Planets?", 1828
"Life on Other Planets---What are the Possibilities?", 1952
"Life on Other Worlds", 1737
"Life on the Planet Venus now Seems Impossible", 2080
"Life on the Planets", 1743, 1762
"Life on Venus", 2137, 2256
Life Sciences and Space Research XI..., 1730
"Life Search Turns up More Clues in Space", 1976
"Life-supporting Planets", 1808
"Life Without end", 89
"Life's Spread Through the Universe", 1885
Lighthall, W. D., 74
"Like Nothing on Earth", 2531
Liljegren, Anders, 939

"Limits of Organic Life in our Solar System", 1741
Lindsay, Gordon, 1066, 3015, 3209, 3772
Lipman, C. B., 1771
Liskowitz, J.W., 2565, 2656
Liss, Jeffrey G., 360, 2364
"A List of Sightings by Astronomers", 684
"Little Inhabited Stars", 1213
"Little Men From Mars", 1378
Littman, Arnold, 2976
Litynski, Z., 1871
Lleget Colomer, Mario, 3057
Lloyd, Dan, 399, 490, 2697, 3688
Lob, Jacques, 792, 3210
Lobsack, Theo, 3678
Locke, L. J., 556
Lockyer, J. N., 23
Loftin, H., 1820
Loftin, Robert, 3097
Long, A., 3348
Lonnqvist, C., 996
"A Look at Alleged 1897 UFO Crash in Aurora, Texas", 678
"Look! It's Flying Discs Again", 2792
"Looking for Martians (Design for an Automated Biological Laboratory, ABL)", 2653
Lore, Gordon I. R., 3098, 3128, 3129
Lorenzen, Coral, 166, 345, 366, 367, 382, 404, 548, 605, 760, 829, 2877, 2982, 3436, 3652
Lorenzen, Jim, 166, 760
"Lost Cause: Condon Report", 3686
Lovell, A. C. B., 34, 1549, 2773, 2809

Lovelock, James E., 1973, 2671, 2692
Lovitch, A., 1373
Lowell, P., 7, 71, 2009, 2011, 2013, 2014, 2015, 2052, 2069
"The Lowestoft Sighting: Object Observed for an Hour", 353
Lozina-Lozinsky, L. K., et al, 2288
Lucchesi, Dominick, 290
Luce, C. B., 3545
Lucero, Daniel P., et al, 2714
Luchitsky, I., 1996
"Luciano Galli's Contact Claim", 844
Luckiesch, Matthew, 175
Lucretius, Carus T., 18
"Lueurs Singulières dans l'Ardèche...", 622
Lugo, Francisco A., 1018
Luna, Walter F., 767
"Lunar Orbiter 1 Photo of Earth Taken in August 1966 Suggests...That Venus Might Support Life...", 2216
"Lust for Life: Search for Life on Other Worlds", 2693
Lynch, J., 2534
Lyon, Bill, 549
Lyustiberg, Villen, 3649, 3654

M

"M. Berger on Societies and Publications in U.S., Devoted to Saucer News", 2405
"M. Faye on Interplanetary air", 1732
"'M.O.C.' et Leurs Occupants vus au sol en 1967", 916
Macaluso, Giuseppe, 3130
McCampbell, James M., 3232
McClain, Edward F., 1516
McCooley, G., 2497, 2499
"McColley, Grant, H. W. Miller Saint Bonaventure, Francis Mayron William Vorilong and the Doctrine of a Plurality of Worlds", 1157
McCoy, John, 1011
McDermot, M., pseud., 2755, 2864
McDonald, James E., 564, 565, 623, 936, 3058, 3099-3102, 3131, 3656
"McDonald Observatory Astronomers Report They Have Obtained First 'Absolutely Conclusive Proof' That Water Exists in Martian Atmosphere", 2258
Macdougal, D. T., 1128
MacDougall, C. D., 2959
MacGowan, R. A., 1034, 1067
The McGraw-Hill Encyclopedia of Space, 3059
McGraw, Walter, 3722
McHugh, L. C., 1252, 1254, 1862, 2529
Mackay, E. A. I., 2395
Mackay, Ivar, 2385
McLaughlin, Robert B., 191
Macpherson, H., 1142
Macvey, J. W., 1045, 1285, 1474, 2127, 2779, 2857
McWane, Glenn, 173
Madden, Ruth S., 707
Maddock, D. P., 3723
Magee, Judith, 400, 657
Magnolia, L. R., et al, 130
"Major General Ramey Reports None of 1500 Reports to U.S. Air Force Since 1947...", 3307
"Major Keyhoe's Book, The Flying Saucer Conspiracy...", 3381

Mallan, Lloyd, 1399, 3060, 3586, 3609
Malthaner, Hubert, 642
Maluquer Wahl, J. J., 2784
Mamikunian, Gregg, 1056, 1695
"Man-Made Flying Saucer", 3334
"Man not Alone in Cosmos", 1189
"Man not in Space", 1270
Manas, J. H., 2983
Mandell, Siegfried, 3368
Maney, Charles A., 307, 339, 386, 461, 2198, 2876, 2977, 3424, 3517
Mann, Mary, 262, 1269, 1866
Mannel, Cliff, 1571
"Many Habitable Worlds", 1162
Marais, D., 1179
Marayo Magdaleno, Feliciano, 2131
Marbarger, J. P., editor, 2757
"Marconi Reports That he has no message From Mars Yet", 1484
Marcus, A. H., 2245
Marey, G., 3640
Margaria, R., 1027, 1276, 1298, 2814
"Mariner IV Could not see Life Forms on Mars", 2663
<u>Mariner Mars 1971 Project</u>, 2597
Mark-Age, 798
Markowitz, William, 2375, 3620
Marov, Michael, 2289
"Mars Biological lab Designs are Sought", 2182
"Mars: Drs. B. J. Bok and F. Watson on Possibility of Life and Contact with Earth", 1497
"Mars has Plant Life, but no Canals", 2116
"Mars Radar Contact Made", 1564
"Mars Surface Laboratory", 2243

"Mars Surface Multi-cratered: Mariner IV Photographs", 2230
"Marsh Gas in Michigan: Latest UFO Incident", 420
"Marsh gas on Mars", 2212
Marshall, James S., ed., 735
"Martian Environment", 2168
"Martian Life: a Darkening Wave", 2194
"Martian Vegetation", 1818
"Martian Wolf Trap in Action; Soil Samples to be Examined for Evidence of Living Organisms", 2220
"Martians Over France", 811
Martin, Anthony R., 2746
Martin, C., 1719
Martin, D. M., 2852
Martin, Dan, 2862
Martin, James S., Jr., 2271
Martin Marietta Corp. Research Institute for Advanced Studies, 2583
Martynov, Dmitri, 1583
Mascall, E. L., 2447
"Maser Here. Hello There! Radiation Waves From Hydroxyl Clouds", 1597
Masi, Roberto, 1054
Mason, B., 1908
Mason, E. C., 1735
Mather, S. M., 2092
Mathew, W. D., 1747, 1748
Maude, A. P., 2086
Mauer, E. F., 218
Maunder, E. W., 988
Maxia, V., 3240
Maxim, H. P., 1758
May, J. A., 3335
Mayer, Cornell H., 2632
"Me Earthman, you Gasbag: Theories of Josef Shklovskii and Carl Sagan", 1364

Mead, M., 3827
Meadows, A. J., 1964
Meerloo, Joost A., 2377
"Meeting Extraterrestrials", 819
Meinschein, Warren G., 2556, 2639
The Mel Noel Story, the Inside Story on the U.S. Air Force Secrecy on UFO's, 3016
Melpar, Inc., 1463, 2557, 2562, 2576
Menzel, Donald H., 222, 365, 370, 827, 1019, 1644, 1664, 1696, 2087, 2853, 2913, 2986, 3061, 3315, 3493, 3499
Mercer, D. M. A., 1587
Mercier, A., 2010
Merek, Edward L., 2584, 2698
MERINT Radiotelegraph Procedure, 1465
Merker, Donald, 3653
Merriman, C. C., 67
Merz, E. J., 2823
Mesnard, Joel, 512, 547, 624, 699
"Message From Space", 1577
"Metalaw", 820
"Meteoritic Life Studied", 1945
Michael, Cecil, 2847
Michel, Aimé, 256, 268, 270, 277, 278, 355, 405, 824, 825, 871, 876, 911, 927, 952, 1237, 1990, 2355, 2551, 2943, 2960, 2961, 3017, 3132, 3502, 3515, 3706, 3752, 3812
Michel, Aimé, et al, 771
Michell, John F., 3062
Middlehurst, B. M., 2029
Midwest UFO Conference, 2d, St. Louis (1971), 3183
Midwest UFO Conference. 3d, Quincy, Ill. (1972), 3211
Mikhailov, A. A., et al, 61

Miller, E. C., 843, 1569
Miller, Evelyn, 2304
Miller, H. W., 2499
Miller, James R., 2749
Miller, Max B., 281, 310, 836, 1513, 2868, 2948, 2967, 3389
Miller, R. D., 2931, 3103
Miller, Stanley L., 98, 1277, 2161
Miller, Stewart, 3604
Miller, Will, 2304, 2996
Mills, David M., 1469
Milton, S., 2751
Mingus, Ron, 3539
Misraki, Paul, 1035, 3018
"Missionaries to Space", 2521
"Mr. Cooke Goes to Zomdic", 2871
Mitchell, Berry, 741
Mitchell, Helen, 741
Mitton, Simon, 1432
Moffat, Samuel, 1057, 1697
Moll, Horacio M., 1404, 1989
Mollohan, Hank, 289
Molton, P., 1421, 2282, 2286, 2287
Monti, Adriano, 1090
"Moon Life Seen Possible", 2150
Moore, Patrick, 784, 1641, 1663, 1788, 1790, 1933, 2023, 2050, 2107, 2112, 2275, 2631, 2811
"More About Marliens", 503
"More Flying Saucers", 3278
"More Intensive Scientific Study of Reported Sightings Urged by D. L. Morgan, Jr...", 3602
"More Saucers Over North Korea", 205
Moreux, Théophile, 987, 998
Morgan, Dean, 306
Morison, R., 2972

Moroz, V. I., 60
Morrison, P., 1459, 1515
Morse, E. S., 2012
"Moscow Radio Reports Prof. Tikhov Holds Life Exists on Mars and Vegetation is Blue", 2099
Moseley, E. L., 1150
Moseley, James W., 1187, 2507, 3184, 3480
Moseley, James W., comp., 3063
Motz, Lloyd, 1046, 1566
Mowbray, Lionel, 1563
Moyer, Ernest P., 2475
Mueller, Robert F., 2257
Mugler, Charles, 1002
Mukerji, B., 1850
Mukhin, L. M., 114, 1107
Mulholland, J., 3316
Muller, H. J., 87, 1304, 1631, 1909
Muller, Wolfgang D., 1012, 2759, 2764
Mumford, G. S., 1943, 2205
Mumford, N. W., 1133
Munday, John C., 3633
Mundo, Laura, 752, 1052, 2461
Murchie, G., 59
Murdin, Paul, 1306
Murray, Jacqueline, 1441
Mustapa, Margit, 2326
Mutschall, Vladimir, 1584, 2568
"Mysterious Broadcast From Space Ship?", 1511
"The Mysterious Chunk of Hardware at Ottawa", 3655
"Mysterious Wave of UFO's Reported in Mass Sighting", 694
"Mystery of Life Nearly Solved", 1883
The Mystery of Other Worlds Revealed, 1001, 2306
"The Mystery Satellite: a Study in Confusion", 319

N

"N. J. Berrill Book, Worlds Without End, on Life in Space and Possibly on Other Planets, Reviewed", 1326
"N. S. Kardashev Suggests CTA-21 and CTA-102, Powerful Sources of Radio Emission in Constellations Aries and Pegasus...", 1575
"N. Tesla Predicts Intercommunication", 1494
"NASA Biologists Discover Rare Earth Organism", 2294
"Nasa Studies Planetary Habitation Methods", 2270
"NASA Tests Boost Jupiter Life Possibility", 2687
"Nasa to Send Sampling Device to Mars in 1966 to Find out if Life Exists There", 2636
NASA-University Conference on the Science and Technology of Space Exploration, Chicago, Nov. 1-3 (1962), 1665, 2555
Nash, William B., 214
"National Investigations Commission on Aerial Phenomena Reveals U.S. Air Force Issued Directive in Dec. 1959...", 315
National Investigations Committee on Aerial Phenomena, 123, 2997, 3133
"National Investigations Committee on Aerial Phenomena Organized to Provide 'Honest Information'", 2407

"National Radio Astronomy
  Observatory Reports...
  Clouds of Formaldehyde
  Thought to be Chemical
  Precursor of Life, in
  many Parts of Milky
  Way...", 1975
"Nature's Way", 1999
Nebel, Long John, 745
Nelson, Albert F.J.H.N.,
  24, 1637, 1660, 1642
  2022
Nelson, Buck, 2848, 2859,
  2869
Netto, Paulo Coelho, 146,
  3233
Neumann, Temple W., 2667
Neville, T., 1872
"New Clues to UFO Electrical
  Interference", 3527
"New Debate Over Life on
  Mars", 2264
"New Evidence of Intelligent
  Life on Other Worlds", 1334
"New Flying Saucer Story",
  3816
"New Light on Flying Saucers",
  3601
"New Method may Detect
  Martian Life", 2678
<u>The New Report on Flying
  Saucers</u>, 3064
"New Saucer Epidemic", 211
"The New Zealand 'Flap' of
  1909", 376
Newburn, R. L., 2030
Newcomb, S., 1130, 1736,
  2051
Newell, H., 2629
Newman, J. R., 1570
"The Next 25 Years on Mars:
  Colonies on the Moon, and
  First men on Mars", 2820
Nicolson, Marjorie H., 2756
Nieman, C. W., 1482
Nieman, H. W., 1482
"The Night of September 29",
  298
Nikander, Jukka, 2268

"9 UFO's Sighted in
  Suburbs North of
  Chicago", 670
"1961: This May be the
  Year", 3464
"1966 Tully", 3574
"No Chance of Life on
  Venus...", 2157
"No Evidence for
  Saucers", 3409
"No new Theories,
  Evidence, Saucers,
  Says Pentagon in Lift-
  ing UFO lid", 3353
"No Room for Christian
  Faith", 2510
"No Visitors From
  Space", 3312
Nollet, A. R., 3445
Noonan, Allen, 2887
Norkin, Israel, 2949
Norman, Ernest L., 1007
Norman, Mark A., 2318
Norman, Paul, 2824, 3523
Norman, Samuel, 255
Norman, Susan M., 2787
Normyle, W. J., 2681
Norris, Geoffrey, 3402
North Atlantic Treaty
  Organization. Advisory
  Group for Aerospace
  Research and Develop-
  ment, 1467
"Northwestern University
  Astronomy Dept. Chair-
  man, Dr. Hynek, Urges
  Scientific Study...",
  3558
"Le Nostre Analisi:
  Inchiesta UFO in
  Calabria", 710
"I Nostri Aderenti ci
  Hanno Chiesto il Parere
  Sull' AIAS...", 2418
"Notes and Comment:
  Saucer Flap", 421
Novick, Aaron, 1861, 1873
"Now They're in Italy:
  Astral Intruders", 241

"La Nuit du 17 au 18 Juillet 1967", 527

## O

Oakley, C. O., 840, 1546
Oberbeck, S. K., 1428
Oberth, Hermann, 1181, 2463, 3460, 3478
O'Brien R., 1312
"Observations Etrangères", 381, 388
"Observations Françaises Plus Anciennes", 240
"Observations Françaises Récentes", 467
"Observations Hors Presse", 468
"Observations of 'Mysterium Phenomen' of Radio Emissions From Milky Way Show Emissions Meet Almost all Criteria for Artificial Communications...", 1596
O'Donnell, A. J., 2591
Of the Plurality of Worlds: an Essay; Also Dialogue on the Same Subject, 2428
"Official Air Force Statements on Unidentified Flying Objects", 3412
The Official Guide to UFO's..., 3104
Ogden, Richard C., 2801, 3446
Ogles, George W., 3613
Oliver, B. M., 1472, 1559, 1620, 2593
Oliver, Frederick S., 2298
Olsen, Thomas M., 3065
"On the Flying Saucer Trail", 231, 3349
"On to the Planets", 2800
"One of 3 Kentucky National Guard Planes", 182
"ONIFE Investiga Caso Carcaraña", 971

Oparin, A. I., 29, 37, 97, 104, 1302, 1365, 1656, 1691
Opik, E. J., 38, 1381
"Opposition Flap 1965", 391
Ordway, Frederick I., 125, 1058, 1067, 1347, 1369, 1698
Ordway, Frederick I., et al, 1666, 1881, 2774
"Organic Life Debated", 1946
"Organic Production on Mars", 2273
Orgel, L. E., 2008
"Origin of Life", 118
Orlando, Carlos, 620, 644
Ormond, Ron, 822, 1440
Osorio, Luis E., 2932
Ostlin, M. T., 2455
Ostriker, Alicia, 1430
Osuna Llorente, Manuel, 686
"Other Beings on Other Planets?", 1196
"Other Earth", 1392
"Other-Worldly Faith", 2513
"Other Worlds Than Ours", 1139
Otrotchenko, V. A., 114
Ottesen, Eric A., comp., 134
"Out-of-the-Blue Believers: Civilian Saucer Intelligence of New York", 2411
"Out of This World: Convention of the Amalgamated Flying Saucer Clubs of America", 2416
Ovenden, Michael W., 1036, 1667, 1854
Owen, Tobias, 2232, 2252
Owens, Ted, 757
"Oxygen on Mars as an Indication of Life", 2068
Oyama, Vance I., 2584, 2670, 2698

Ozick, Cynthia, 1422
"Ozmology", 2613

## P

Pace, Anthony R., 3109, 3212
Pack, Warren E., 3437
Page, Henrietta M., 138
Page, Thornton, 516, 3669
Pallmann, Ludwig F., 2782
Palmer, Ray, 198, 436, 454, 1222, 1442, 2910, 3281
Paluzie Borrell, A., 2770
"Panel Appointed by National Sciences Academy at Govt. Request Urges Effort to Land Automated Biological Lab on Mars by 1971...", 2664
"Panel of Four Leading Scientists and one Theologian Agrees on Nov. 19, That Highly Advanced Civilizations Flourish Elsewhere in Universe...", 1412
Pannekoek, A., 1657
Panovkin, B. N., 1401
Parkes, A. S., 1822
Parks, R. J., 2614
Parry, A., 2128
"Pascagoula ou les Pecheurs Pechés: un Drôle d'Occupant", 967,
Pascalis, Bernardino, 1593, 3591
Pastorino, Luiz P., 478
Patrovsky, Venceslav, 3134
Patterson, R. H., 1120
Pauquet, P. P., 2921
Pay, R., 2660, 2677
Pearman, J. P. T., 1047, 1701
Pease, Harry S., 981, 3831
Pederiali, Giuseppe, 3042, 3086
Pedersen, Frank, 3135
Pedrajo, Manuel, 2922

Pelley, William D., 727, 1451, 2305
Pendergast, R., 1321
Pennington, J., 351
Penzias, A. A., 1997
"The People who see 'Flying Saucers'", 2356
Perego, Alberto, 156, 158, 562, 1219, 2962, 3165
Pereira, Flavio A., 1647, 1801, 3019, 3066
Pereira, Jader U., 934, 946
Perez, George R., 2579
Perkins, Elmer Corp. Aerospace Systems, 2588
Perret, Jacques, 220
"Persons, by R. Pucetti (Review, by Bernard Murchland)", 1396
"Perturbations Psychiques", 2373
Pettengill, G. H., et al, 2623
"Philco Labs Developing Protein Detector to Probe for Life on Mars", 2648
Philip, Brother, 2322
Philipp, Franz, 3166
Phillips, Ted, 717, 721
Phillips, Z. T., 592
Philosophical Speculations Concerning Life on Earth and in Outer Space, 2464
"Physical Evidence Landing Reports", 371
"Pi in the sky", 414
Pickering, J. S., 1028
Pickering, W. H., 2017, 2075
Piel, J., 486
Pierce, J. R., 2804
"Pies in the sky in Mexico City", 193
Pikelner, Solomon B., 1682
Pinotti, Roberto, 793, 1357, 1363, 3245
Pirie, N. W., ed., 2772

Pittendrigh, Colin S., 1701
Pittendrigh, Colin S., et al, eds., 2040
Pittenger, Norman W., 2509
The Planet Mercury, 1971..., 2049
Planet Mercury Sends Greetings, 2341
"Planetary Biology Soon", 1824
"Planets", 1764
"Plant Life on Mars", 2096
Plantier, J., 2760
Platillos Volantes, 2998
"Plenty Going on in the Skies", 229
"Plenty of Lebensraum in Other Parts of Universe", 1774
"Plurality of Worlds", 2489
"Plurality of Worlds Inhabited", 2490
Pokrovskii, Giorgii I., 2747
"Police Chase low Flying UFO", 413
Pollard, William G., 2277
Ponnamperuma, Cyril, 55, 105, 113, 117, 1944, 2199, 2287, 2720, 2733
Ponnamperuma, Cyril, ed., 1725
Pono, W., 39
"Populating Other Planets", 2791
Posin, Dan Q., 313, 1325, 1668, 1928
"Poursuites Dans le Ciel", 484
Powell, Conley, 1419
Powers, William T., 517, 521
Pratt, Bob, 691
"Pravda i Vynsysel ob UFO", 3583
"Pre-Life on Jupiter?", 2219
"Pres. Eisenhower Reports U.S. Air Force has Assured him Saucers are not Invading Earth...", 3359

"Prevalence of Planets and the Probability of Life", 1927
Price, George R., 2606
"Priest Astronomer's Report: Observations of an Argentine Astronomer", 536
Priestly, Lee, 2410
"Primitive Life on Mars?", 2276
Prince, A. E., et al, 1653
Pritchett, E. B., 2337
"The Problem of Witness Reliability", 3755
"Problems Common to the Fields of Astronomy and Biology: a Symposium", 1797
Proceedings of the Eastern UFO Symposium, Jan. 23, 1971, 3185
Proctor, R. A., 1124
"Prof. A. Arrhenius Declares Life Possible on Planet", 2059
"Prof. E. Campbell on Chance of Vegetation and Life", 1752
"Prof. F. Ziogel of Moscow Aviation Institute Urges 'Joint Effort' of World Scientists to Determine Nature of UFO's...", 2424
"Prof. H. N. Russell Holds Many (Planets) May Be Inhabited", 1778
"Professor Hermann Oberth Defends the Flying Saucer", 3483
"Prof. J. Strong Sees Discovery of Water Vapor in Atmosphere Increasing Chances of Existence of Life (on Venus)", 2178

"Prof. Lovell's Article on Probability of Sentient Life Existing Elsewhere in Cosmos...", 1275
"Prof. McDonald Urges NASA, National Science Foundation or Like Organization Without Vested Interest Open Study of Reported Sightings...", 3570
"Prof. P. Morrison Sees Jupiter, Mars, and Venus Capable of Supporting 'Something Like Life'", 1823
"Prof. S. Arrhenius Predicts That Venus Will Carry Culture When our World Dies", 1146
"Professor Teller Calls UFO's Miracles on TV Interview", 3536
"Prof. W. W. Howells Speculates Beings on Other Planets Resemble Mythical Centaurs", 2138
"Prof. Wald Sees Intelligent Life as Culmination of Cosmic Evolution...", 1355
The Project A Report, 2914
Project Cyclops: A Design Study of a System for Detecting Extraterrestrial Intelligent Life, 2598
"Project Ozma", 2608
"Project Ozma and Other Efforts to Determine if Life Exists on Other Worlds Discussed", 2619
"Project Ozma, Designed to try to Pick up Intelligible Signals That Might Have Been Sent From Other Worlds...", 2607
"Project Ozma Off", 2617
"Project Viking", 2752
"A Propos du Symposium de Quimper: Deux Declarations d'Astronomes", 3798

"Proposal for Coordinated International Effort to Deal with Sightings...", 2425
"Proteins Could Give Clue to Unearthly Life on Mars", 2153
Pryor, H., 103, 2680
Prytz, John, 615, 920, 961, 1411, 1612, 1966, 3561, 3703, 3709, 3724, 3747, 3749, 3767
"Psychoanalyzing the Flying Saucers", 3264, 3273
Puccetti, Roland, 768
"Puerto Velaz: Plato Volador", 722
Pursglove, S. D., 2633
Pye, Fred, 619

Q

"Quarantine for Space Travelers?", 1941
Quarnström, Gunnar, 154, 2978
"Quarter's Polls on Flying Saucers", 3263
"Que S'est-il Passé à Marliens", 496
"Que S'est-il Passé à Plestan?", 403
Quimby, Freeman H., 1954, 2563, 2564
Quiroga, R. A., 2812

R

"R. Heflin Releases 3 Photos of Saucer-Shaped Flying Object Taken Aug. 3 near El Toro Marine Airfield, Calif...", 397
"Rabbi N. Lamm Says Existence of Rational Beings on Other Planets is Compatible with Jewish Theology...", 1356

"Race of Flying Men", 1117
"Radar Detection of Cosmic Radio Waves Revealed...", 1499
"Radio Uganda Reports Pres. Amin Sees UFO Near Kampala", 662
Radmer, Richard, 2729
Radunskaya, Irina, 2207
Raible, D. C., 1249, 1255
Ramo, S., ed., 2031
Rampa, T. L., 2860
Rand Corporation, 1669, 1683, 2999, 3105
Randolph, James R., 1217, 1792
"Randolph School's Test Project Proves Life can Exist on Mars", 2117
Rankow, Ralph, 462, 522
Rapp, Daniel J., 1407
Rasool, S. I., ed., 64
Rathbun, Mabel, 406
Ratliff, Buffard, 932
Raynes, Brent, 559, 567
Rea, D. G., 1310, 2173, 2177, 2684, 2709
Redon, Père, 713, 976, 2394
Rees, M. J., 1608
Reeve, Bryant, 1466, 2050
Reeve, Helen, 2050
Rehn, K. G., 3020, 3067, 3136
Remaley, Sally, 552
Rember, Winthrop A., 2849
"Remember the Flying Saucers?", 3256
Renshaw, Jr., Charles C., 1415
"Reporting Unidentified Flying Objects", 3503
"Reports Finding Evidence of Outer-space Life in Meteorite", 1867
"Reports From Everywhere", 233
"Retired Rear Admiral Fahrney Says Reports Indicate High-Speed, Directed Objects are Entering Earth's Atmosphere", 259

"Retour sur Attigneville, L'incident de Xertigney", 889
"Retour sur Valensole, les Conclusions de Notre Enquêteur", 872
Revelle, R., 1895
"Rev. D. C. Raible Sees Possibility of Existence of Rational Beings With Cultures and Civilizations Superior to Those of Earth...", 1248
Reynolds, O. E., 1896, 1954, 2581
Rhodes, George, 1815
Ribera, Antonio, 155, 168, 302, 333, 341, 350, 359, 425, 509, 513, 528, 568, 851, 1443, 1445, 2979, 3000, 3004, 3137, 3246, 3524
Ribes, Jean C., 1093, 1105, 1402
Rice, R. V., 2144
Rich, Valentin, 1241
Richards, H. M. S., 1053
Richards, Sam, 2512
Richardson, R. S., 2101
Rifat, Alain, 504
Rimes, Nigel, 940, 3701
"Ring of Lights set up in Texas as UFO Bait", 689
"Riverside, Calif. UFO Said to be Meteor", 672
"Robert Jastrow", 1433
Roberts, A. W., 1188
Roberts, August C., 269, 290, 543
Roberts, Keith, 859
Robertson, J. G., 1619
Robey, Donald H., 3439, 3441
Robinson, G. S., 925
Robinson, Jack, 882
Robinson, L., 1132
Robinson, Mary, 882
Rocha, H., 1015, 1016
Rocky Mountain Bioengineering Symposium, 1692
Rocquet, R., 1635
Rodgers, Philip, 1508

Rodriguez, Vincente C., 3186
Rogers, Warren, 487, 3596
Romaniuk, Pedro, 794, 3138, 3802
Rorvik, David M., 1420
Rosenberg, Paul, 1557
Ross, H. E., 1322
Ross, John C., 185, 226, 234, 282, 316, 325, 344, 1212, 3350, 3430, 3698
Rossi, Dora Nelly V. de, 785
Rothberg, Gerald, 3725
Rougeron, Camille, 2403, 3267
Roulet, Alfred, 1114
Rousseau, P., 43
Roussel, Robert, 429
Rowe, Kelvin, 1453
Rubenchik, L. I., 1708
Rublowsky, J., 1037, 1273, 1281, 2143
Rudaux, L., 2019
Ruddy, John, 3636
Ruggieri, Guido, 32
Ruppelt, Edward J., 3351, 3385, 3392
Rush, J. H., 1645
Russell, H. N., 13, 1148, 1161, 1763, 1767
Russen, D., 2754
"Russian Report: Is There Life on Mars?", 1238
"Russian UFO's", 540
"Russians say That Flying Saucers Exist", 330
Ruzic, Neil P., 1335
"Rymdbesok--Eller vad?", 3641
Rynin, Nikolai A., 2783

S

"S. Nixon, NICAP Researcher, Eyes UFO's Seen by Astronauts", 679
Saalsaa, Geneva, 554
Sable, Martin H., 136

Sacksteder, Fred V., 455
Saenz, Manuel, 761
Sagan, Carl, 56, 95, 106, 115, 1059, 1068, 1305, 1423, 1621, 1658, 1670, 1684, 1693, 1731, 1920, 1965, 2002, 2004, 2028, 2032, 2133, 2146, 2154, 2162, 2226, 2228, 2229, 2750, 3585
Sagan, Carl, ed., 1475
Sagan, Carl, et al, 1886, 2248, 2736
Salisbury, Frank B., 174, 1185, 1791, 1887, 1921, 2113, 2197, 3247, 3587
Salisbury, John W., 2249
Salisbury, John W., ed., 139
Sall, T., 2657
Salmon, C. F., 1143
Sanctillean, 2906
Sanderson, Ivan T., 409, 1437, 3139
Sanderson, Jay, 553
Sanford, Ray, 1503
Sanger, E., 2762
Santesson, Hans S., ed., 3107
Santos, Maurice, 3167
Sarrantonio, Albert, 3728, 3794
Satterthwaite, Gilbert E., 1506
"Saucer Blue Book", 3378
"Saucer Craze Continues Despite Facts", 3526
"Saucer Diehards", 3742
"The Saucer Question", 2501
"Saucer Season", 209
"Saucer Session for Spaceship Sighters...", 2408
"Saucer Sightings Fall Sharply", 228
"Saucer Sightings Rise During Year", 463
"Saucer Sorcery: Cartoons", 3361

"A Saucer, two men, and 'Little Creatures'", 831
"Saucerens: UFO Watchers Report Eerie Events", 688
"Saucer-Eyed Dragons", 3272
"Saucers and the Iron Curtain: a Report From Czechoslovakia", 320
"Saucers Down to Earth", 3668
"Saucers Downed (Air Force Report on)", 3375
"Saucers' End: Condon Report", 3681
"Saucers Explained", 3453
"Saucers in the News", 291, 293
"Saucers? No, Skyhooks", 3286
"Saucers Over Paris", 264
"Saucers, Pancakes and Such", 3472
"Saucers Under Glass", 3309
Saunders, Alex, 300, 937, 941, 954, 962, 1591, 2391, 2392, 2549, 3737
Saunders, David R., 600, 616, 625, 626, 3106, 3140
Saunders, Joseph F., 1974
Saunders, Richard, 3679
"I saw a Flying Saucer", 267
Scamehorn, Howard, 2765
Schalin, Sven, 646
Schang, Casimiro A., 441
Schatzman, Evry, 1048
Schiaparelli, G. V., 1699
Schindler, Charles A., et al, 2152
Schmidt, Reinhold, 744, 832
Schmidt, Robert A., 942
Schneider, Adolf, 795
Schoenherr, Luis, 379, 566, 2362, 2366, 2369, 3736
Schopf, J. W., 2046
Schopfer, Siegried, 2933
Schroeder, W., 3416
Schuessler, John F., 3732
Schurmeier, M. H., 2722
Schuster, O. J., 2435

Schütze, Alfred, 3642
Schwartz, Alan W., 1091, 1985
Schwartz, Alan W., ed., 1723, 1726, 3213
Schwartz, Berthold E., 590, 618, 928, 929, 978, 2388, 3643, 3761, 3782, 3786
Schwartz, R. N., 1541
"Science, Cups or Saucers?", 3405
"Scientific Opinion Divided About Life on Jupiter", 2295
"Scientists at International Congress on Astronautics...", 3317
"Scientists Detect Radio Waves From Planet Venus", 1504
Sclanders, I., 1303
Scorneaux, Jacques, 979, 980
Scott, David H., 3616
Scully, Frank, 2907
Seaman, E. A., 469
"Search for Civilizations", 2621
"Search for Life in Space", 2647
"Search for Martian Life", 2665
Searle, G., 2491, 2495
Seckbach, Joseph, 2269
"Secret Conference Sponsored by National Sciences Academy at Green Bank, W. Va., in November 1961 on Possibilities and Problems of Communicating With Other Worlds Revealed and Discussed", 1553
See, T. J. J., 1740
"Seeing Things", 275
"Seeking Extraterrestrial Civilizations...", 1409
"Seen any Flying Saucers Lately? The UFO Controversy", 3248

Seevior, Peter M., 385
Sendy, Jean, 2465, 2473, 2480
Senelier, Jean, 514
Serdobolskii, V. I., 1709
"Serious Flaws in AF Special Report, 14 Revealed by NICAP Analysis", 3413
Serpas, Paul F., 244
"Service Holds Ground on Flying Saucers", 3440
Serviss, Garrett P., 1628, 1734
Sesma, Fernando, 772
Settimo, Gianni, 3648
Seybold, Paul G., 1685
Shalett, Sidney, 3260
Shanklin, H. A., 251
Shapley, Harlow, 80, 1638, 1648, 1786, 1838, 1841, 1851, 1890, 1907, 2527
Shapley, Harlow, ed., 88
Sharp, Alan W., 906
Sharp, Peter F., 2620, 3461, 3498, 3715
Shaver, Richard S., 3777, 3795
Shaw, G., 1982
Sheath, P., 1099
"Sheffield's Sensational Week", 348
Sheldon, Jean, 897
Sheridan, H., 215
Sherman, Carl, 3634
Sherwood, John C., 165, 3068
Shipley, M., 990
Shklovskiy, I. S., 1060, 1068, 1460
Shneour, Elie A., 134, 1697
"Shooting Down the UFO's: Condon Report", 3685
"Should we Advertise?", 3774
Shritenour, Joan, 1448
Shuldiner, Herbert, 3622
Shuttlewood, Arthur, 439, 762, 769, 3069, 3187
Sibol, R. F., 593
"Sickles in the Sky: Communist UFO Observations", 3636

Siegel, S. M., 1987
Siegel, S. M., et al, 2164, 2165
Sievers, E., 2934
Siforov, V. I., 1330
"Sightings by Scientists", 265
"Sightings in Britain", 638, 659
"Sightings in Finland", 627
"Sightings Spur Review; UFO Probe Methods, Findings Studied", 419
Sigma, Rho, 3789
"Signalisation Interplanétaire", 1495
"Signals From Outer Space", 1603
"Signals From Space?", 1545
"The Significant Report From France", 401
"Signs of Life on Mars: Life on Mars Hard to see", 2180
"Signs, Portents, and Flying Saucers", 176
Siguret, B., 1344
Silva, R. I. da., 1069
Silvano, Ceccarelli, 845
Silver, Brent W., 2837
Silvers, Samula M. R., 3773
Simões, Auriphebo Berrance, 2968
Simons, Howard, 3572
Simons, Rodger L., 3254
Simpson, G. G., 1313
Sinclair, A. C. E., 2265
"Singulière Observation Russe en 1663", 585
Sinton, William M., 1830, 1902, 2119, 2126, 2163
"Sir Bernard Lovell, Dr. H. S. Brown, Others Discuss Possibility of Life on Other Worlds and of Radio Contact With Them...", 1554
Sitter, W. de, 12

"Situation Report, What is the Unidentified Flying Object Situation These Days?", 3597
"6 Scientists Testify Before House of Representatives Committee That Objects Merit Serious Study...", 3672
Skalrewitz, Norman, 287
Sky and Telescope (periodical), 57
The sky is Haunted", 1163
Slaboda, Emil, 3399
Slate, B. A., 645, 666, 2892, 3778
Slater, Alan E., 1195, 1283, 1311
Slater, J., 1882
Slater, Robert M., 2838
Slater, Tim, 3805
"Slim Chance for Saucer Sightings", 283
Slipher, E. C., 2066
Sloan, Eugene A., 1832, 1842
Sloan, Richard K., 2691
Smart, R., 1942
Smith, A. U., 1822
Smith, Bernard, 1507
Smith, E. R., 3471
Smith, J. L., 1569
Smith, Robert G., 706, 714, 1567
Smith, S. L., 107, 440, 448, 1352, 1359
Smith, Warren, 502, 3053
Smith, Wilbert B., 773, 2685
Smoluchowski, R., 1937, 2238
Sneath, Peter H. A., 1720
"Snova 'Letaiushchie Tarelki'?", 3650
Soffen, G. A., 2271, 2735
"Some 'Saucers' may be Electrical", 442
"Somebody out There?", 1226
"Something in the air", 210
Sondy, Dominic, 266

La 'Soucoupe' Carrée de Bolazec ou le Tracteur Volant", 408
"Soucoupes Carrées Avant Bolazec", 430
Soule, Gardner, 3070
Soviet Bloc Research in Geophysics, Astronomy and Space; #275, 65
"Soviet Scientist E. Krinov Allays Fears of Invasion of USSR by Mars, Radio Broadcast", 1186
"Space and Bugs", 1834
Space, Gravity, and the Flying Saucer, 2758
"Space Life on Earth...", 2135
"Space-Life Study Lags", 1906
"Space men Predicted", 1265
"Space Radio Noise Picked up", 1590
"Space Ships--and the People They Visit", 3429
Space Theology", 2506
"Space Visitors: Examples of Mysterious and Well-authenticated Unidentified Flying Objects, by 'Theorist'", 818
"Space Visitors: From Which Planets do They Originate, and is There Life on Them?", 1190
"Spacious Talk", 1517
"Speaking of Pictures: a Rash of Flying Disks Breaks out Over the U.S.", 177
"Special Article by C. Flammarion on Mars, Possibility of Life and of Communications Discussed", 1488
"A Speecy by Wilbert B. Smith", 3494
Speer, Herbert V., 2951

Spencer, Dwain F., 2830
Spencer, John W., 1438
Spencer, S. M., 2666
Spiegler, Paul E., 128
Spitzer, L., 2793
Sponsler, G. C., 2695
Spraggett, Allen, 2883
Sprinkle, R. L., 2343, 2383, 3071, 3108
Stamey, Dennis, 935
Stanford, Ray, 2963
Stanford, Rex G., 518, 2963, 3644
Stanford University. Design Div., 2572
Stanford University. School of Medicine, 2561
Stanley, Neil, 3259
Stanley, R. R., 1427
Stanton, L. J., 3021, 3598
Stanway, Roger H., 3109, 3212
Stapledon, O., 2789
"Star Search: new Instruments Search the Heavens for Intelligent Signals From Other Worlds", 1556
"Stardust and Moonshine: UFO Sightings", 692
Stecklong, Fred, 3141
Steen, J., 1893
"Steep Rock Flying Saucer", 801
Steiger, Brad, 470, 912, 191, 1448, 3022, 3072, 3110, 3665
Steiner, G., 1833
Steiner, R., 219
Steinhauser, Gerhard R., 2484
Sternfeld, A. J., 1769
Stetson, H. T., 11
Stevens, C., 2543, 2547
Stevens, Stuart, 3548
Stewall, Frank, 2432
Stewart, Edward A., 796
Stewart-Gordon, James, 1450
Stimson, T. E., 1160
Stine, G. H., 3599

Stokes, Bill, 968
Straight, M., 3344
Straiser, J. A., 1923
"Strange Effects From EM Waves", 3528
"Strange Shapes Seen in the sky", 253
Stranges, Frank E., 763, 1024, 2374, 2969, 2973, 3023
Strauch, Arthur A., 431
Strentz, Herbert, 3024, 3552
Stringfield, Leonard H., 2952
"The Struggle for Existence on the Planet Mars", 1742
Strughold, Hubertus, 1169, 1804, 1821, 1849, 1903, 2021, 2095, 2098, 2102, 2145, 2215, 2798
Struve, O., 19, 91, 1789
Stuart, Jerry L., 2650
"Student, Ben Baron, Takes Picture of Piedmont Mo., UFO", 669
"Studies Show Martian Life", 2120
"Study Mars With Photos", 2675
"Study of Radio Noises Generated in Space Outside the Earth Discussed", 1498
"Study Radar 'Ghosts'", 3306
Stuhlinger, Ernst, 1809, 1810
Stumbough, Virginia, 2351
"The Substance and the Shadow", 3775
Sukhotin, B. V., 1473
Sullivan, Walter, 1070, 1350, 1387, 1610, 2193
Sullivan, Walter, et al, 1700
Sumner, F. W., 2312
"Sur les Effets Biologiques des Champs Magnétiques Intenses", 2370

"Sur les Solitudes Glacées de l'Antarctique", 396
A Survey of Life-Detection Experiments for Mars, 2599
"Suspect Human Life on Millions of Planets", 1203, 1205
Sutherly, Curtis K., 3758
Swart, Hans, 1331
Swedenborg, E., 2436, 2438-2440
Swezey, K., 212
Sykes, Egerton, 764, 1076, 3481
Sylvester, J., 2953
Sytinskaia, N. N., 1636
Szachnowski, Antoni, 2412

T

Tacker, Lawrence J., 2974, 3463
"Taking no Chances", 3477
Tarade, Guy, 2348, 3142, 3188
Tass Report: Mars: Complex Investigations..., 2047
"Tass Reports Astronomer G. Sholomitsky Finds CTA-102 Radio Source Emits 'Flickering' Radio Waves Every 100 Days...", 1579
Tassi, D., 44
Taylor, H. J., 196
Techter, David, 687, 969
Techtran Corp., Glen Burnie, Md., 1710
Tel Aviv University. Dept. of Environmental Sciences, 1712
Tello, Luis R., 1853, 2642
Temm, P. N., 3173
"Temperature Inversions Cause 'Flying Saucers!'", 217
Templin, K. W., 578
Terblanche, Le Roux, 2984

"Texas Woman Recalls Crash of 'Spaceship' in 1897", 675
Teyfel, V. G., ed., 1727
"That Prospective Communication With Another Planet", 1479
"The Theology of Saucers", 2500
"Theory of God as Extra-Terrestrial Visitor Presented", 2548
"The Theory of Life on Mars is Dealt a Blow by Spectra", 2500
"There's Intelligent Life on the Moon", 1214
"Things That go Whiz; Flying Saucers", 3261
Thomas, Andrew, 1368
Thomas, Dorothy, 2315
Thomas, Franklin, 734
Thomas, Jane, 643
Thomas, P., 3025
Thompson, T. A., 407
Thompson, William C., 808
Thomsen, D. E., 2290, 2726
Thomson, J. R., 2674
Thomson, M., 3327
"Three Amazing Cases of UFO Healing", 3826
Three Undiscovered Planets, 2041
Thurgood, Peter A., 3796
Tikhov, G. A., 1029, 1038, 1184, 1826
"Tiny Instrument to see if Life on Mars", 2626
Tobias, C., 1882
Tobin, Michael, 1386
Tocquet, Robert, 45
Todd, D., 1492
Toft, Ron, 3750
Tomas, Andrew, 1115, 3234
Tombaugh, Clyde W., 1227, 1278
Tomorrow Show, 3235, 3832

Tomorrow Show (Devoted to UFO's), 3249
Tompkins, Daniel N., 2689
Toogood, Granville, 720
Toulet, François, 497, 1991
Towner, Cliff R., 260
Towner, Larry E., 235
Townes, C. H., 1541
Townsend, L. T., 2433
Translations From Priroda #6, 1973, 3236
"The Transmission of Life to a Dead World Despite the Ultra-violet ray", 1744
"Tribal Memories of the Flying Saucers", 190
Trofimov, Alexei, et al, 2742
Troitskii, V., 2715
The True Report on Flying Saucers, 3073
"The Truth About the Book. 'The Report on Unidentified Flying Objects' by Edward J. Ruppelt", 3428
Tsung, Thomas, 1049
Tsvetikob, A. N., 1839
Tucci, Eduardo A., 3143
Tufty, Barbara, 2643, 3495
"Two Classic Sightings", 356
"Two USSR Newspapers Score Russians Believing Reports of Flying Saucers", 326
"Two USSR Writers Hypothesize That Light Signals Have Been Received on Earth 3 Times From Habitants of Planet 61 Cygni", 1573
Tyler, Steven, 1080, 3111
"The Type of Mind That Believes in Life on Other Worlds", 1144
Tyrode, J., 965
Tzonis, Konstantin, 2577

## U

"UFO Activity in Brazil During 1965", 491
"Gli UFO Debattuti al' ONU...", 2420
UFO Flight. Visit to Planet Selo, 2863
UFO: Flying Saucers, 3112
"UFO Global Landings Show Dramatic Increase, E-M Effects Noted", 660
"UFO Inquiries set a Record", 3588
"UFO News 1973", 3804
UFO, Past, Present and Future, 3250
"UFO Photographs, Anymore?", 3615
"UFO Problem: Call for Security Against Space War", 3776
"UFO Report", 3821
"UFO Report Rejects Nonterrestrial Origin", 3687
"UFO Sighting Reports Over the British Isles", 628
"UFO Sightings at sea", 569
"UFO Skepticism: or Objectivity?", 3704
"UFO Study Credibility Cloud?", 3666
UFO Terminology, 3113
UFO'er-Flyvende Cigarer..., 3168
"Ufologian Tila ja Puharich", 3813
"Ufology: New Report Debunks Belief That Unidentified Flying Objects are Buzzing the Earth", 3492
UFO's; a Scientific Debate, 3214
"UFO's Again?", 696, 3605

"UFO's and the Laws of Physics", 3624
"Ufo's and UFOnauts, Landings More Frequent in 1973", 676
UFO's: do you Believe?, 3251
"UFO's for Real?", 3565
"UFO's Grounded", 3683
"UFO's: Military Watergate of Hell", 3823
"UFO's? No! Lens Flare? Yes!!, 380
"UFO's not From Mars", 3555
"UFO's or Kugelblitz?", 3553
"UFO's Over Washington: Call for Renewed Scientific and Governmental Saucer Research", 544
"UFO's Real, Scientist Says", 3807
"UFO's: Scientific Debate, edited by Sagan and Page, Reviewed", 3779
"UFO's to be Probed...", 3571
"UFO's Visiting From Space, Goldwater Says", 3808
Underwood, R. S., 1159, 1296
Unesco, 145
Unger, George, 2448
"Unidentified Aerial Phenomena", 560
Unidentified Flying Object Research Committee, 151, 152
"Unidentified Flying Objects, Motion Picture Documentary on Flying Saucers...", 3386
"Unidentified Flying Objects (UFO's)", 3567
"Unidentified Objects?", 3387
"USSR Academy of Sciences Astronomical Council Pres. Mustel Urges International Astronomical Union 1967 General Assembly Discuss International Effort to Locate Signals From Other Worlds...", 1589

"USSR Astronomers Call for 5-Year Cooperative Effort by Radio-Astronomy Centers Around the World...", 1351
"...USSR on October 16 Reports Detection of Radio Signals that may have Originated with Another Civilization...", 1622
"USSR Prof. Vorontsov-Velyaminov Sees Infinite Number of Worlds Inhabited by Intelligent Beings", 1233
"USSR Scientists Study Mars Plant Life", 1780
United Press International, 3074
U. S. Air Force, 2399, 2898, 2900, 3027, 3114
U. S. Air Force. Air Materiel Command, 2901
U. S. Air Force. Air Technical Intelligence Center, 2397, 2398, 2935
"U. S. Air Force, Announcing Development of Flying-Saucer Type Aircraft and Releasing Sketch of Such Craft Being Built...", 3372
"U. S. Air Force Calls Objects Natural Phenomena...", 3304
U. S. Air Force Cambridge Research Laboratories..., 137, 142
"U. S. Air Force Cites 24-Hour Readiness to Challenge Unidentified Objects...", 3303
"U. S. Air Force Consultant J. Allen Hynek Comments on UFO's", 3790
"U. S. Air Force Describes Saucer Reports as Hoaxes", 274

"U. S. Air Force Ends its Investigation of Unidentified Flying Objects", 3710
"U. S. Air Force Ends Probe, Discounts Reports", 2601
"U. S. Air Force Intelligence Center Reports...", 272
"The U.S. Air Force News Release", 3479
"U. S. Air Force Releases Details of Talk Between jet Pilot and Airport Control Tower Just Before he Crashed While Chasing Unidentified Object...", 3310
"U. S. Air Force Reports no Evidence of Flying Saucers Found in 483 Investigations", 352
"U. S. Air Force...Reports (no UFO's)...", 3513
"U. S. Air Force Reports Sightings off Sharply Since 1952...", 238
U. S. Air Force Systems Command, Kirtland AFB, New Mexico, 126
"U. S. Air Force Warns Weather Balloons Might be Mistaken for Saucer", 3294
"U. S. and USSR Scientists Study Plant Life on Mars...", 2091
"U. S. Army Air Force Drops Inquiry into Reports of Strange Objects...", 178
U. S. Congress. House. Committee on Armed Services, 3028
U. S. Congress. House. Committee on Science & Astronautics, 3115, 3144
U. S. Congress. House of Representatives, 1671

"U. S. Earth Satellite Program Held not Responsible for Saucer Rumors", 247
"U. S. Federal Government Proposes Plan to Build Mammoth Radio Telescope...", 2723
U. S. Geological Survey, 62
U. S. Joint Publications Research Service, 66, 1686
U. S. Kennedy Space Center, 3075
U. S. National Academy of Sciences- National Research Council, 2578
U. S. National Academy of Sciences- National Research Council. National Academy of Sciences Panel, 3145
U. S. National Academy of Sciences- National Research Council. Space Sciences Board, 2585
"U. S. National Academy of Sciences Survey Committee...", 2740
U. S. National Aeronautics & Space Administration, 132, 133, 1659, 1687, 1688, 1703, 1711, 1714, 2558, 2569, 2594
U. S. National Aeronautics & Space Administration. Ames Research Center, 50
U. S. National Aeronautics & Space Administration. Scientific and Technical Information Facility, 131
"U. S. National Bureau of Standards Reports Radio Signals Emitted With Energy of 2000 Tons of TNT About Once a Second Indicate Disturbances on or in Planet With Energies

of 100 Million Tons of TNT...", 1524
"U. S. Navy Reports First Conclusive Radio Signals Picked up by Michigan University Radio Telescope From Saturn", 1532
"U. S. News & World Report Claims Saucers are Revolutionary U. S. Navy Craft...", 3269
"United States Report: Is There Life on Mars?", 1240
U. S. Scientific Advisory Panel on Unidentified Flying Objects, 3029
"U. S. Scientists, Drs. C. Ponnamperuma, I. R. Kaplan, and C. Moore Report They Have Discovered 17 Amino Acids...", 1992
"U. S. Space Plans Offer Clue to UFO Problem", 3486
"U. S. to Accept Existence of Space Visitors?", 3824
"U. S. Unmanned Pioneer 10 Spacecraft...Will Contain Message in Scientific Symbols Stating who Sent it and Where They Live...", 1609
"Universal Decoding Plan for Interstellar Messages: Project Ozma", 1533
"University Professors Support Increased UFO Research", 3700
"...Unmanned U. S. Mariner 9 Flight to Mars May Have Been Most Fruitful Scientific Experiment Ever Conducted in Space...", 2725
Uphof, J. C. T., 1768
"Upper Atmosphere and Space Research", 2727
Uranus, Lover of man, Speaks, 2342

Urey, H. C., 21, 1654, 1795, 1947
Uriondo, Oscar A., 170, 3005, 3116, 3189, 3237
Usdin, Earl, 2579

V

Valentine, Tom, 1434
Vallée, Jacques, 337, 362, 372, 373, 510, 587, 591, 594, 602, 603, 606, 613, 617, 629, 786, 850, 957, 2422, 3006, 3030, 3146, 3484, 3529, 3533, 3559, 3607, 3699
Vallée, Janine, 3030, 3484, 3607
Vance, Adrian, 665
Van den Berg, Basil, 2875
Vanquelef, G., 949
Van Sommers, T., 2413
Van Tassel, George, 738, 2954
Varner, Joseph E., 2683
Varsavsky, Carlos M., 1081
"The Vaudriat Sighting", 357
"Le Vaudriat II", 466
Vaughn, D., 68
Vaughn, S. K., 1650, 1836
Veestraeten, Door J., 415
"Vegetation on Mars?", 2115
Velt, Karl, 746, 2423, 2987
Vendégek a Világűrből? Mirol vall a "Repülő' Csézealjak" es a Paleo-asztronautika Irodalma, 3215
"Venus, a Hellhole", 2231
"Venus is Dead and too hot", 2221
"Venus Lives?", 2240
"Venus, Near Neighbor of the Earth", 2103

"The Venus Probes, Venera-5 and Venera-6", 2712
Verplaetse, Juliaan, 930
Vertregt, M., 1581
Vesco, Renato, 3117, 3147, 3190, 3216
Veverka, Joseph, 2226
Vezina, Allan K., 539
Victoria University, Wellington, New Zealand, 1672
Vidal, Franco, 770
Villela, Rubens Junqueira, 677
Viney, Basil, 2508
Vinther, L. W., 3291
Vishniac, Wolf, 1701, 1844
Vishniac, Wolf, ed., 1728
"Visitors From Venus; Flying Saucer Yarn", 799
Vliegende Schotels. Fantasie of Werkelijkheld?, 2909
Vnezemnye Tsivilizatsii. Problemy Mezhzvezdnoi Sviazi, 1468
Vogt, Cristian, 2944, 3511
Volkman, Frank, 252
Von Braun, Werner, 1912, 2170
von Cues, N., 1630
Von Däniken, Erich, 2485
Von Hoerner, Sebastian, 1547, 1548, 2817, 2818
"Vonkeviczky e gli UFO all'UNU...", 2421
Von Krueger, Frederick, 787, 1108
"Voorname Kronologie", 1360
Voronin, M. A., 1314
Vries, T. E. de, 40, 1025
Vsesoyuznove Soveshchaniye No Probleme Vnezemnykh-Tsivilizatsily, Byurakan, 20-23 May, (1965), 1061

## W

"W. L. Laurence on Dr. C. Sagan Discussion of Possibility That Life Exists on Other Planets", 2183
"W. L. Laurence on House of Representatives Committee Hearings", 3476
"W. L. Laurence on Possibility of Sentient Life on Other Worlds as Indicated by Prof. Calvin...", 1290
"W. L. Laurence on Theories That Intelligent Life Exists in Distant Solar Systems and may be Signalling Earth...", 1124
"W. Marconi Says Queer Signals Occur in London and New York Simultaneously", 1480
"W. Sullivan on Rash of... Sightings of UFO's over Various Areas of U.S....", 690
"W. Sullivan Article on Past and Planned Efforts to Find Life on Mars", 2669
"W. Sullivan Discusses Dr. Lederberg's Views...", 1338
"W. Sullivan Discusses UFO Studies in U.S.", 3635
"W. Sullivan on Pulsar Sources: Views That They are Manifestations of Intelligent Life in Other Civilizations...", 1602
"W. Sullivan's Book, We are not Alone: The Search for Intelligent Life on Other Worlds, Reviewed", 3508

Wagner, Bernard M., 1390, 2582
Waithman, Robert, 192, 3270
"Waiting for the Little men", 815
Wald, G., 77, 99, 1829
Walker, J. C. G., 1616
Walker, Sydney, 542
Wallace, A. R., 2014
Walling, Theodore L., 257
Walosin, Frank, 1388
"Walter Sullivan, on Dr. Bracewell's Suggestion, Published in Nature, That Some Civilized Galactic Community may be Trying to Communicate with our Solar System...", 1528
Ward, Bernie, 3825
Warder, George W., 985
"The Warminster Phenomenon", 864
Warren, Donald I., 574, 577
"Washington's Blips", 207
Wassilko-Serecki, Zoe, 3369
Wat zijn UFO's?, 3217
"Water on Mars Might not Support Life", 2006
Watkins, H. D., 2174, 2196
Watkins, Keith, 489, 501
Watson, Charles E., 2703
Watson, S. H., 2533
Watson, W. H., 511
Webb, J. A., 1535, 1561
Webb, J. W., 1461
Webb, W. H., 1461
Webster, G., 1796, 2511
Webster, Robert N., 202, 3290, 3341, 3430
Wegner, Willy, 147, 1100
"The Weight of man on the Planets", 1738
Weiss, Sara, 2844
"Well-witnessed Invasion, by Something: Australia to Michigan...", 417
Wellman, Wade, 1910, 2365, 3475, 3525
Wells, Herbert G., 986
Wells, Leslie E., 2867

Wells, Jeff, 705
Wells, R. A., 2267
Wend, Richard E., 2237
Wentworth, Jim, 947
Weor, Samael A., 3031
Weston, Charles R., 2201, 2651
Westphal, Peter G., 3118
Wetmore, W. C., 2184
Weverbergh, Julian, 3218
Weverbergh, Louis J., 172, 3191
Whalley, Paul, 1604
"What Goes on up There?", 1784
"What the Air Force Believes About Flying Saucers", 3262
"What the British Press Reports on Flying Saucers", 437
"What Were the Flying Saucers?", 3293
"Whatever Happened to UFO's?", 3741
"What's Going on in the Skies", 208
"What's Stirring on Mars", 1953
"What's up There in Space?", 2704
"Where There's Hope There may be Life: Mars", 2200
"Where There's Life...", 1938
"Whewell on Plurality of Worlds", 1118
Whipple, F. L., 2104, 2105
White, George S., 17, 994
"White House is Hiding the Truth About Flying Saucers", 3663
Whittington, George, 3268
"Who Sees Flying Saucers?", 3733
Whritenour, Joan, 912, 3072, 3110, 1448
"Why are we Keen About Mars?", 1753

Wiersema, G. S., 658
Wiley, John P., Jr., 1395, 1413, 2743
Wilhelm, John, 122, 1625
Wilkins, H. P., 2024, 2308
Wilkins, Harold T., 1171, 2923-2924, 3276
Wilks, Willard, 2677
Willems, Louis, 1361, 3564
Williams, Edward D., 3809
Williams, Frances (Mathis), 683
Williams, H. S., 1140
Williamson, A. A., 2502
Williamson, George H., 729, 1020, 1452, 1462, 2316, 2964
Wills, J., 1119, 1123
Wilson, A. G., 81
Wilson, Clifford, 1100
Wilson, Harlan, 236, 284, 323
Wilson, Harvey, 3674
Wilson, H. C., 1136, 1137
Wilson, R. H., Jr., 1044
Wilson, Richard, 3287
"The Wind is up in Kansas", 213
Winder, R. B. H., 523, 902, 3576, 3629
Wolf, R., 3720
Wolfgang, Richard, 1986
Wood, A. T., Jr., et al, 2251
Wood, J. A., 100
Wood, R. H., 3283, 3292
Woodley, Morris, 248
Woodruff, Michael, 3705
Woolen, R. W., 2823
Wooster, H., 1868
Wooster, H., et al, 1586
"Worlds in the Sky", 1122
Wright, Eric, 541
Wright, T., 3, 3119
Wylie, C. C., 3300, 3338
Wylie, Philip, 3600
Wynne, E. S., 1852

X

"X-15 Pilot Shows his Film", 342

Y

Yagoda, Herman, 2036
Young, A. T., 2735
Young, Louis B., ed., 1904
Young, Mort, 3076, 3612
Young, R. S., 120, 1660, 1673, 1702, 1993, 2710
Young, Richard Stuart, et al, 2570
Younghusband, F., 991, 992, 1149
"Youth Tell of Seeing 'Glowing Sphere' at Pt. Dume, Calif.", 664

Z

Zayin, Karl, 3760
Zaitsev, Vyacheslav, 1371, 1391
Zarco de Egea, J., 1266
Zemkla, Interplanetary Avatar, 2861
Zenabi, J., 2915
Zerpa, Fabio, 972
Zigel, F., 1267, 3667
Zinsstag, Lou, 150, 2854, 3491
"Znaniye-Sila (publication) Reports USSR Develops Saucer", 3407
Zubek, T. J., 2530
Zulli, Alphonse, 443
Zullo, Allan, 704, 719, 2553, 3814
Zungri, Giuseppe, 3238

## SUBJECT INDEX

Note:   Numbers immediately following topics, sub-topics, places and individuals treated as subjects refer <u>not</u> to pages, but rather to individual, numbered bibliographic <u>items</u>.

### A

Andromeda (planet), 1292
Animal life on other planets in general (excluding intelligent life, see Meteors and meteorites
Animals, injury by humanoids, see Humanoids, injury to animals
Anthropology (Earth) and Flying saucers, see Flying saucers and humanoids, relationships with Earth's anthropology and history
Associations, space research and UFO, see Organizations concerned with UFO's; Space research, etc.
Astrogeology, see Planets, physical, geological, chemical and biological aspects of; Space research
Astrogeophysics, see Planets, physical, geological, chemical and biological; Planets, physical, geological, chemical and biological aspects of; Space research
Astronautics, see Space flight; Flying saucers
Astronomy, 1, 4, 8, 10, 26, 31, 39, 41, 42, 44, 48, 53, 61, 62, 66, 81, 116, 1657, 1869, 2437, 2460

### B

Bases, flying saucer in oceans, including Bermuda Triangle, see Flying saucer bases in oceans, including Bermuda Triangle
Bases, flying saucer on and under Earth's surface, see Flying saucer bases on and under Earth's surface
Bibliographies of flying saucers, see Flying saucers, bibliographies
Bibliographies of life on other planets, see Life on other planets, bibliographies
Biochemistry of planets and of space, see Planets, physical, geological, chemical and biological aspects of; Space Research
Biological aspects of planets, see Planets, physical, geological, chemical and biological aspects of; Space research
Biophysics of planets and of space, see Planets, physical, geological, chemical and biological; Planets, physical, geological, chemical and biological, aspects of; Space research

312

Blue Book, Project, 3013, 3029, 3175, 3712

## C

Chemical aspects of planets, see Planets, physical, geological, chemical and biological aspects of; Space research
Condon Report, 3085, 3106, 3145, 3560, 3566, 3568, 3571, 3573, 3581, 3589, 3639, 3708, 3710, 3711, 3731
CTA-102 (planet), 1575, 1579
CTA-21 (planet), 1575, 1576

## D

Damage from flying saucers, see Flying saucers, danger from
Detection of flying saucers and extraterrestrial life, see Flying saucers and extraterrestrial life, detection
Doubts concerning existence of flying saucers, see Flying saucers, alternate explanations

## E

Evolution of planets, see Planets, origin of; Planets, physical, geological, chemical and biological aspects of
Exobiology, see Life on planets

ESP contact, see Psychic contacts with life on other planets and with humanoids, including mental telepathy, ESP, etc.
Extraterrestrial life, see Life, intelligent, on other planets; Life on other planets (excluding intelligent life)

## F

Flying saucer bases in oceans, including Bermuda Triangle, 696, 1437, 1438, 1439, 1441, 1443, 1444, 1445, 1448, 1450
Flying saucer bases on and under Earth's surface, 1435-1450, 3659
Flying saucers, 2899-3824
Flying saucers, alternate explanations, 175, 181, 201, 210, 217, 221, 227, 253, 272, 276, 283, 315, 317, 352, 380, 420, 442, 444, 681, 2959, 2982, 3095, 3107, 3240, 3242, 3244, 3254, 3269, 3280, 3283, 3286, 3287, 3294, 3304, 3306, 3307, 3311, 3312, 3315-3319, 3325, 3328, 3329, 3336, 3347, 3353, 3359, 3363, 3370, 3372, 3378, 3382-3385, 3396-3398, 3409, 3412, 3426, 3440-3442, 3445, 3448, 3451, 3457, 3463, 3473, 3492, 3493, 3449, 3503, 3520, 3521, 3526, 3531, 3555, 3561-3563, 3595, 3610, 3613, 3625, 3634, 3654, 3657, 3663, 3666, 3680-3682, 3685, 3686, 3689-3693, 3698, 3735, 3747, 3748, 3750, 3795

Flying saucers and extr-
  terrestrial life, 407,
  516, 1059, 1085, 1295,
  1296, 1297, 1298, 1314,
  1320, 1384, 1409, 1417,
  1503, 1511, 1688, 1694,
  2199, 2554-2753
Flying saucers and human-
  oids, relationships with
  Earth's anthropology and
  history, 1078, 1094, 1101,
  1102, 1104, 1106, 1111,
  1112, 1115, 1116, 1151,
  1171, 1175, 1187, 1284,
  1299, 1317, 1323, 1343,
  1354, 1357, 1363, 1372,
  1385, 1389, 1393, 1397,
  1403, 1423, 1426, 1434,
  2470, 2474, 2483, 2485,
  2507, 2541, 3025, 3043,
  3045, 3077, 3177, 3197,
  3215, 3234, 3246, 3298,
  3302, 3706, 3767, 3791,
  3812
Flying saucers, bibli-
  ographies, 123, 135, 136,
  138, 141, 143, 144, 146
  147
Flying saucers, cartoons,
  3087, 3361
Flying saucers, children's
  books, 3120
Flying saucers, conference
  proceedings, 3176, 3178,
  3183, 3185, 3189, 3211,
  3718, 3798
Flying saucers, danger
  from, 465, 485, 922,
  936, 1186, 1191, 1258,
  1274, 1388, 2361, 2368,
  2370, 2373, 2374, 2380,
  2386, 2388, 2924, 2970,
  2973, 3048, 3072, 3111,
  3128, 3203, 3221, 3403,
  3427, 3469, 3523, 3528,
  3556, 3627, 3630, 3645,
  3676, 3730, 3739, 3776,
  3778, 3783
Flying saucers, landings,
  149, 158, 160, 170, 189,
  284, 387, 443, 450, 452,
  496, 503, 504, 514, 515,
  526, 529, 530, 541, 567,
  579, 590, 592, 594, 597,
  600, 601, 606, 613, 617,
  621, 626, 641, 642, 648,
  658, 666, 674, 676, 687,
  691, 697, 715, 716, 717,
  720, 721, 816, 860, 861,
  2912, 3164, 3171, 3699,
  3701, 3703
Flying saucers, photo-
  graphs, 195, 199, 219,
  303, 323, 336, 342, 349,
  375, 397, 431, 436, 474,
  475, 481, 506, 521, 522,
  526, 543, 618, 619, 665,
  669, 790, 2400, 3049,
  3050, 3073, 3074, 3079,
  3090, 3323, 3386, 3395,
  3615, 3669, 3794
Flying saucers, propulsion,
  2778, 2794-2797, 2799,
  2802, 2807, 2826-2838
Flying saucers, sightings,
  150, 151-159, 161-169,
  171-188, 148-723, 2792,
  3065, 3754, 3755, 3757,
  3765, 3792, 3800, 3802,
  3813, 3818

G

Galaxy, see Universe
Geological aspects of
  planets, see Planets,
  physical, geological,
  chemical and biological
  aspects of; Space
  research
Grudge, Project, 2900
Gulliver, Project, 2633

H

History (Earth) and
  flying saucers, see
  Flying saucers and

314

humanoids, relationships with Earth's anthropology and history

Humanoids and Earth people, possible result of contact between, 3257, 3258, 3825

Humanoids and life on other planets, contact with, see Psychic contact with life on other planets and with humanoids, including mental telepathy, ESP, etc.

Humanoids, contacts (including psychic), with Earth inhabitants, 332, 538, 724-981, 1006, 1071, 1412, 2441, 2856, 2861, 3198, 3222, 3824

Humanoids, injury to animals, 878, 959, 3621

Humanoids, kidnapping of Earth inhabitants, 567, 885, 893, 897, 899, 2867, 2873, 2877-2879, 2883-2885, 3223, 3648

Humanoids, linguistic communication with, 743, 784, 804, 839, 840, 933, 941, 1536, 1582

Humanoids, medical treatment of Earth inhabitants, 904, 2384, 3826

Humanoids, physiology, 779, 958, 960, 977

Humanoids, religious aspects, 957, 1050, 1073, 1109, 1356, 1423, 1426, 1428

I

Inhabitants of Mars, see Mars, inhabitants of; Humanoids...

Intelligent life on other planets, see Life, intelligent on other planets, possibilities for; Humanoids...; Men in black

Interstellar communication, see Space communication (by TV, radar, radio, radiotelescope, artificial satellite or travel)

J

Jupiter, possibilities for life (excluding intelligent), on, 1594, 2066, 2132, 2139, 2141, 2219, 2222, 2223, 2232, 2234, 2237, 2241, 2252, 2261, 2282, 2287, 2292, 2293, 2295, 2687, 2746

K

Kidnapping of Earth inhabitants by humanoids, see Humanoids, kidnapping of Earth inhabitants

L

Landings of flying saucers, see Flying saucers, landings

Lanulos (planet), 780

Laws and regulations concerning Earth inhabitants and humanoids, see Metalaw

Life in space, see Space research; Planets, physical, geological, chemical and biological aspects of

Life, intelligent, on other planets, 17, 18, 32, 45, 47, 54, 56, 57, 60, 70, 71, 72, 73, 81, 89, 91, 95, 96, 102, 103, 104, 106, 107, 114, 115, 116, 121, 122, 148, 395, 982-1434, 1555, 2798, 2808, 2923, 2925, 3142, 3205, 3252

Life, intelligent, on other planets, possibilities for, see Triton, Pluto, Venus, Mercury, Mars, Jupiter, Moon, Sun, CTA-102, 61 Cygni, Andromeda, Lanulos; Men in black; Humanoids; Flying saucers and extraterrestrial life, detection; Mars, inhabitants of; Intelligent life on other planets; Life on other planets in general (excluding intelligent life); Life on other planets, bibliographies

Life on other planets, bibliographies, 123-147

Life on other planets in general (excluding intelligent life), 1628-2295, 2554-2753

Life, origin of, 11, 17, 18, 24, 29, 30, 33, 35, 36, 43, 44, 46, 48, 49, 50, 52, 56, 73, 74, 77, 78, 79, 80, 83, 84, 85, 86, 87, 90, 91, 93, 95, 98, 99, 103, 104, 105, 106, 107, 111, 113, 117, 118, 120, 122, 1062, 1355, 1649, 1691, 1732, 1795, 2801

Linguistic communication with humanoids, see Humanoids, linguistic communication with

## M

Magnet, Project, 2908

Mariner, Project, 2666, 2679, 2713, 2716, 2725, 2750, 3608

Marks on ground from flying saucers, see Flying saucers, landings

Mars, inhabitants of, 140, 735, 748, 811, 824, 937, 968, 1007, 1041, 1042, 1133, 1134, 1135, 1143, 1158, 1185, 1186, 1217, 1222, 1238, 1240, 1250, 1261, 1267, 1278, 1280, 1308, 1310, 1319, 1332, 1339, 1340, 1344, 1378, 1407, 1791, 2925, 3482, 3484

Mars, possibilities for life (excluding intelligent), on, 1753, 1779, 1804, 1818, 1830, 1887, 1893, 1900, 1902, 1903, 1912, 1923, 1933, 1936, 1937, 1953, 1978, 1979, 1995, 2000, 2010-2017, 2020-2028, 2033-2040, 2042, 2045, 2047-2048, 2050-2052, 2054-2075, 2067-2069, 2072-2074, 2076-2078, 2082, 2084-2087, 2089-2091, 2095-2102, 2104-2106, 2109-2110, 2113-2120, 2123-2124, 2126, 2128-2131, 2134, 2136, 2140, 2142, 2144-2145, 2147-2148, 2151, 2153-2156, 2158, 2165, 2167-2177, 2180-2197, 2200-2203, 2209-2212, 2215, 2217, 2220, 2226, 2228, 2230, 2233, 2238, 2239, 2243, 2245, 2246, 2248, 2249, 2254, 2255, 2258-2260, 2262-2264, 2267, 2271-2273, 2275, 2276, 2278, 2281, 2283-2285, 2288,

2580, 2597, 2599, 2600,
2611, 2624-2626, 2634,
2636, 2637, 2648, 2654,
2661-2666, 2669, 2675-
2678, 2684, 2689, 2707-
2709, 2716-2720, 2722,
2724, 2726, 2729, 2730,
2734-2738, 2742, 2750-
2753
Medical treatment of Earth
  inhabitants by humanoids,
  see Humanoids, medical
  treatment of Earth
  inhabitants
Medicine, space, 2757, 2772,
  2819
Men in black, 766, 778,
  917, 923, 953
Mental telepathy, ESP, etc.,
  see Psychic contacts
  with other planets and
  with humanoids, includ-
  ing mental telepathy,
  ESP, etc.
Mercury, possibilities for
  life (excluding intel-
  ligent), on, 2009, 2030,
  2042, 2049, 2092, 2107,
  2132, 2242
Metalaw (regulations
  involving humanoids
  and Earth inhabitants),
  749, 820, 920, 925, 930,
  1193, 1390
Meteors and meteorites,
  life in, 83, 236, 239,
  672, 1650, 1732, 1771,
  1795, 1856, 1867, 1872,
  1867, 1913, 1915, 1922,
  1930, 1945, 1947, 1962,
  1982, 1992, 2001, 2135,
  3510, 3512, 3513
Moon, intelligent life on
  the, 986, 1207, 1214,
  1273, 1335, 1365, 1380,
  1825, 2923
Moon, possibilities for
  life (excluding intel-
  ligent), on, 1892, 1954,
  2019, 2025, 2032, 2045,
  2075, 2079, 2108, 2111,
  2121, 2122, 2133, 2143,
  2149, 2150, 2204, 2208,
  2214, 2578, 2629, 2632
Multivator, Project, 2638,
  2641

O

Organizations concerned
  with UFO's, space
  research, etc., 2397-
  2427
Origin of life, see Life,
  origin of; Planets, origin of
Origin of planets, see
  Planets, origin of
Ozma, Project, 1533, 1539,
  2607, 2608, 2612, 2613,
  2615-2617, 2628, 2640

P

Philosophical and reli-
  gious aspects concern-
  ing UFO's, intelligent
  life on other planets,
  etc., see Religious
  and philosophical
  aspects
Photos of flying saucers,
  see Flying saucers,
  photographs
Physical appearance of
  humanoids, see Humanoids,
  physiology
Physical aspects of
  planets, see Planets,
  physical, geological,
  chemical and biological
  aspects of; Space
  research
Pioneer, Project, 2743
Planets, physical, geo-
  logical, chemical and
  biological, aspects of,
  50, 55, 60, 62, 75, 81,

82, 86, 87, 104, 105,
  110, 113, 114, 117, 133,
  146, 1056, 1169, 1229,
  1290, 1634, 1638, 1639,
  1643, 1646, 1647, 1653,
  1658, 1659, 1660, 1662,
  1666, 1669, 1675, 1681,
  1701, 1702, 1704, 1705-
  1711, 1716-1731, 1733,
  1738, 1750-1777, 1779-
  1790, 1795-2295, 2213,
  2567, 2569, 2575, 2584
Plant life on other planets,
  see life on other planets
  in general (excluding
  intelligent life); Meteors
  and meteorites
Planets, origin of, 4, 9,
  11, 15, 16, 19, 21, 23,
  25, 32, 35, 40, 41, 47,
  51, 60, 67, 68, 82, 84,
  100, 101, 109, 119, 121,
  1355
Pluto, possibilities for
  life (excluding intel-
  ligent), on, 2070, 2094,
  2127
Psychic contacts with life
  on other planets and
  with humanoids, includ-
  ing mental telepathy,
  ESP, etc., 2296-2396,
  2954, 3199

## R

Religious and philosophical
  aspects, 2327, 2332,
  2333, 2344, 2357, 2375,
  2428-2553, 2983, 2992,
  2993, 3197, 3491, 3670
Religious aspects of humanoids, see Humanoids,
  religious aspects

## S

Search for flying saucers
  and extraterrestrial
  life, see Flying saucers
  and extraterrestrial life;
  Flying saucers and extraterrestrial life, detection; Space communica-
  tion (by TV, radar, radio,
  radiotelescope, artificial
  satellite or space travel)
Second Story, Project, 2908
Sightings of flying
  saucers, see Flying
  saucers, sightings
Sign, Project, 2901
61 Cygni (planet), 1573
Sociological contacts
  between Earth inhabitants
  and humanoids, see
  Metalaw; Humanoids,
  contacts (including
  psychic), with Earth
  inhabitants
Solar system, see Universe
Space communication (by TV,
  radar, radio, radiotele-
  scopes, artificial satellite or travel), 145,
  1103, 1224, 1225, 1226,
  1242, 1244, 1246, 1275,
  1292, 1305, 1324, 1327,
  1351, 1451-1627, 2603,
  2605-2753, 2783
Space flight, 1852, 1873,
  2754-2843, 2941, 2983
Space medicine, see Med-
  icine, space
Space probes, see Pioneer,
  Viking, Mariner, Wolftrap,
  Gulliver, Ozma Projects;
  Multivator
Space research, 48, 66, 102,
  116, 125-127, 2773, 2809,
  2810
Sun, intelligent life on
  the, 985, 2125

## T

Teleportation (travel by flying saucer), 2844-2897, 3198
Teleportation (travel by flying saucer), to Alpha Centauri, 2857
Teleportation (travel by flying saucer), to Mars, 2844, 2848, 2849, 2855
Teleportation (travel by flying saucer) to planet Selo, 2863
Teleportation (travel by flying saucer) to the Moon, 2859, 2864
Teleportation (travel by flying saucer), to Venus, 2848
Triton, possibilities for life (excluding intelligent), on, 2094

## U

UFO's approaching Earth, see Blue Book, Second Story, Sign, Grudge, Magnet Projects
UFO's, see Flying saucers
Unidentified flying objects (UFO's), see Flying saucers
Universe, 2, 3, 5, 6, 7, 9, 10, 12, 13, 14, 15, 20, 23, 24, 26, 30, 31, 33, 34, 36, 37, 38, 39, 40, 43, 51, 53, 57, 59, 64, 67, 74, 75, 78, 79, 80, 84, 89, 94, 97, 101, 112, 119, 148, 1072, 1096, 1142

## V

Valensole, France, humanoid contact in, 867, 868, 871, 872, 876, 911, 1446
Venus, beings from, 733, 737, 740, 750, 777, 784, 792, 799, 800, 879, 989, 1133, 1146, 1281, 2866
Venus, possibilities for life (excluding intelligent), on, 2042, 2053, 2061, 2062, 2065, 2071, 2072, 2080, 2081, 2083, 2088, 2097, 2103, 2112, 2137, 2142, 2152, 2157, 2166, 2178, 2179, 2198, 2206, 2207, 2216, 2221, 2225, 2227, 2229, 2231, 2235, 2236, 2240, 2244, 2247, 2250, 2251, 2255-2257, 2260, 2265, 2266, 2268, 2269, 2289, 2618, 2631, 2681, 2712
Viking, Project, 2734, 2735, 2738, 2752, 2841

## W

Warminster, mystery, 3069
White, Barney and Betty, meeting with humanoids, 755, 870, 881, 885, 905, 981, 2858, 2882
Wolftrap, Project, 2641, 2651

## Z

Zeta Reticuli (planet), 2843

INDEX TO DIRECTORY SECTION

I. **ORGANIZATIONS CONCERNED WITH SPACE RESEARCH AND FLYING SAUCERS**

    Argentina.......................................3833-3841
    Australia.......................................3842-3844
    Belgium.........................................3845-3849
    Canada..........................................3850
    Denmark.........................................3851
    England.........................................3852-3856
    France..........................................3857-3858
    Italy...........................................3859-3860
    Spain...........................................3861
    Sweden..........................................3862
    Switzerland.....................................3863
    Turkey..........................................3864

    United States of America

        Alabama.....................................3865
        Arizona.....................................3866-3867
        California..................................3868-3873
        District of Columbia........................3874
        Illinois....................................3875-3878
        Kentucky....................................3879
        Maryland....................................3880-3881
        Michigan....................................3882-3883
        New Jersey..................................3884-3888
        New York....................................3889
        North Carolina..............................3890-3891
        Ohio........................................3892-3894
        Oklahoma....................................3895
        Oregon......................................3896
        Pennsylvania................................3897
        South Carolina..............................3898
        Tennessee...................................3899
        Texas.......................................3900
        Virginia....................................3901
        Washington..................................3902

II. **PERIODICALS CONCERNED WITH SPACE RESEARCH AND FLYING SAUCERS**

    Argentina.......................................3903-3904
    Belgium.........................................3905
    Canada..........................................3906-3908
    Denmark.........................................3909
    England.........................................3910-3917
    Finland.........................................3918

France..........................................3919-3921
Netherlands.....................................3922-3924
New Zealand.....................................3925
Spain...........................................3926
United State of America.........................3927-3968

ABOUT THE AUTHOR


MARTIN HOWARD SABLE received A.B. and M.A. degrees from
Boston University; M.S. in Library Science from Simmons
College; and earned his doctorate at Universidad Nacional
Autonoma de Mexico.

Dr. Sable is a Professor in the School of Library Science
of the University of Wisconsin-Milwaukee. His area of
expertise is the teaching of general and specialized
reference courses, among which is "Reference Materials
and Sources in Science and Technology." Professional
experiences include being Staff Librarian and Bibliographer
at Northeastern University, Boston, Massachusetts; Language
Librarian at California State College at Los Angeles;
Reference Librarian at Los Angeles County Library, Hawthorne,
California; Assistant Research Professor at Latin American
Center, University of California-Los Angeles.

Professor Sable holds membership in Institute of International Education, Library Association of Colombia (honorary),
American Library Association, Conference on Latin American
History, Latin American Studies Association, American
Association of University Professors.

His published works include:

A Selective Bibliography in Science and Engineering (G.K. Hall, 1964);
Master Directory for Latin America (Latin American Center, University of California, Los Angeles, 1965);
Periodicals for Latin American Economic Development, Trade and Finance: An Annotated Bibliography (Latin American Center, University of California, Los Angeles, 1965);
(Contributor) Jose Rubia Barcia and M.A. Zeitlin, editors, Unamuno, Creator and Creation (University of California Press, 1967);
A Guide to Latin American Studies, 2 vols. (Latin American Center, University of California, Los Angeles, 1967);
UFO Guide: 1947-1967 (Rainbow Press, 1967);
Communism in Latin America, an International Bibliography: 1900-1945 (Latin American Center, University of California, Los Angeles, 1968);
A Bio-Bibliography of the Kennedy Family (Scarecrow, 1969);

*Urbanization Research, with Special Reference to Latin America:
An Inventory* (Center for Latin American Studies, University
of Wisconsin, Milwaukee, 1969);
*Latin American Agriculture: A Bibliography* (Center for Latin
American Studies, University of Wisconsin, Milwaukee, 1970);
*Latin American Studies in the Non-Western World and Eastern
Europe* (Scarecrow, 1970);
*Latin American Urbanization: A Guide to the Literature,
Organizations and Personnel* (Scarecrow, 1971);
*International and Area Studies in Librarianship: Case Studies*
(Scarecrow, 1973)

Contributor to many periodicals, including,
*New England Modern Language Association Bulletin*;
*Current History*;
*American Documentation*;
*International Library Review*;
*RQ*;
*International Educational and Cultural Exchange*;
*Wisconsin Library Bulletin*

Since Sable speaks fluent Spanish and French, and is
competent in Portuguese, German, Italian, Latin, and Hebrew,
this bibliography profits from an international approach to
the literature of exobiology.